Avoiding Inelastic Strains in Solder Joint Interconnections of IC Devices

Avoiding Inelastic Strains in Solder Joint Interconnections of IC Devices

Authored by

Ephraim Suhir

CRC Press
Taylor & Francis Group
Boca Raton London New York

CRC Press is an imprint of the
Taylor & Francis Group, an **informa** business

First edition published 2021
by CRC Press
6000 Broken Sound Parkway NW, Suite 300, Boca Raton, FL 33487-2742
and by CRC Press
2 Park Square, Milton Park, Abingdon, Oxon, OX14 4RN

A catalog record has been requested for this book.

ISBN: 978-1-138-62473-3 (hbk)
ISBN: 978-0-429-46047-0 (ebk)

Typeset in Times LT Std
by KnowledgeWorks Global Ltd.

Dedication

To the Suhirs: my wife Raisa, my late parents Leonid and Betya, my children Elena and Gene, and my grandchildren Scarlett and Fiona

Contents

Preface

"Do not follow where the path may lead, but go where no path is—and leave a trail"

—Chinese saying

All the design-for-reliability (DfR) problems addressed in this book were solved using analytical ("mathematical") modeling. Such modeling occupies a special place in the predictive modeling effort: it enables not only coming up with simple relationships that clearly indicate "what affects what," but, even more importantly, can often explain the physics of phenomena much better than FEA modeling, or even experimentation, can. In addition, analytical and FEA models are based, as a rule, on different assumptions, and if these two modeling approaches result, in the problem of interest, in close computed data, then there is good reason to believe that the obtained information is both accurate and trustworthy.

The majority of the problems considered in this book are based on the method of interfacial compliance. The development of this method (see the author's 1986 *ASME Journal of Applied Mechanics* paper "Stresses in bi-metal thermostats") enabled obtaining closed form engineering solutions for problems that otherwise could be addressed only using FEA. I hope that you, the reader, will be able/willing to apply the solutions suggested in this book to better understand the physics of failure, to consider these solutions when developing your FEA preprocessing models and, hopefully, even develop analytical models of your own.

Here are some problems envisioned and questions asked in connection with today's practices concerning reliability evaluations and assurances of electronic, photonic, MEMS, and MOEMS (optical MEMS) devices, products, and systems:

- Solder joint interconnections are, as is known, the reliability bottleneck of electronic and photonic packaging structures. Could the induced stresses in the solder material can be relieved by appropriate DfR measures, even, perhaps, to an extent that the inelastic strains in the solder material could be avoided, thereby resulting in a significant increase in the materials fatigue lifetime?
- Electronic and photonic products that underwent highly accelerated life testing (HALT), passed the existing qualification tests (QT) and survived burn-in testing (BIT) often exhibit nonetheless premature field failures. Are today's methodologies and practices adequate? Could these practices be improved to an extent that if the product passed the QTs and BIT, it will satisfactorily perform in the field?
- In many applications, such as aerospace, military, long-haul communications, and medical, high reliability of electronics and photonic materials and products is particularly imperative. Could the operational reliability of such products be assured if it is not predicted, that is, not quantified?

- And if such quantification is found to be necessary (in many situations it is the case), could it be done on the deterministic, that is, on a nonprobabilistic basis, or, considering that, in effect, the probability of failure is never zero, PDfR and probabilistic risk analyses are a must?
- Should electronic product manufacturers keep shooting for an unpredictable and, perhaps, unachievable, very long, such as, say, 20 years or so, product lifetime or, considering that every 5 years a new generation of devices appear on the market and that 20-year predictions are rather shaky, should electronic and photonic products manufacturers settle for a shorter, but well substantiated, predictable, and assured lifetime, with a high (actually, adequate) probability of nonfailure?
- And how should such a lifetime be related to the acceptable (specified) probability of nonfailure for a particular product and application?
- The best engineering product is, as is known, the best compromise between the requirements for its reliability, cost-effectiveness and (short-as-possible) time-to-market; it goes without saying that, in order to make any kind of optimization possible, the reliability of such product should be quantified, but how could such a quantification and optimization be done? Effective modeling is certainly required.
- The bathtub curve (BTC), the experimental "reliability passport" of a mass-fabricated electronic product, reflects the inputs of two critical irreversible processes: the statistical failure rate (SFR) process that results in a reduced failure rate with time and the physics-of-failure (aging, degradation) process that leads to an increased failure rate. Could these two critical processes be separated and what could possibly be done to improve them?
- BIT is a costly undertaking: early failures are avoided and the infant mortality portion of the BTC is supposedly eliminated at the expense of the reduced yield. But what is even worse is that the elevated BIT stressors might not only eliminate "freaks," but could cause permanent damage to the main population of the "healthy" products. It is not even clear if BIT is needed at all, so the first question is "to BIT or not to BIT?" and if "to BIT," how to optimize and predict the outcome of this process?
- In many reliability problems, human factors play an important role: equipment and instrumentation reliability (both hard- and software) and human performance often contribute jointly to the outcome of a particular mission or an extraordinary situation. How could the interaction and possible integration of a human and a system be considered and quantified?

Some of these practically important questions are addressed and, to an extent possible, answered in this book.

Ephraim Suhir,
IEEE, ASME, SPIE, IMAPS Life Fellow;
APS, IoP (UK), SPE Fellow; AIAA Associate Fellow
Los Altos, California, USA, June 2020

Foreword

I am truly honored to write this Foreword since it is an opportunity to pay tribute to Prof. Suhir's many seminal contributions to the field of microelectronics packaging reliability. I have always regarded Prof. Suhir as a role model since my graduate school days at Lehigh University (1987–1992). In the late 1980s, concerns about the reliability of microelectronic packages, due to thermo-mechanical stresses induced during processing and in application environments, had started to dominate many research conversations in academic and industrial circles. Distinguished researchers such as Profs. Evans, Hutchinson, Barnett, and Suhir (in those days he was Dr. Suhir from Bell Labs), and my PhD advisor, Prof. Fazil Erdogan, along with (the then) upcoming thought leaders such as Huajian Gao, Herman Nied, Ahmet Kaya, Zhigang Suo, and Jianmin Qu, were extending and reinterpreting stress models for anisotropic multilayer composites. Bell Labs, IBM, and GE were places of great intellectual ferment, and leading thinkers like Prof. Suhir were applying the ideas of classical mechanics (for example, Prof. Timoshenko's work) to understand and solve cracking and delamination problems in packaging. Prof. Suhir's contributions are, in effect, extensions of the Timoshenko theory for assemblies comprised of dissimilar materials and subjected to the change in temperature, such as for assemblies typical in electronics packaging. I remember listening to Prof. Suhir for the first time when he came to Lehigh to deliver a graduate seminar on stress modeling of microelectronics packages as multilayer composites. What struck me then and continues to resonate most about Prof. Suhir's work is how well he combines technical rigor with a robust pragmatism toward applied solutions. To me, this is his enduring contribution to the field. By describing the problems and developing solutions in coherent, practical terms, he continues to inspire many generations of engineers, in addition to personally participating in driving relevant change. It is no surprise then that from those days to today, over as span of so many decades, Prof. Suhir has had an immensely significant role in shaping the field of microelectronics reliability.

I left Lehigh in the early 1990s and joined Intel, initially as a stress analyst responsible for developing models that helped ensure package reliability. Prof. Suhir's advice to be pragmatic and yet retain technical depth continued to guide my thinking as we collaborated with some of the best minds in academia and industry to build our modeling team, tool capabilities, and a laboratory where we could validate our models. Our efforts in this area continue on to this day and they reflect a broader consensus in the industry about the crucial need for predictive modeling. In this context, I have had the privilege of staying in touch with Prof Suhir's work and have been fortunate to be a coauthor on two well-received papers:

- E. Suhir, R. Mahajan, A. Lucero, L. Bechou, "Probabilistic Design for Reliability (PDfR) and a Novel Approach to Qualification Testing (QT)," 2012 IEEE/AIAA Aerospace Conference, Big Sky, Montana, 2012.
- E. Suhir, R. Mahajan, "Are Current Qualification Practices Adequate?" *Circuit Assembly*, Apr. 2011.

As my career at Intel has evolved, I have moved on to help define new heterogeneous package architectures. Today, the importance of heterogeneous integration to drive the performance of semiconductor devices is increasingly and broadly recognized. There is great need for predictive modeling to guide the definition of heterogeneous architectures and hence the importance of the work in this book. As Prof. Suhir points out in his preface and in Chapter 1, predictive models are a major constituent of making the right design and materials decisions in defining robust packaging architectures. Predictive modeling is essential to reduce development cycle times, development costs, and in elucidating complex interactions in the multimaterial composite package structures. Analytical modeling is an excellent complement to finite element analysis (FEA) for a mechanistic understanding of physical phenomenon. This book reflects Prof. Suhir's technical depth and pragmatism as he develops different models that offer excellent examples of how analytical models can be developed to understand the elastic and inelastic behavior of interfaces and solder joints. Different, widely used package architectures are comprehensively covered. Additionally, the influences of geometry, material inhomogeneity, and different types of loading conditions are excellently addressed in this book. I would particularly draw attention to the development of failure-oriented-accelerated-testing (Chapter 10), a topic of considerable interest to practitioners in the field and the discussion on probabilistic design for reliability. The latter topic is both timely and relevant as we encounter increasing complexity in packaging structures moving forward. Overall, I continue to be guided by Prof. Suhir's thinking on the value of analytical thinking on understanding package reliability and am pleased to be a part of this book. It is an excellent reference that I am convinced will be useful for many generations of engineers.

Ravi Mahajan,
Intel Fellow, High Density Interconnect Pathfinding,
ASME Fellow, IEEE Fellow, VP of Publications & Managing
Editor-in-Chief of the IEEE Transactions of the CPMT,
Assembly Technology, Intel Corporation,
June 2020

Foreword

The book's title *Avoiding Inelastic Strains in Solder Joint Interconnections of IC Devices* is, in a way, misleading because the title provides an impression that the book is focused on a rather narrow, although certainly important, aspect of the reliability of electronic packaging materials and structures, while, in effect, it covers a number of critical problems in this field of engineering and even beyond this field. Quite a few critical messages of this groundbreaking book will make, I am sure, a difference in several aspects of electronic and photonic packaging engineering: new designs and technologies, applied materials, reliability evaluations and predictions, solder joint interconnections, accelerated testing and burn-ins, cost-effectiveness, as well as new applications.

First, a few words about the author. I've known Ephraim, both personally and based on his publications, for many years since I joined, about 30 years ago, the electronic packaging field. Ephraim's name was, and still is, well known to everyone involved or interested in this field as the one, who, as his 2004 ASME Worcester Warner Medal Award citation says, "made outstanding contributions to the permanent literature of engineering through a series of papers in Mechanical, Microelectronic, and Optoelectronic Engineering, which established a new discipline known as the Structural Analysis of Microelectronic and Photonic Systems." And here is what his 2019 IEEE EPS Field award (as a matter of fact, it was I who nominated him) says of Ephraim's numerous contributions: "With over 40 years of pioneering work in modeling and reliability engineering, Ephraim Suhir has enabled electronic packaging engineers to accurately predict stress in advanced packaged components for the design of more reliable devices. He was one of the earliest researchers to introduce the use of rigorous mechanics principles in electronic systems. His closed-form solutions have provided the electronics industry with invaluable tools for ensuring reliability and cost savings during the design process by eliminating errors early in the design process. He has applied his techniques to advanced components and packaged structures such as microelectronics, photonics, photo-voltaic, and thermo-electronic modules. Every serious mechanics practitioner and researcher in the electronics packaging field has been influenced by Suhir's groundbreaking contributions."

Now, about this well-written and highly informative CRC book. In 1991, Van-Nostrand Reinhold published Ephraim's monograph *Structural Analysis in Microelectronic and Fiber Optic Systems, vol.1, Basic Principles of Engineering Elasticity and Fundamentals of Structural Analysis* and in 1997, McGraw-Hill published his monograph *Applied Probability for Engineers and Scientists.* Both books were written during Ephraim's approx. 20-year long tenure with the Basic Research area, Bell Labs, Murray-Hill, New Jersey, and the copyrights belong to Bell Labs. Being familiar with these two excellent books, I would say that the present 2020 CRC monograph is, in effect, a logical and a fruitful continuation of his 1991 Van-Nostrand book (it provides numerous practical examples that use methods and approaches addressed in this book and because of that, with some stretching, perhaps, could be viewed as the never-published vol. 2 of this book), with an impact of his 1997 McGraw-Hill monograph. As Ephraim's 2017 IEEE EPS Exceptional Technical Achievement Award citation says: "for the development of numerous probabilistic design concepts that enable effective and rapid assessment of the probability of failure of electronic products."

Here are, in my opinion, the major, both pioneering and practically important, messages of this CRC book.

1. The author has advanced and effectively applied classical analytical methods of structural analysis and elasticity theory to various critical electronic and photonic packaging problems, with an emphasis, of course, on the mechanical behavior and performance of solder joint interconnections. I particularly appreciate this contribution because I am professor of computational mechanics and reliability at the University of Greenwich, London, UK, because my PhD degree is in computational modeling and also because, before joining Greenwich, I worked for 3 years at Carnegie Mellon University, USA, as a research fellow in materials engineering. As Dr. Suhir has indicated, analytical ("mathematical") modeling occupies a special place in the predictive modeling effort in the field of electronic packaging mechanics and materials. In today's world, when computational, and, particularly, FEA methods dominate the modeling approaches in electronics and photonics, analytical methods successfully complement these approaches, and, because the numerical and analytical models are based, as a rule, on different assumptions, it is always advisable, as Ephraim has indicated, to use both approaches in every material's reliability of importance.
2. The author has suggested, advanced, and implemented the "method of interfacial compliance," and demonstrated how this analytical method could be successfully used in many electronic packaging reliability problems, including those concerning solder joint interconnections. Using this method, he demonstrated, particularly, that the maximum strains in them could be minimized by an appropriate physical design of the solder material structure, sometimes even to an extent that inelastic strains that lead to low-cycle fatigue conditions could be avoided, but if not, there is always a way to predict and to minimize the size of the inelastic zones at the ends of a soldered assembly. The shorter these zones, the longer is the fatigue life of the solder material.
3. The author has considerably advanced the general field of electronics and photonics reliability by indicating and demonstrating that when the operational reliability is imperative (this is typically the case for aerospace, long-haul communication, military, or medical devices and systems), it has to be quantified to be assured, and that a highly focused and highly cost-effective failure-oriented-accelerated-testing (FOAT), geared to a physically meaningful, powerful, and flexible predictive model, such as the multiparametric kinetic Boltzmann–Arrhenius–Zhurkov (BAZ) model that he has suggested, can be employed to do that. Ephraim has demonstrated in his book, using numerous practical examples, how this could be done.

4. The author has established a new direction in the design-for-reliability of electronic and photonic systems—the probabilistic design for reliability (PDfR)—and has shown how this concept could be used in many practical electronic and photonic packaging problems, and even beyond this field. Particularly, he has demonstrated that the never-zero probability of failure is closely connected with the projected/expected lifetime of a material or a device and has shown how these could be evaluated using the FOAT approach and the PDfR concept.
5. Using a simple example, the author has shown how the total cost of reliability could be minimized and that such a minimization has to do with the availability of the product, which is, as is known, the probability that the product is available to the user, when needed.
6. A significant contribution has been made by the author to understanding the underlying physics and, to some extent, to predicting the outcome of a burn-in testing (BIT) effort, and, first of all, to answer the fundamental "to BIT or not to BIT?" question.
7. Since human factors play an important role in the behavior and performance of electronic and photonic systems for some applications (say, for aerospace missions or extraordinary situations), the author has indicated that the developed methods could also be applied in, for example, "human-in-the-loop" problems. Such a possibility has been indicated in his 2018 CRC book *Human-in-the-Loop: Probabilistic Modeling of the Outcome of an Aerospace Mission,* but is certainly beyond the scope of this book.

I read this book with significant interest, will use its findings and recommendations in my work, and have no doubt in my mind that many of its readers will do the same.

Christopher Bailey,
IEEE EPS President,
Greenwich, London, UK

Author Biography

Ephraim Suhir is on the faculty of the Portland State University (PSU), Portland, Oregon, USA, Technical University, Vienna, Austria, and James Cook University, Queensland, Australia. He is also CEO of a Small Business Innovative Research (SBIR) ERS Co. in Los Altos, California, USA. Ephraim is Foreign Full Member of the National Academy of Engineering, Ukraine (he was born in that country), and Life Fellow or Fellow of several leading professional societies. He has authored about 450 publications (patents, technical papers, book chapters, books), presented numerous keynote and invited talks worldwide, and received many professional awards, including the 1996 Bell Labs Distinguished Member of Technical Staff (DMTS) Award for developing effective methods for predicting the reliability of complex structures used in AT&T and Lucent Technologies products; the 2004 ASME Worcester Read Warner Medal for outstanding contributions to the permanent literature of engineering and laying the foundation of a new discipline "Structural Analysis of Electronic Systems" (Ephraim is the third "Russian American," after S. Timoshenko and I. Sikorsky, to receive this prestigious award); the 2019 IEEE Electronic Packaging Society (EPS) Field award for seminal contributions to mechanical reliability engineering and modeling of electronic and photonic packages and systems; and also the 2019 International Microelectronic Packaging Society's (IMAPS) Lifetime Achievement award for making exceptional, visible, and sustained impact on the microelectronics packaging industry and technology. His current Linked-in (Research Gate) data are: Profile strength: "all-star," RG score: 41.40 (higher than 97.5% of RG members), over 22,000 downloads ("reads"), and about 5,000 citations.

Here is the shortened list of Suhir's previously published books:

E. Suhir, *Human-in-the-Loop: Probabilistic Modeling of an Aerospace Mission Outcome*, CRC Press, 2018.

E. Suhir, D. Steinberg, T. Yi, (eds.), *Dynamic Response of Electronic and Photonic Systems to Shocks and Vibrations*, John Wiley, 2011.

X. Fan, E. Suhir, *Moisture Sensitive Plastic Packages of IC Devices*, Springer, 2010.

E. Suhir, C.P. Wong, Y.C. Lee, (eds.), "Micro- and Opto-Electronic Materials and Structures: Physics," in *Mechanics, Design, Packaging, Reliability*, 2 volumes, Springer, 2008.

E. Suhir, M. Fukuda, C.R. Kurkjian, (eds.), "Reliability of Photonic Materials and Structures," *Materials Research Society Symposia Proceedings*, vol. 531, 1998.

E. Suhir, *Applied Probability for Engineers and Scientists*, McGraw-Hill, 1997.

E. Suhir, R.C. Cammarata, D.D.L. Chung, M. Jono, (eds.), "Mechanical Behavior of Materials and Structures in Microelectronics," Materials Research Society Symposia Proceedings, vol. 226, 1991.

E. Suhir, *Structural Analysis in Microelectronic and Fiber Optic Systems*, Van Nostrand Reinhold, 1991.

1 Analytical Modeling, Its Role and Significance

"There are things in this world, far more important than the most splendid discoveries—it is the methods by which they were made."

—Gottfried Leibnitz, German mathematician

Modeling is the major approach of any science, whether pure or applied. Engineering models could be experimental or theoretical. Experimental models (test specimens) are usually of the same physical nature and, in electronics and photonics engineering, even on the same scale as the actual objects. Theoretical models use abstract notions. Their ultimate goal is to reveal nonobvious, latent, even sometimes paradoxical, relationships hidden in the input information. No theoretical model can provide results that are not contained in the input data and in the taken assumptions. Experimental models, on the other hand, can occasionally lead to new and unexpected results. A famous example is the surprise discovery of radiation by Henri Becquerel in 1896, which gave birth to nuclear chemistry.

Theoretical models can be either analytical or numerical (computational). Analytical models [1–12] employ mathematical methods of analysis. Today's numerical models are mostly computer-aided. The most widespread models in the stress–strain evaluations and physical design for reliability (DfR) of electronic and photonic materials and systems use finite element analysis (FEA) (see, for example, [13–20]). FEA strongly depends on the use of computers and went a long way from its initial application in the 1960s to avionic structures (by J.H. Argyris and his associates at the University of Stuttgart in Germany) to broad applications in a wide variety of engineering disciplines and efforts, including material science and reliability physics.

Experimental and theoretical models should be viewed as equally important for the design of a viable, reliable, cost-effective, and timely product. In electronic engineering problems, experimental models are, however, perceived as more significant than in many other areas of engineering, such as aerospace, civil, ocean, and so on, and qualification testing is usually the main way to make a viable electronic or a photonic component into a reasonably reliable and marketable product. There are several reasons for that. Let us indicate some of them:

1. Experiments could be carried out with full autonomy without necessarily requiring theoretical support;
2. Unlike theory, testing can be used as a final proof of the viability and reliability of a product and is, indeed, essential requirement of various specifications for electronic or photonic products;

3. Experiments in the high-tech field, expensive as they might be, are considerably less costly than those in "macro-engineering," such as naval architecture, or aerospace engineering, where "specimens" might cost millions of dollars;
4. High-tech experimentations are much easier to design, organize, and conduct than in the macro-engineering world;
5. Materials, whose properties are not completely known, are often and successfully employed in high-tech engineering products; such a practice would be completely inappropriate in macro-engineering systems;
6. Lack of information about the material properties is often viewed as an obstacle for implementing theoretical modeling, including FEA, but not for implementing the materials themselves, provided that their performance is proven experimentally;
7. Many leading specialists in high-tech engineering, such as experimental physicists, materials scientists, chemists, and chemical engineers, traditionally use experimental methods as their major research tool. It is not just a coincidence that 11 out of 12 Bell Labs Nobel laureates were experimentalists.

At the same time, experimental investigations, unlike theoretical modeling, require, as a rule, considerable time and significant expense. What is even more important is that experimental data inevitably reflect the effect of the combined action of a variety of factors affecting the material, phenomenon, or the product of interest. Because of that, experimental evaluations, important as they are, are, in effect, a sort of a "black box," and are often insufficient to understand the underlying physics of the behavior and performance of a material or a device. Such a lack of insight leads to tedious, time-consuming, and costly efforts. In addition, experimental data cannot be simply extended to new situations or new designs that are different from those already tested. It is always easy to recognize purely empirical relationships obtained by formal processing of experimental data: these relationships often contain fractional exponents and coefficients, odd units, and so on. Although such relationships have a certain practical value, the very fact of their existence should be attributed to the lack of knowledge in the given area of applied science. Typical examples are power law, such as the one used in proof testing of optical fibers, when the time to delayed fracture, "static fatigue," is evaluated, and an inverse power law, such as numerous relationships of Coffin–Manson type, when evaluating the lifetime of solder joint interconnections in electronic products.

A good theoretical and physically substantiated and meaningful model could be as practical as the most thoroughly conducted experimentation. Here are some considerations on what could be gained by using theoretical modeling in electronics and photonics materials science and engineering:

1. Predictive modeling, both analytical and simulation-based, is able, unlike experimentation, to shed light on the role of each particular parameter that affects the behavior and performance of the material or the product of interest;

2. Although testing can reveal insufficiently robust elements, it is usually incapable of detecting superfluously robust ones; overengineered products may have excessive weight and be more costly than necessary: in mass-produced expensive devices, superfluous reliability may entail substantial and unnecessary additional costs;
3. Theoretical modeling can often predict the outcome of an experiment in less time and at considerably lower expense than the actual experiment can;
4. In many cases, theory serves to discourage wasting time on useless experiments; a classical example is the numerous attempts to build impossible heat engines that have been prevented by a study of the theoretical laws of thermodynamics; while this is, of course, a classical and an outstanding example of the triumph of a theory, there are also numerous, although less famous, examples, when plenty of time and expense was saved because of prior theoretical modeling of a problem of interest;
5. In the majority of research and engineering projects, a preliminary theoretical analysis enables to obtain valuable information about the phenomenon or the object, and gives an experimentalist an opportunity to decide what and how should be tested or measured, and in what direction success might be expected;
6. By shedding light on "what affects what," theoretical predictive modeling often serves to suggest new useful experiments; as is known, theoretical analyses of stresses in bi- tri- and multi-material assemblies and in semiconductor thin films triggered numerous experimental investigations aimed at the rational physical design of semiconductor crystal grown systems;
7. Theory can be used to interpret empirical results, to bridge the gap between different experiments, and to extend the existing experience on new materials, components, and structures;
8. One cannot do without a good theory when developing rational (optimal) designs; the best engineering design is, in effect, the best compromise between reliability, cost-effectiveness, and time-to-market (to completion); the idea of optimization of structures, materials, and costs has penetrated many areas of modern engineering; no progress in this direction could be achieved, of course, without application of theoretical methods that would enable to quantify the role of these critical factors.

Analytical modeling occupies a special place in the predictive modeling effort. Such modeling is able not only to come up with simple relationships that clearly indicate "what affects what" in the structure or design of interest, but, more importantly, can often explain the physics of phenomena much better than the FEA modeling, or even experimentation, can. Since the mid-1950s, FEA modeling has become the major modeling tool for theoretical stress–strain evaluations in materials, mechanical, structural, aerospace, maritime, and other areas of engineering and applied science. Even Maxwell and Navier–Stokes equations are solved today using FEA. This should be attributed, first of all, to the availability of powerful and flexible computer programs that enable to obtain, within a reasonable time, a solution to almost any stress–strain or other external loading–related problem, but partially also to the illusion that FEA is the ultimate and indispensable tool for solving any design or stress-analysis problem. However, the truth of the matter is that FEA and broad application of computers has by no means made analytical solutions unnecessary or even less important, whether exact, approximate, or asymptotic. Simple and physically meaningful analytical relationships have invaluable advantages, because of the compactness of the information and clear indication on the role of various factors affecting the given phenomenon or a device performance. These advantages are especially significant when the parameter under investigation depends on more than one variable, which is typically the case. As to the asymptotic analytical techniques, they can be successful in many cases when there are difficulties in the application of computational methods, such as in problems containing singularities. Such problems are often encountered in high-tech materials engineering, where assemblies comprised of dissimilar materials are widely employed. But even when application of FEA encounters no difficulties, it is always advisable to investigate the problem analytically before carrying out FEA. Such a preliminary investigation helps to reduce computer time and expense, develop the most feasible and effective preprocessing model and, in many cases, avoid fundamental errors. It is noteworthy in this connection that FEA has been originally developed for structures with complicated geometry and/or with complicated boundary conditions (such as avionics, maritime, or some civil engineering structures) when it might be difficult to apply analytical approaches. As a consequence, FEA has been especially widely used in those areas of engineering in which structures of complex configuration are typical, such as aerospace, maritime, and offshore structures or some civil engineering structures. In contrast, electronic and photonic structures are usually characterized by simple geometries and can be easily idealized as beams, flexible rods, rectangular or circular plates, composite structures of relatively simple geometry, and so on. Therefore, there is an obvious incentive for a broad application of analytical modeling in electronics and photonics materials science and engineering. Additional incentive is due to the fact that adjacent structural elements in electronics materials engineering often have dimensions that differ by orders of magnitude. Examples are multilayer thin film structures fabricated on thick substrates or adhesively bonded assemblies, in which the bonding layer is, as a rule, significantly thinner than the bonded components. Since the mesh elements in a FEA model must be compatible, FEA of such structures often becomes a problem of itself, especially in regions of high stress concentration. Such a problem does not occur, however, with an analytical approach. Another consideration in favor of analytical modeling is associated, as previously mentioned, with an illusion of simplicity in applying FEA procedures and trust in the FEA data. Many users of FEA programs sincerely believe that the "black box" they deal with will automatically provide the right answer as long as they push the right key on the computer keyboard. At times, a hasty, thoughtless, and incompetent application of computers can result in more harm than good by creating an impression that a solution has been obtained, when, in effect, this "solution" is simply wrong. Although it is usually easy to obtain *a* FEA solution, especially with today's user-friendly and powerful software, it might be quite difficult to obtain *the* right one. And how would one know that they obtained the correct and trustworthy solution if there is nothing to compare it with? FEA and analytical models are typically based on different assumptions: FEA is a numerical continuum mechanics tool, while the available close-form analytical solutions use mostly approximate structural analysis and strength-of-materials methods. Clearly, if the FEA data are in good agreement with the results of an analytical modeling then there is a good reason to believe that the obtained solution is accurate and trustworthy.

A crucial requirement for an effective analytical model is simplicity and clear physical meaning. A good analytical model should be based on physically meaningful considerations and produce simple relationships, clearly indicating the role of the major factors affecting the phenomenon or the object of interest. A good example is the Arrhenius equation widely used in electronics reliability problems. One authority in applied physics remarked, perhaps only partly in jest, that the degree of understanding of a phenomenon is inversely proportional to the number of variables used for its description. It takes a lot of imagination, intuition, appropriate assumptions, and effort to come up with a meaningful analytical expression, while it is typically and merely skill that is needed for the application of FEA simulation.

While an experimental approach, unsupported by a meaningful theory, is blind, theory, not supported by an experiment, is dead. An experiment forms a basis and provides input data for a theoretical model, and determines its viability, accuracy, and limits of application. It is also noteworthy that limitations of a theoretical model are different in different problems, particularly when structural analyses, not elasticity theory approaches, are used. It is the experimental modeling that is often the "supreme and ultimate judge" of a theoretical model. The limitation of a particular theoretical model could be also assessed based on a more general theoretical model: limitations of a linear approach could be determined on the basis of a more comprehensive nonlinear model. Experiment can often be rationally included into a theoretical solution to a problem of interest. Even when some relationships and structural characteristics lend themselves, in principle, to theoretical evaluation, it is sometimes simpler to determine these relationships empirically. For example, the spring constant of an elastic foundation provided by the primary coating of an optical fiber could be evaluated experimentally and then included into the analytical or numerical predictive model.

To conclude this chapter, I would like to quote the late Vladimir Lefebvre, a mathematical psychologist at the University of California, Irvine: "Mathematics is just a manifestation of the laziness of human beings. Instead of using many, many words and implications, sometimes we can use a few formulae instead."

REFERENCES

1. E. Suhir, "Analytical Modeling in Structural Analysis for Electronic Packaging: Its Merits, Shortcomings and Interaction with Experimental and Numerical Techniques," *ASME Journal of Electronic Packaging*, vol. 111, No. 2, June 1989.
2. E. Suhir, "Axisymmetric Elastic Deformations of a Finite Circular Cylinder with Application to Low Temperature Strains and Stresses in Solder Joints," *ASME Journal of Applied Mechanics*, vol. 56, No. 2, 1989.
3. E. Suhir, "Analytical Thermal Stress Modeling in Electronic and Photonic Systems," *ASME Applied Mechanics Reviews*, vol. 62, No. 4, 2009.

4. E. Suhir, "Thermal Stress Failures: Predictive Modeling Explains the Reliability Physics Behind Them," *IMAPS Advanced Microelectronics*, vol. 38, No. 4, Jul.–Aug. 2011.
5. E. Suhir, "Predictive Modeling is a Powerful Means to Prevent Thermal Stress Failures in Electronics and Photonics," *Chip Scale Reviews*, vol. 15, No. 4, Jul.–Aug. 2011.
6. E. Suhir, "Thermal Stress Failures in Electronics and Photonics: Physics, Modeling. Prevention," *Journal of Thermal Stresses*, June 3, 2013.
7. E. Suhir, "Analytical Modeling Occupies a Special Place in the Modeling Effort," *Short Communications Journal of Physical Mathematics*, vol. 7, No. 1, 2016.
8. E. Suhir, "Analytical Modeling Enables Explanation of Paradoxical Behaviors of Electronic and Optical Materials and Assemblies," *Advances in Materials Research*, vol. 6, No. 2, 2017.
9. E. Suhir, "Solder Joint Interconnections in Automotive Electronics: Design-for-Reliability and Accelerated Testing," Abstracts Proceedings, SIITME, Jassy, Romania, 2018.
10. E. Suhir, "Analytical Thermal Stress Model for a Typical Flip-Chip Package Design," *Journal of Materials Science: Materials in Electronics*, vol. 29, No. 4, 2018.
11. E. Suhir, "Low-Cycle-Fatigue Failures of Solder Material in Electronics: Analytical Modeling Enables to Predict and Possibly Prevent Them-Review," *Journal of Aerospace Engineering and Mechanics (JAEM)*, vol. 2, No. 1, 2018.
12. E. Suhir, "Interfacial Shearing Stresses and Warpage of Flip-Chip Packages: Application of Analytical Thermal Stress Modeling," *Scholar Journal of Applied Sciences and Research (SJASR)*, 2020, in review.
13. K.J. Bathe, *Finite Element Problems in Engineering*, Prentice Hall, 1982.
14. L.L. Meekisho, "Thermal Stress Analysis of Contact Problems by the Finite Element Method," Ph.D. Dissertation, Carleton University, Canada, 1988.
15. J.N. Reddy, D.K. Gartling, *The Finite Element Method in Heat Transfer and Fluid Dynamics*, CRC Press, 1994.
16. K.H. Huebner, F.A. Thornton, T.G. Byrom, *The Finite Element Method for Engineers*, 3rd ed., John Wiley & Sons, 1995.
17. D. Roylance, *Finite Element Analysis*, Massachusetts Institute of Technology, Cambridge, 2001.
18. A.I. Southpointe, *Guide to ANSYS User Programmable Features*, SAS IP, Inc, Canonsburg, 2004.
19. V.B.C. Tan, M.X. Tong, K.M. Lim, C.T. Lim, "Finite Element Modeling of Electronic Packages Subjected to Drop Impact," *IEEE CPMT Transactions*, vol. 28, No. 3, Oct. 2005.
20. M. Li, R. Torah, J. Liu, et al., "Finite Element Analysis (FEA) Modeling and Experimental Verification to Optimize Flexible Electronic Packaging for E-Textiles," *Microsystems Technologies*, 2020.

2 Method of Interfacial Compliance

"There is nothing more practical than a good theory."

—Kurt Zadek Lewin, German American psychologist

2.1 INTRODUCTION

Thermal loads arise during manufacturing, testing, and operation of electronic, optoelectronic, and photonic devices, assemblies, packages, and systems. Stresses and deformations caused by thermal loads are the major contributors to their finite service life. High thermal stresses and strains can lead not only to physical (mechanical) damage, but also to a functional failure of the electronic or photonic product. If the heat produced by the chip cannot readily escape then the highly localized thermal stresses in the integrated circuit (IC) can lead to the PN junction failure. Thermally induced warpage is a serious problem in today's manufacturing of electronic packages. Low-temperature microbending in dual-coated optical fibers, although it might not be high enough to lead to appreciable bending stresses, could result nonetheless in significant added transmission losses. Thermal stresses are responsible for "curling" of optical fibers during drawing. Substantial loss in optical coupling efficiency occurs when the lateral displacement in the gap between two light-guides becomes too large because of thermal stress-related phenomena (elevated deformations, stress relaxation in laser welds, and so on). The ability to predict (model) and, if necessary, minimize the adverse consequences of thermal stresses and deformations is of obvious practical importance in electronics and photonics reliability engineering.

Pioneering work on thermal stress modeling belongs to Timoshenko [1]. He applied a structural analysis (strength of materials) approach to evaluate the stresses in, and the bow of, bimetal thermostat strips. Timoshenko did not address interfacial shearing and peeling stresses in his paper, but indicated that these stresses cannot be determined using engineering methods of structural analysis, but could be evaluated, if necessary, using methods of the elasticity theory. This was done by Aleck [2] who employed theory-of-elasticity to model thermal stresses in a rectangular plate clamped along the edge. Both approaches were extended later on by numerous investigators, including the field of electronics and photonics engineering (see, for example, [3–22]). Suhir [7–10] managed to come up with a structural analysis (strength of materials) engineering solution for interfacial shearing and peeling thermally induced stresses in thermostat-like assemblies of finite size, such as for assemblies typical in electronics and optical engineering. This has been done by introducing a concept of interfacial compliance. This concept enables to separate the roles of loading, not even necessarily thermal, and structural characteristics of the assembly, and, owing to that, obtain simple and physically meaningful closed-form solutions for the stresses and strains, including interfacial ones. No singularities appear in the obtained expressions. This is because the concept of the interfacial compliance, unlike the elasticity theory, whether analytical or finite element analysis (FEA)–based, is an approximate structural analysis (strength of materials) method and does not require that all the theory-of-elasticity equations and conditions of this theory are fulfilled in every point of the body under stress. Some of the rather numerous problems successfully solved using the method of interfacial compliance include thermal stresses in thin films [23, 24], in trimaterial assemblies [25–27], in coated optical fibers [28–31], and in assemblies with seal glass bonds, whose coefficient of thermal expansions (CTEs) have to be treated as random variables [32]; problems, in which the probabilistic design for reliability (PDfR) concept is employed [33]; assemblies with power cores sandwiched between dissimilar insulated metal substrates [34]; and delamination (interfacial fracture) problems [35]. Mishkevich and Suhir [36] established good agreement between the analytical modeling data and FEA predictions in various thermally induced stress problems, and have shown that although the interfacial shearing and peeling stresses are coupled, a simplified approach can be used, when analytical model is employed: the shearing stress can be evaluated in an approximate analysis without considering its coupling with peeling, and that the peeling stresses can be evaluated with sufficient accuracy (confirmed by FEA modeling) from the computed shearing stress. It has been shown, particularly, that the longitudinal gradient of the shearing stress plays the role of the excitation force when the peeling stress is evaluated.

An overview of the substance, attributes, and applications of the method of the interfacial compliance was recently published in a special issue dedicated to the 90th birthday of Prof. Richard Hetnarski, editor of the Taylor and Francis Group *Journal of Thermal Stresses* [37]. A modification of the method of interfacial compliance is set forth below. In addition to the general theory of thermal stresses in bimaterial and trimaterial assemblies, we also show, having in mind soldered assemblies, how the concept of interfacial compliance could be used in the interfacial fracture (delamination) problem [35].

2.2 STRESSES IN THE MIDPORTION OF A MULTIMATERIAL BODY SUBJECTED TO A CHANGE IN TEMPERATURE

Let a multimaterial assembly (body) consisting of n components be fabricated at an elevated temperature and subsequently cooled down to a low (room or testing) temperature. The thermally induced strain, ε, at the state of equilibrium should be the same for all the components, and could be evaluated for the ith component as follows:

$$\varepsilon = -\alpha_i \Delta t + \lambda_i T_i^0, \quad i = 1, 2, 3, \ldots, n. \tag{2.1}$$

Here, Δt is the change in temperature, α_i is the CTE of the ith component's material, T_i is the thermally induced force acting in the component's cross sections,

$$\lambda_i = \frac{1}{E_i^0 h_i}, \quad i = 1,2,3,...,n \tag{2.2}$$

is the longitudinal (axial) compliance of the component, $E_i^0 = \frac{E_i}{1-\nu_i}$ is the effective Young's modulus of the material, E_i is its Young's modulus, and ν_i is Poisson's ratio. The condition

$$\sum_{i=1}^{n} T_i^0 = 0 \tag{2.3}$$

of equilibrium results in the following formula for the induced strain:

$$\varepsilon = -\alpha_e \Delta t, \tag{2.4}$$

where

$$\alpha_e = \frac{\sum_{i=1}^{n} \alpha_i E_i^0 h_i}{\sum_{i=1}^{n} E_i^0 h_i} \tag{2.5}$$

is an effective CTE of the assembly. The first term in formula (2.1) is unrestricted thermal contraction and the second term is the elastic extension evaluated in accordance with Hooke's law.

Formulas (2.1) and (2.4) lead to the following expression for the induced forces T_i acting in the materials' cross sections:

$$T_i^0 = E_i^0 h_i (\alpha_i - \alpha_e) \Delta t. \tag{2.6}$$

The corresponding stresses are

$$\sigma_i^0 = \frac{T_i^0}{h_i} = E_i^0 (\alpha_i - \alpha_e) \Delta t \tag{2.7}$$

and are component thickness independent. This stress depends on the generalized Young's modulus of the material (i.e., Young's modulus with consideration of its Poisson's ratio), the thermal mismatch of the given material with the effective CTE of the assembly, and the change in temperature.

In a situation when one of the components of the assembly is significantly stiffer (thicker, and/or has a much higher Young's modulus) than the other components (such as, say, substrate in a thin film structure), then, as follows from formula (2.5), the effective CTE, α_e, of the assembly is simply the CTE of the stiff component's material. In this case, it is only the mismatch of the given material with this material that determines the induced force and stress in the material, and not its mismatch with the adjacent materials.

2.3 BIMATERIAL ASSEMBLY: INTERFACIAL SHEARING STRESSES

In the case of a bimaterial assembly, formulas (2.5) and (2.6) yield:

$$T_1^0 = -T_2^0 = (\alpha_1 - \alpha_2)\frac{\Delta t}{\lambda}, \tag{2.8}$$

where

$$\lambda = \lambda_1 + \lambda_2 = \frac{1}{E_1^0 h_1} + \frac{1}{E_2^0 h_2} \tag{2.9}$$

is the total longitudinal (axial) compliance of the assembly.

To determine how the thermally induced forces are distributed along an assembly of finite length, we seek these forces in the form

$$T_{1,2}(x) = T_{1,2}^0 \chi(x). \tag{2.10}$$

The function $\chi(x)$ has to be determined. To do that, we use the following approximate formulas for the interfacial longitudinal displacements:

$$u_1(x) = -\alpha_1 \Delta t x + \lambda_1 \int_0^x T_1(\xi) d\xi - \kappa_1 \tau(x),$$

$$u_2(x) = -\alpha_2 \Delta t x + \lambda_2 \int_0^x T_2(\xi) d\xi + \kappa_2 \tau(x). \tag{2.11}$$

The origin of the coordinate x is in the mid-cross section of the assembly.

The first terms in formulas (2.11) are stress-free thermal displacements. The second terms are determined using Hooke's law, are due to the thermally induced forces, and reflect an assumption that the displacements of all the points in the given cross section are the same. The third terms are, in effect, corrections to this assumption and account for the fact that the interfacial displacements are somewhat larger than the displacements of the inner points of the given cross section. The structure of these terms reflects an assumption that sought corrections can be computed as a product of the interfacial compliance, κ_1 or κ_2, of the corresponding component and the interfacial shearing stress, $\tau(x)$, in the given cross section. Formulas for the evaluation of the interfacial compliances have been suggested in [7–9]. These formulas are based on the theory of elasticity solution of the Ribiére problem for a long-and-narrow strip and depend on how the strip is loaded [7–9, 18]. If the strip is loaded on its long edges by equal distributed forces applied in an antisymmetric fashion (in- or outward), its interfacial compliance is $\kappa = \frac{h}{G}$, where h is the height (thickness) of the strip, and G is the shear modulus of its material. This is typically the adhesive/solder layer. If the strip is loaded only on one of its long edges in an antisymmetric fashion, while the opposite long edge is loading free, its interfacial compliance is only $\kappa = \frac{h}{3G}$. These are typically the adherends in thermally mismatched assemblies.

The interfacial shearing stress $\tau(x)$ is related to the forces $T_{1,2}(x)$ acting in the cross sections of the bonded components as follows:

$$T_{1,2}(x) = \pm \int_{-a}^{x} \tau(\xi)d\xi, \tag{2.12}$$

where a is half the assembly length. The condition of the compatibility of the displacements (2.11) can be written as

$$u_1(x) = u_2(x) + \kappa_0 \tau(x), \tag{2.13}$$

where κ_0 is the interfacial compliance of the bonding layer, if any. If no separate material is used as the bonding layer (e.g., when thermo-compression bonding is employed), then the interfacial compliance of the assembly is due only to the bonded components themselves, and the second term in the right part of formula (2.13) should be omitted.

Introducing formula (2.11) for the interfacial displacements into the condition (2.13) and considering formulas (2.10) and (2.12) for the induced forces, we obtain the following equation for the sought function $\chi(x)$ that accounts for the nonuniform distribution of the forces acting in the cross sections of the assembly components:

$$\chi'(x) - k^2 \int_{0}^{x} \chi(\xi)d\xi = -k^2 x. \tag{2.14}$$

Here,

$$k = \sqrt{\frac{\lambda}{\kappa}} \tag{2.15}$$

is the parameter of the interfacial shearing stress, and

$$\kappa = \kappa_0 + \kappa_1 + \kappa_2 \tag{2.16}$$

is the total interfacial compliance of the assembly. In typical adhesively bonded or soldered assemblies, in which the thickness and Young's modulus of the bonding layer are significantly smaller than those of the bonded components, the axial compliance λ of the assembly is due primarily to the bonded components. Its interfacial compliance, however, is due to both the adherend and the bonding materials. This is because these compliances are defined as ratios of the component thicknesses to their shear moduli, and these ratios might be very well comparable for the adherends and the adhesive.

The integral equation (2.14) can be reduced, by differentiation, to the following differential equation:

$$\chi''(x) - k^2\chi(x) = -k^2. \tag{2.17}$$

Its solution

$$\chi(x) = 1 - \frac{\cosh kx}{\cosh ka} \tag{2.18}$$

indicates that the thermally induced forces $T_{1,2}(x)$ are symmetric with respect to the mid-cross section of the assembly.

The zero boundary conditions $T_{1,2}(\pm a) = 0$ for the sought forces lead to the conditions $\chi(\pm a) = 0$ for the function $\chi(x)$. From (2.10), (2.12), and (2.18) we obtain the following simple formula for the thermally induced interfacial shearing stress:

$$\tau(x) = T_1'(x) = T_1^0\chi'(x) = -kT_1^0\frac{\sinh kx}{\cosh ka} = -k(\alpha_1 - \alpha_2)\frac{\Delta t}{\lambda}\frac{\sinh kx}{\cosh ka}. \tag{2.19}$$

The maximum interfacial shearing stresses

$$\tau_{\max} = \tau(a) = -k\frac{\Delta\alpha\Delta t}{\lambda}\tanh ka \tag{2.20}$$

occur at the end cross sections $x = \pm a$. and, as evident from (2.20), change from

$$\tau_{\max} = \tau(a) = -k^2\frac{\Delta\alpha\Delta t}{\lambda}a = -\frac{\Delta\alpha\Delta t}{\kappa}a \tag{2.21}$$

in the case of a short assembly to

$$\tau_{\max} = \tau(a) = -k\frac{\Delta\alpha\Delta t}{\lambda} = -\frac{\Delta\alpha\Delta t}{\sqrt{\lambda\kappa}} \tag{2.22}$$

in the case of a sufficiently long assembly, when $\tanh ka$ can be put equal to one. As evident from equations (2.21) and (2.22), the maximum shearing stress $\tau_{\max}$ at the ends of a bimaterial assembly subjected to the change in temperature is proportional to the assembly length, in the case of a short assembly, and becomes assembly length independent for assemblies characterized by high ka values. Equations (2.21) and (2.22) also show that this stress is inversely proportional to the interfacial compliance of the assembly and is independent of its axial compliance, while, for a long and stiff assembly, it is inversely proportional to the square root of the axial and interfacial compliances.

2.4 BIMATERIAL ASSEMBLY: INTERFACIAL PEELING STRESSES

To determine the interfacial peeling stress $p(x)$, we proceed from the following equations of equilibrium:

$$\int_{-a}^{x}\int_{-a}^{\xi} p(\xi)d\xi dx = D_1 w_1''(x) + \frac{h_1}{2}T_1(x) = -D_2 w_2''(x) + \frac{h_2}{2}T_2(x). \tag{2.23}$$

Here $w_1(x)$ and $w_2(x)$ are the deflections of the assembly components,

$$D_1 = \frac{E_1 h_1^3}{12\left(1-\nu_1^2\right)}, \quad D_2 = \frac{E_2 h_2^3}{12\left(1-\nu_2^2\right)} \tag{2.24}$$

are their flexural rigidities (the components are treated here as elongated rectangular thin plates), and $T_{1,2}(x)$ are the thermally induced forces acting in their cross sections and related to the interfacial shearing stress by formula (2.12). The left part of equation (2.20) is the bending moment caused by the peeling load $p(x)$. The first terms in the right parts of these equations are bending moments caused by the elastic bending forces. The second terms in the right parts are the moments caused by the thermally induced forces $T_1(x)$ and $T_2(x)$ acting in the cross sections of the assembly components.

We assume that the peeling stress $p(x)$ is related to the deflection $w_1(x)$ and $w_2(x)$ as

$$p(x) = K[w_2(x) - w_1(x)], \tag{2.25}$$

where K is the through-thickness interfacial spring constant. In an approximate analysis, this spring constant can be evaluated by the formula:

$$K = \frac{1}{C_1 + C_2}, \tag{2.26}$$

where

$$C_1 = \frac{1-\nu_1}{3E_1}h_1, \quad C_2 = \frac{1-\nu_2}{3E_2}h_2 \tag{2.27}$$

are the through-thickness compliances of the assembly components. Formula (2.25) indicates that no peeling stress can possibly occur in a cross section, where both components of the assembly have the same deflections. From (2.23) and (2.25) we obtain the following differential equation for the peeling stress function, $p(x)$:

$$p^{IV}(x) + 4\beta^4 p(x) = 2\beta^4 h_{12}\tau_1'(x), \tag{2.28}$$

where

$$\beta = \sqrt[4]{K\frac{D_1 + D_2}{4D_1D_2}} \tag{2.29}$$

is the parameter of the interfacial peeling stress,

$$h_{12} = \frac{h_1D_2 + h_2D_1}{D_1 + D_2} \tag{2.30}$$

is the effective thickness of the assembly, and the derivative $\tau_1'(x)$ of the interfacial shearing stress function, $\tau_1(x)$, is expressed as

$$\tau_1'(x) = -k^2T_1^0\frac{\cosh kx}{\cosh ka}. \tag{2.31}$$

In the taken approach, the interfacial shearing stress is evaluated assuming that it is not affected by the interfacial peeling stress and that the peeling stress can be evaluated based on the determined magnitude and the distribution of the shearing stress. The validity of this assumption has been confirmed by the FEA carried out for assemblies with comparative thickness of its components.

The equation of the type (2.28) is known in the engineering theory of beams lying on continuous elastic foundations (see, for example, [18, 38]), where a similar equation was obtained, however, for the deflection functions, and not for the interfacial peeling stress.

The particular solution to the inhomogeneous equation (2.28) can be sought in the form

$$p_{p.s.}(x) = -A\frac{\cosh kx}{\cosh ka}. \tag{2.32}$$

Introducing this solution into equation (2.28) we find

$$A = \frac{\varsigma^4}{1+\varsigma^4}k^2T_1^0\frac{h_{12}}{2}, \tag{2.33}$$

where the ratio

$$\varsigma = \frac{\beta\sqrt{2}}{k} \tag{2.34}$$

considers the relative role of the parameters of the interfacial shearing and peeling stresses. The general solution to equation (2.28) can be sought as follows:

$$p(x) = C_0 V_0(\beta x) + C_2 V_2(\beta x) - A\frac{\cosh kx}{\cosh ka}, \tag{2.35}$$

where C_0 and C_2 are constants of integration, and the functions $V_0(\beta x)$ and $V_2(\beta x)$ are expressed as

$$V_0(\beta x) = \cosh\beta x\cos\beta x, \quad V_2(\beta x) = \sinh\beta x\sin\beta x. \tag{2.36}$$

These functions obey the following simple rules of differentiation:

$$\begin{aligned} V_0'(\beta x) &= -\beta\sqrt{2}V_3(\beta x), \quad & V_1'(\beta x) &= \beta\sqrt{2}V_0(\beta x), \\ V_2'(\beta x) &= \beta\sqrt{2}V_1(\beta x), \quad & V_3'(\beta x) &= \beta\sqrt{2}V_2(\beta x). \end{aligned} \tag{2.37}$$

The functions $V_1(\beta x)$ and $V_3(\beta x)$ are

$$V_{1,3}(\beta x) = \frac{1}{\sqrt{2}}(\cosh\beta x\sin\beta x \pm \sinh\beta x\cos\beta x) \tag{2.38}$$

The first two terms in solution (2.35) represent the general solution to the homogeneous equation corresponding to equation (2.28).

Since there are no external forces acting on the assembly, the peeling stress function, $p(x)$, must be self-equilibrated, and, since this function should be also symmetric with respect to the origin, the following conditions of equilibrium should be fulfilled:

$$\int_0^a\int_0^x p(\xi)d\xi dx = 0, \quad \int_0^a p(x)dx = 0 \tag{2.39}$$

The first condition indicates that the bending moment at the ends of the assembly should be zero, and the second condition indicates that the shearing force acting in the through-thickness direction of the assembly should be zero as well. These conditions result in the following equations for the constants of integration:

$$V_1(\beta a)C_0 + V_3(\beta a)C_2 = \varsigma A\tanh ka,$$

$$V_2(\beta a)C_0 - [V_0(\beta a) - 1]C_2 = \varsigma^2 A\left(1 - \frac{1}{\cosh ka}\right) \tag{2.40}$$

These equations yield:

$$C_0 = \varsigma A \frac{[V_0(\beta a)-1]\tanh ka + \varsigma V_3(\beta a)\left(1-\frac{1}{\cosh ka}\right)}{V_1(\beta a)[V_0(\beta a)-1]+V_2(\beta a)V_3(\beta a)},$$

$$C_2 = \varsigma A \frac{V_2(\beta a)\tanh ka - \varsigma V_1(\beta a)\left(1-\frac{1}{\cosh ka}\right)}{V_1(\beta a)[V_0(\beta a)-1]+V_2(\beta a)V_3(\beta a)}. \tag{2.41}$$

The peeling stress is therefore expressed as follows:

$$p(x) = \varsigma A \frac{[V_0(\beta a)-1]\tanh ka + \varsigma V_3(\beta a)\left(1-\frac{1}{\cosh ka}\right)}{V_1(\beta a)[V_0(\beta a)-1]+V_2(\beta a)V_3(\beta a)} V_0(\beta x) +$$
$$+\varsigma A \frac{V_2(\beta a)\tanh ka - \varsigma V_1(\beta a)\left(1-\frac{1}{\cosh ka}\right)}{V_1(\beta a)[V_0(\beta a)-1]+V_2(\beta a)V_3(\beta a)} V_2(\beta x) - A\frac{\cosh kx}{\cosh ka} \tag{2.42}$$

For sufficiently large βa values, such as for long assemblies with stiff (in the through-thickness direction) interfaces (say, $\beta a \triangleright 2.5$), formulas (2.41) can be simplified as

$$C_0 = 2\varsigma A e^{-\beta a}(\sqrt{2}\cos\beta a + \varsigma(\sin\beta a - \cos\beta a)),$$
$$C_2 = 2\varsigma A e^{-\beta a}(\sqrt{2}\sin\beta a - \varsigma(\sin\beta a + \cos\beta a)) \tag{2.43}$$

and solution (2.39) can be simplified as

$$p(x) = \varsigma A e^{-\beta(a-x)}[(\sqrt{2}-\varsigma)\cos\beta(a-x) + \varsigma\sin\beta(a-x)] - Ae^{-k(a-x)} \tag{2.44}$$

Hence, the maximum value of the peeling stress that occurs at the ends of a long assembly with a stiff interface is

$$p(a) = A[\varsigma(\sqrt{2}-\varsigma)-1] = \frac{\varsigma^4}{1+\varsigma^4}[\varsigma(\sqrt{2}-\varsigma)-1]k^2 T_1^0 \frac{h_{12}}{2}. \tag{2.45}$$

2.5 TRIMATERIAL ASSEMBLY: INTERFACIAL SHEARING STRESSES

In the case of a trimaterial assembly, formulas (2.5) and (2.6) yield:

$$T_1^0 = \Delta t \frac{\alpha_1(\lambda_2+\lambda_3) - \alpha_2\lambda_3 - \alpha_3\lambda_2}{\lambda_1\lambda_2+\lambda_2\lambda_3+\lambda_3\lambda_1} = \Delta t \frac{\lambda_2(\alpha_1-\alpha_3)+\lambda_3(\alpha_1-\alpha_2)}{\lambda_1\lambda_2+\lambda_2\lambda_3+\lambda_3\lambda_1},$$

$$T_2^0 = \Delta t \frac{\alpha_2(\lambda_1+\lambda_3)-\alpha_1\lambda_3-\alpha_3\lambda_1}{\lambda_1\lambda_2+\lambda_2\lambda_3+\lambda_3\lambda_1} = \Delta t \frac{\lambda_1(\alpha_2-\alpha_3)+\lambda_3(\alpha_2-\alpha_1)}{\lambda_1\lambda_2+\lambda_2\lambda_3+\lambda_3\lambda_1},$$

$$T_3^0 = \Delta t \frac{\alpha_3(\lambda_1+\lambda_2)-\alpha_1\lambda_2-\alpha_2\lambda_1}{\lambda_1\lambda_2+\lambda_2\lambda_3+\lambda_3\lambda_1} = \Delta t \frac{\lambda_1(\alpha_3-\alpha_2)+\lambda_2(\alpha_3-\alpha_1)}{\lambda_1\lambda_2+\lambda_2\lambda_3+\lambda_3\lambda_1}, \quad (2.46)$$

and the interfacial displacements can be sought in the following approximate forms:

$$u_1(x) = -\alpha_1\Delta tx + \lambda_1 \int_0^x T_1(\xi)d\xi - \kappa_1\tau_1(x),$$

$$u_{21}(x) = -\alpha_2\Delta tx - \lambda_2 \int_0^x [T_1(\xi)+T_3(\xi)]d\xi + \kappa_2\tau_1(x)$$

$$u_{23}(x) = -\alpha_2\Delta tx - \lambda_2 \int_0^x [T_1(\xi)+T_3(\xi)]d\xi + \kappa_2\tau_2(x),$$

$$u_3(x) = -\alpha_3\Delta tx + \lambda_3 \int_0^x T_3(\xi)d\xi - \kappa_3\tau_2(x) \quad (2.47)$$

Here, $u_1(x)$ are the interfacial displacements of the component #1 at its boundary with the intermediate component #2; $u_{21}(x)$ are the interfacial displacements of the component #2 at its boundary with the component #1; $u_{23}(x)$ are the interfacial displacements of the component #2 at its boundary with the component #3; $u_3(x)$ are the interfacial displacements of the component #3 at its boundary with the component #2; $T_1(x), T_2(x)$, and $T_3(x)$ are the thermally induced forces acting in the cross sections of the components #1, #2, and #3, respectively; $\tau_1(x)$ is the interfacial shearing stress at the boundary between the components #1 and #2; $\tau_2(x)$ is the interfacial shearing stress at the boundary between the components #2 and #3; and κ_1, κ_2, and κ_3 are the interfacial compliances of the components. In these expressions for the interfacial displacements of the component #2, we have considered that the three thermally induced forces have to be in equilibrium, and therefore the condition

$$T_2(x) = -[T_1(x)+T_3(x)] \quad (2.48)$$

must take place. We seek the forces $T_1(x), T_2(x)$, and $T_3(x)$ in the form of equation (2.10), such as

$$T_1(x) = T_1^0\chi(x), \quad T_2(x) = T_2^0\chi(x), \quad T_3(x) = T_3^0\chi(x), \quad (2.49)$$

and require that the conditions

$$u_{21}(x) = u_1(x), \quad u_{23}(x) = u_3(x) \quad (2.50)$$

of the displacement compatibility are fulfilled. If there are compliant layers employed at the interfaces between the assembly components, then the conditions (2.50) should be replaced by the conditions of the condition (2.16) type.

The conditions (2.50) of the displacement compatibility require, considering the relationships (2.18) and (2.19), that the following homogeneous equations for the forces T_1^0 and T_3^0 are fulfilled:

$$[(\kappa_1+\kappa_2)k^2-(\lambda_1+\lambda_2)]T_1^0-\lambda_2 T_3^0=0,$$
$$-\lambda_2 T_1^0+[(\kappa_2+\kappa_3)k^2-(\lambda_2+\lambda_3)]T_3^0=0 \tag{2.51}$$

The forces T_1^0 and T_3^0 cannot be zero, and therefore the determinant of equations (2.51) must be zero. This leads to the following biquadratic equation for the parameter k of the interfacial shearing stress:

$$k^4-(k_{12}^2+k_{23}^2)k^2+\eta k_{12}^2 k_{23}^2=0, \tag{2.52}$$

where the following notation is used:

$$k_{12}=\sqrt{\frac{\lambda_1+\lambda_2}{\kappa_1+\kappa_2}},\quad k_{23}=\sqrt{\frac{\lambda_2+\lambda_3}{\kappa_2+\kappa_3}},\quad \eta=\frac{\lambda_1\lambda_2+\lambda_2\lambda_3+\lambda_3\lambda_1}{(\lambda_1+\lambda_2)(\lambda_2+\lambda_3)} \tag{2.53}$$

Equation (2.52) results in the following expression for the interfacial shearing stress parameter:

$$k=\sqrt{\frac{k_{12}^2+k_{23}^2}{2}\left[1+\sqrt{1-\eta\left(\frac{2k_{12}k_{23}}{k_{12}^2+k_{23}^2}\right)^2}\right]} \tag{2.54}$$

When $k_{23}=0$ (the interface #2 is infinitely compliant, i.e., simply does not exist), $k=k_{12}$. When $k_{12}=0$ (the interface #2 is infinitely compliant, i.e., does not exist), $k=k_{23}$. When all the assembly components are identical (only the CTEs might be different), formula (2.54) yields: $k=k_{12}\sqrt{\frac{3}{2}}=1.2247k_{12}$. Thus, trimaterial assemblies are characterized by higher parameters of the interfacial shearing stress than bimaterial assemblies.

After the parameter k of the interfacial shearing stresses is determined, the interfacial shearing stresses $\tau_1(x)$ and $\tau_2(x)$ can be evaluated as

$$\tau_1(x)=T_1'(x)=T_1^0\chi'(x)=-kT_1^0\frac{\sinh kx}{\cosh ka},$$
$$\tau_2(x)=T_3'(x)=T_3^0\chi'(x)=-kT_3^0\frac{\sinh kx}{\cosh ka} \tag{2.55}$$

2.6 TRIMATERIAL ASSEMBLY: INTERFACIAL PEELING STRESSES

In the case of a trimaterial assembly, the equations of equilibrium for the peeling stresses $p_1(x)$ and $p_2(x)$ acting in the interfaces #1 and #2, are as follows:

$$\int_{-a}^{x}\int_{-a}^{\xi} p_1(\xi)d\xi dx = D_1 w_1''(x) + \frac{h_1}{2}T_1(x),$$

$$-\int_{-a}^{x}\int_{-a}^{\xi} p_1(\xi)d\xi dx + \int_{-a}^{x}\int_{-a}^{\xi} p_2(\xi)d\xi dx = D_2 w_{2,}''(x) - \frac{h_2}{2}[T_1(x) + T_3(x)],$$

$$\int_{-a}^{x}\int_{-a}^{\xi} p_2(\xi)d\xi dx = D_3 w_{3.}''(x) + \frac{h_3}{2}T_3(x). \tag{2.56}$$

Note that the curvatures of the assembly components depend on both the interfacial peeling stresses and the axial forces, such as on the interfacial shearing stresses. Even if the peeling stresses $p_1(x)$ and $p_2(x)$ are the same, the curvature of the component #2 might not be zero, but will depend on the level of the axial forces (and interfacial shearing stresses).

We assume that the peeling stresses $p_1(x)$ and $p_2(x)$ are related to the deflection functions $w_1(x)$, $w_2(x)$, and $w_3(x)$ of the assembly components by the equations:

$$p_1(x) = K_1[w_2(x) - w_1(x)], \quad p_2(x) = K_2[w_2(x) - w_3(x)] \tag{2.57}$$

where K_1 and K_2 are spring constants of the interfaces in the through-thickness direction. Solving equations (2.56) for the corresponding deflection functions, substituting the obtained expressions into the relationships (2.57) and differentiating the obtained relationships twice with respect to the coordinate x, we obtain the following equations for the peeling stress functions acting at the interfaces #1 and #2:

$$p_1^{IV}(x) + 4\beta_1^4 p_1(x) - \frac{K_1}{D_2}p_2(x) = -k^6\left(\varsigma_1^4\frac{h_{12}}{2}T_1^0 + \xi_1^4\frac{h_2}{2}T_3^0\right)\frac{\cosh kx}{\cosh ka}$$

$$p_2^{IV}(x) + 4\beta_2^4 p_2(x) - \frac{K_2}{D_2}p_1(x) = -k^6\left(\varsigma_2^4\frac{h_{23}}{2}T_3^0 + \xi_2^4\frac{h_2}{2}T_1^0\right)\frac{\cosh kx}{\cosh ka} \tag{2.58}$$

where

$$\beta_1 = \sqrt[4]{K_1\frac{D_1 + D_2}{4D_1D_2}}, \quad \beta_2 = \sqrt[4]{K_2\frac{D_2 + D_3}{4D_2D_3}} \tag{2.59}$$

are the parameters of the interfacial peeling stresses at the interfaces #1 and #2, respectively, and the following notation is used:

$$\varsigma_{1,2} = \frac{\beta_{1,2}\sqrt{2}}{k}, \quad \xi_{1,2} = \frac{1}{k}\sqrt[4]{\frac{K_{1,2}}{D_2}}, \quad h_{23} = \frac{h_2D_3 + h_3D_2}{D_2 + D_3} \tag{2.60}$$

We seek the particular solutions to equations (2.58) as

$$p_{1,p.s.}(x) = A_1 \frac{\cosh kx}{\cosh ka}, \quad p_{2,p.s.}(x) = A_2 \frac{\cosh kx}{\cosh ka}. \tag{2.61}$$

Introducing these solutions into equations (2.55), we obtain the following equations for the constants A_1 and A_2

$$\begin{aligned} (1+\varsigma_1^4)A_1 - \xi_1^4 A_2 &= -k^2\left(\varsigma_1^4 \frac{h_{12}}{2} T_1^0 + \xi_1^4 \frac{h_2}{2} T_3^0\right), \\ -\xi_2^4 A_1 + (1+\varsigma_2^4)A_2 &= -k^2\left(\varsigma_2^4 \frac{h_{23}}{2} T_3^0 + \xi_2^4 \frac{h_2}{2} T_1^0\right). \end{aligned} \tag{2.62}$$

Equations (2.62) yield:

$$\begin{aligned} A_1 &= -k^2 \frac{\left[\varsigma_1^4(1+\varsigma_2^4)\frac{h_{12}}{2} + \xi_1^4\xi_2^4 \frac{h_2}{2}\right]T_1^0 + \xi_1^4\left[\varsigma_2^4 \frac{h_{23}}{2} + (1+\varsigma_2^4)\frac{h_2}{2}\right]T_3^0}{(1+\varsigma_1^4)(1+\varsigma_2^4) - \xi_1^4\xi_2^4} \\ A_2 &= -k^2 \frac{\xi_2^4\left[\varsigma_1^4 \frac{h_{12}}{2} + (1+\varsigma_1^4)\frac{h_2}{2}\right]T_1^0 + \left[\varsigma_2^4(1+\varsigma_1^4)\frac{h_{23}}{2} + \xi_1^4\xi_2^4 \frac{h_2}{2}\right]T_3^0}{(1+\varsigma_1^4)(1+\varsigma_2^4) - \xi_1^4\xi_2^4} \end{aligned} \tag{2.63}$$

We seek the solutions to equations (2.58) as follows:

$$\begin{aligned} p_1(x) &= C_{01}\cosh\gamma_1 x \cos\gamma_2 x + C_{21}\sinh\gamma_1 x \sin\gamma_2 x - A_1 \frac{\cosh kx}{\cosh ka} \\ p_2(x) &= C_{02}\cosh\gamma_1 x \cos\gamma_2 x + C_{22}\sinh\gamma_1 x \sin\gamma_2 x - A_2 \frac{\cosh kx}{\cosh ka} \end{aligned} \tag{2.64}$$

Introducing the homogeneous parts of these solutions into equations (2.58) and requiring that the constants of integration in these solutions are nonzero, we obtain the following formulas for the parameters γ_1 and γ_2:

$$\gamma_{1,2} = \sqrt{\frac{\sqrt{\mu_1^2 + \mu_2^2} \pm \mu_2}{2}}, \tag{2.65}$$

where

$$\mu_{1,2} = \sqrt{(\beta_1^4 + \beta_2^4) \pm 2\beta_1^2\beta_2^2\sqrt{1+\delta^2}}. \tag{2.66}$$

and

$$\delta = \sqrt{\frac{D_1 D_3}{(D_1 + D_2)(D_2 + D_3)}} \tag{2.67}$$

is the parameter of the flexural rigidities. Note that in the case of a bimaterial assembly ($D_1 = 0$ or $D_3 = 0$) the parameter δ is zero. In the case when one of the components (say, #1 or #3) is significantly more rigid than the other two (this is the case of a thin film system fabricated on a thick substrate, or a substrate/IC system attached to a heat sink), the parameter δ becomes independent of the rigid component's flexural rigidity. This parameter is either $\delta = \sqrt{\frac{D_1}{D_1 + D_2}}$ or $\delta = \sqrt{\frac{D_3}{D_2 + D_3}}$. If the flexural rigidity of all the assembly components is the same, then $\delta = \frac{1}{2}$. Note also that, for a bimaterial assembly, when the parameter δ is zero, $\mu_1 = 2\beta^2$, $\mu_2 = 0$, and $\gamma_1 = \gamma_2 = \beta$, as it is supposed to be.

The constants of integration in solutions (2.64) can be determined from the self-equilibrium conditions for the peeling stress functions. We use the following formulas that can be easily obtained by integration by parts:

$$\int \cosh\gamma_1 x \cos\gamma_2 x dx = \frac{\gamma_1 \sinh\gamma_1 x \cos\gamma_2 x + \gamma_2 \cosh\gamma_1 x \sin\gamma_2 x}{\gamma_1^2 + \gamma_2^2}$$

$$\int \sinh\gamma_1 x \sin\gamma_2 x dx = \frac{\gamma_1 \cosh\gamma_1 x \sin\gamma_2 x - \gamma_2 \sinh\gamma_1 x \cos\gamma_2 x}{\gamma_1^2 + \gamma_2^2}$$

$$\int \sinh\gamma_1 x \cos\gamma_2 x dx = \frac{\gamma_1 \cosh\gamma_1 x \cos\gamma_2 x + \gamma_2 \sinh\gamma_1 x \sin\gamma_2 x}{\gamma_1^2 + \gamma_2^2}$$

$$\int \cosh\gamma_1 x \sin\gamma_2 x dx = \frac{\gamma_1 \sinh\gamma_1 x \sin\gamma_2 x - \gamma_2 \cosh\gamma_1 x \cos\gamma_2 x}{\gamma_1^2 + \gamma_2^2}$$

$$\int\int \cosh\gamma_1 x \cos\gamma_2 x dx dx = \frac{(\gamma_1^2 - \gamma_2^2)\cosh\gamma_1 x \cos\gamma_2 x + 2\gamma_1\gamma_2 \sinh\gamma_1 x \sin\gamma_2 x}{(\gamma_1^2 + \gamma_2^2)^2}$$

$$\int\int \sinh\gamma_1 x \sin\gamma_2 x dx dx = \frac{(\gamma_1^2 - \gamma_2^2)\sinh\gamma_1 x \sin\gamma_2 x - 2\gamma_1\gamma_2 \cosh\gamma_1 x \cos\gamma_2 x}{(\gamma_1^2 + \gamma_2^2)^2} \qquad (2.68)$$

Applying these formulas to the solution for the peeling stress function, $p_1(x)$, in (2.64), we obtain the following equations for the constants C_{01} and C_{21} of integration:

$$\begin{aligned}
&(\gamma_1 \sinh u_1 \cos u_2 + \gamma_2 \cosh u_1 \sin u_2)C_{01} + \\
&+ (\gamma_1 \cosh u_1 \sin u_2 - \gamma_2 \sinh u_1 \cos u_2)C_{21} = kA_1\varsigma_1\varsigma_2\sqrt[4]{1+\delta^2}, \\
&[(\gamma_1^2 - \gamma_2^2)(\cosh u_1 \cos u_2 - 1) + 2\gamma_1\gamma_2 \sinh u_1 \sin u_2]C_{01} + \\
&+ [(\gamma_1^2 - \gamma_2^2)\sinh u_1 \sin u_2 - 2\gamma_1\gamma_2(\cosh u_1 \cos u_2 - 1)]C_{21} = k^2 A_1\varsigma_1^2\varsigma_2^2\sqrt{1+\delta^2}.
\end{aligned} \qquad (2.69)$$

Here $u_1 = \gamma_1 a$ and $u_1 = \gamma_2 a$. The determinant of equations (2.69) is

$$\begin{aligned}
\Delta &= (\gamma_1^2 + \gamma_2^2)(\cosh u_1 - \cos u_2)(\gamma_1 \sin u_2 - \gamma_2 \sinh u_1) = \\
&= -\sqrt{\mu_1^2 + \mu_2^2}(\cosh u_1 - \cos u_2)(\gamma_2 \sinh u_1 - \gamma_1 \sin u_2)
\end{aligned} \qquad (2.70)$$

Then equations (2.69) yield:

$$C_{01} = -\frac{kA_1\varsigma_1\varsigma_2\sqrt[4]{1+\delta^2}}{\Delta}[\mu_1(\cosh u_1 \cos u_2 - 1) - \mu_2 \sinh u_1 \sin u_2 +$$
$$+ k\varsigma_1\varsigma_2\sqrt[4]{1+\delta^2}(\gamma_1 \cosh u_1 \sin u_2 - \gamma_2 \sinh u_1 \cos u_2)]$$
$$C_{21} = -\frac{kA_1\varsigma_1\varsigma_2\sqrt[4]{1+\delta^2}}{\Delta}[\mu_1 \sinh u_1 \sin u_2 + \mu_2(\cosh u_1 \cos u_2 - 1) -$$
$$- k\varsigma_1\varsigma_2\sqrt[4]{1+\delta^2}(\gamma_1 \sinh u_1 \cos u_2 + \gamma_2 \cosh u_1 \sin u_2)] \quad (2.71)$$

Similar expressions can be obtained for the constants C_{02} and C_{22} of integration in the expression for the peeling stress $p_2(x)$: the factor A_1 should be simply replaced in formulas (2.69) by the factor A_2.

For sufficiently long (large a values) and/or stiff (large k and γ_1 values), formulas (2.69) and solutions (2.65) can be simplified, and these solutions yield:

$$p_{1,2}(x) = A_{1,2}\left[\frac{\eta}{\gamma_2\sqrt{\mu_1^2+\mu_2^2}}e^{-\gamma_1(a-x)}[(\mu_1 \cos(\gamma_2(a-x) - \mu_2 \sin(\gamma_2(a-x)) + \right.$$
$$\left. + \eta(\gamma_1 \sin(\gamma_2(a-x) - \gamma_2 \cos(\gamma_2(a-x))] - e^{-k(a-x)}\right], \quad (2.72)$$

where

$$\eta = k\varsigma_1\varsigma_2\sqrt[4]{1+\delta^2}. \quad (2.73)$$

The maximum stresses act at the assembly ends and are as follows:

$$p_{1,2}(a) = A_{1,2}\left[\frac{\eta(\mu_1 - \eta\gamma_2)}{\gamma_2\sqrt{\mu_1^2+\mu_2^2}} - 1\right]. \quad (2.74)$$

2.7 NUMERICAL EXAMPLE

Input Data:

Assembly length: $L = 2a = 20$ mm; Temperature change: $\Delta t = 275°C$;

Component	CTE, α, 1/°C	Young's Modulus, E, kg/mm/sq.	Poisson's Ratio, ν	Thickness, h, mm
#1	2.4E-6	12,500	0.24	0.5
#2	7.0E-6	36,000	0.33	1.0
0	–	12,000	0.33	0.05
#3	16.5E-6	13,000	0.34	2.5

Computed Data:

Axial compliances:

$$\lambda_1 = \frac{1-\nu_1}{E_1 h_1} = \frac{0.76}{12500 \times 0.5} = 1.2160 \times 10^{-4} \text{ mm/kg};$$

$$\lambda_2 = \frac{1-\nu_2}{E_2 h_2} = \frac{0.67}{36000 \times 1.0} = 0.1861 \times 10^{-4} \text{ mm/kg};$$

$$\lambda_3 = \frac{1-\nu_3}{E_3 h_3} = \frac{0.66}{13000 \times 2.5} = 0.2031 \times 10^{-4} \text{ mm/kg};$$

$$\lambda_1\lambda_2 + \lambda_2\lambda_3 + \lambda_3\lambda_1 = 0.2263 \times 10^{-8} + 0.0378 \times 10^{-8} + 0.2470 \times 10^{-8}$$
$$= 0.5111 \times 10^{-8} \text{ mm}^2/\text{kg}^2$$

Parameter of axial compliances

$$\eta = \frac{\lambda_1\lambda_2 + \lambda_2\lambda_3 + \lambda_3\lambda_1}{(\lambda_1 + \lambda_2)(\lambda_2 + \lambda_3)} = \frac{0.5111 \times 10^{-8}}{1.4021 \times 10^{-4} \times 0.3892 \times 10^{-4}} = 0.9366$$

Shear moduli:

$$G_0 = \frac{E_0}{2(1+\nu_0)} = \frac{12000}{2 \times 1.33} = 4511 \text{ kg/mm}^2$$

$$G_1 = \frac{E_1}{2(1+\nu_1)} = \frac{12500}{2 \times 1.24} = 5040 \text{ kg/mm}^2$$

$$G_2 = \frac{E_2}{2(1+\nu_2)} = \frac{36000}{2 \times 1.33} = 13534 \text{ kg/mm}^2$$

$$G_3 = \frac{E_3}{2(1+\nu_3)} = \frac{13000}{2 \times 1.34} = 4851 \text{ kg/mm}^2$$

Shearing compliances:

$$\kappa_1 = \frac{h_1}{3G_1} = \frac{0.5}{3 \times 5040} = 0.3307 \times 10^{-4} \text{ mm}^3/\text{kg}$$

$$\kappa_2 = \frac{h_2}{3G_2} = \frac{1.0}{3 \times 13534} = 0.2463 \times 10^{-4} \text{ mm}^3/\text{kg in a bimaterial assembly}$$

$$\kappa_2 = \frac{h_2}{G_2} = \frac{1.0}{13534} = 0.7389 \times 10^{-4} \text{ mm}^3/\text{kg in a trimaterial assembly}$$

$$\kappa_0 = \frac{h_0}{G_0} = \frac{0.05}{4511} = 0.1108 \times 10^{-4} \text{ mm}^3/\text{kg}$$

$$\kappa_3 = \frac{h_3}{3G_3} = \frac{2.5}{3 \times 4851} = 1.7179. \times 10^{-4} \text{ mm}^3/\text{kg}$$

Total interfacial compliance for a bimaterial assembly is:

$$\kappa = \kappa_1 + \kappa_2 = 0.3307\times10^{-4} + 0.2463\times10^{-4} = 0.5770\times10^{-4}\ \text{mm}^3/\text{kg}$$

For a trimaterial assembly, the longitudinal interfacial compliances are

$$\kappa_{12} = \kappa_1 + \kappa_2 = 0.3307\times10^{-4} + 0.2463\times10^{-4} = 0.5770\times10^{-4}\ \text{mm}^3/\text{kg}$$

$$\kappa_{23} = \kappa_0 + \kappa_2 + \kappa_3 = 0.1108\times10^{-4} + 0.2463\times10^{-4} + 1.7179\times10^{-4} = 2.0750\times10^{-4}\ \text{mm}^3/\text{kg}$$

The parameter of the interfacial shearing stresses is

$$k = \sqrt{\frac{\lambda_1+\lambda_2}{\kappa}} = \sqrt{\frac{1.4021\times10^{-4}}{0.5770\times10^{-4}}} = 1.5588\ \text{mm}^{-1}$$

for the bimaterial assembly. For a trimaterial assembly, we find:

$$k_{12} = \sqrt{\frac{\lambda_1+\lambda_2}{\kappa_{12}}} = \sqrt{\frac{1.4021\times10^{-4}}{0.5770\times10^{-4}}} = 1.5588\ \text{mm}^{-1},$$

$$k_{23} = \sqrt{\frac{\lambda_2+\lambda_3}{\kappa_{23}}} = \sqrt{\frac{0.3892\times10^{-4}}{2.0750\times10^{-4}}} = 0.4331\ \text{mm}^{-1},$$

and the parameter of the interfacial shearing stresses computed for a trimaterial assembly in accordance with formula (2.54) is

$$k = \sqrt{\frac{k_{12}^2+k_{23}^2}{2}\left[1+\sqrt{1-\eta\left(\frac{2k_{12}k_{23}}{k_{12}^2+k_{23}^2}\right)^2}\right]}$$

$$= \sqrt{1.3087\left[1+\sqrt{1+0.9366\times\left(\frac{0.6751}{1.3087}\right)^2}\right]} = 1.6648\ \text{mm}^{-1}$$

This parameter is by 6.8% larger than the parameter of the interfacial shearing stress in a bimaterial assembly.

Through-thickness compliances evaluated for a bimaterial assembly in accordance with formulas (2.27) are:

$$C_1 = \frac{1-\nu_1}{3E_1}h_1 = \frac{0.76\times0.5}{3\times12500} = 0.1013.\times10^{-4}\ \text{mm}^3/\text{kg},$$

$$C_2 = \frac{1-\nu_2}{3E_2}h_2 = \frac{0.67\times1.0}{3\times36000} = 0.0620.\times10^{-4}\ \text{mm}^3/\text{kg}$$

For the trimaterial assembly the through-thickness compliances are:

$$C_2 = \frac{1-\nu_2}{E_2}h_2 = \frac{0.67\times 1.0}{36000} = 0.1861\times 10^{-4}\ \text{mm}^3/\text{kg}$$
$$C_0 = \frac{1-\nu_0}{E_0}h_0 = \frac{0.67\times 0.05}{12000} = 0.0279.\times 10^{-4}\ \text{mm}^3/\text{kg}$$
$$C_3 = \frac{1-\nu_3}{3E_3}h_3 = \frac{0.66\times 2.5}{13000} = 0.4231.\times 10^{-4}\ \text{mm}^3/\text{kg}$$

Through-thickness stiffness for a bimaterial assembly is:

$$K = \frac{1}{C_1+C_2} = \frac{1}{0.1013.\times 10^{-4} + 0.0620.\times 10^{-4}} = 34789.331\ \text{kg/mm}^3$$

Through-thickness stiffnesses for a trimaterial assembly are:

$$K_1 = \frac{1}{C_1+C_2} = \frac{1}{9.8717\times 10^{-4} + 16.1290\times 10^{-4}} = 1.5159\times 10^4\ \text{kg/mm}^3$$
$$K_2 = \frac{1}{C_0+C_2+C_3} = \frac{1}{0.0279\times 10^{-4} + 0.1861\times 10^{-4} + 0.4231\times 10^{-4}}$$
$$= 1.5696\times 10^4\ \text{kg/mm}^3$$

Flexural rigidities:

$$D_1 = \frac{E_1 h_1^3}{12(1-\nu_1^2)} = \frac{12500\times 0.5^3}{12\times 0.942} = 138.2\ \text{kgmm}$$
$$D_2 = \frac{E_2 h_2^3}{12(1-\nu_2^2)} = \frac{36000\times 1.0^3}{12\times 0.8911} = 3366.6\ \text{kgmm}$$
$$D_3 = \frac{E_3 h_3^3}{12(1-\nu_3^2)} = \frac{12000\times 2.5^3}{12\times 0.8844} = 17667.3\ \text{kgmm}$$

Parameter of the flexural rigidities in trimaterial assemblies evaluated by formula (2.67) is as follows:

$$\delta = \sqrt{\frac{D_1 D_3}{(D_1+D_2)(D_2+D_3)}} = \frac{1}{\sqrt{\left(1+\frac{D_2}{D_1}\right)\left(1+\frac{D_2}{D_3}\right)}} = \frac{1}{\sqrt{25.3603\times 1.1906}} = 0.1820$$

If all the assembly components have the same flexural rigidities, then this parameter would be equal to 0.5. If the component #2 were significantly more rigid than the other two, then this parameter would be very small.

Parameters of the peeling stresses in a bimaterial assembly computed by formulas (2.29) and (2.34) are:

$$\beta = \sqrt[4]{K\frac{D_1 + D_2}{4D_1D_2}} = \sqrt[4]{3.4789\times 10^4\frac{3504.8}{4x138.2\times 3366.6}} = 2.8452$$

and

$$\varsigma = \frac{\beta\sqrt{2}}{k} = \frac{4.0237}{1.5588} = 2.5811.$$

For a trimaterial assembly, formulas (2.56) and (2.57) yield:

$$\beta_1 = \sqrt[4]{K_1\frac{D_1 + D_2}{4D_1D_2}} = \sqrt[4]{1.5159\times 10^4\frac{3504.8}{4\times 138.2\times 3366.6}} = 2.3116$$

$$\beta_2 = \sqrt[4]{K_2\frac{D_2 + D_3}{4D_2D_3}} = \sqrt[4]{1.5696\times 10^4\frac{21033.9}{4\times 3366.6\times 17667.3}} = 1.08535$$

$$\varsigma_1 = \frac{\beta_1\sqrt{2}}{k} = \frac{3.2691}{1.6648} = 1.9637, \quad \varsigma_2 = \frac{\beta_2\sqrt{2}}{k} = \frac{1.5349}{1.6648} = 0.9220,$$

$$\xi_1 = \frac{1}{k}\sqrt[4]{\frac{K_1}{D_2}} = 0.6007\sqrt[4]{\frac{1.5159\times 10^4}{3366.6}} = 0.8750,$$

$$\xi_2 = \frac{1}{k}\sqrt[4]{\frac{K_2}{D_2}} = 0.6007\sqrt[4]{\frac{1.5696\times 10^4}{3366.6}} = 1.9611.$$

$$h_{12} = \frac{h_1D_2 + h_2D_1}{D_1 + D_2} = \frac{1683.30 + 138.17}{138.17 + 3366.60} = 0.5197,$$

$$h_{23} = \frac{h_2D_3 + h_3D_2}{D_2 + D_3} = \frac{17667.3 + 8416.5}{3366.60 + 17667.3} = 1.2401.$$

Auxiliary parameters for the peeling stress in a trimaterial assembly predicted by formulas (2.66) are:

$$\mu_1 = \sqrt{(\beta_1^4 + \beta_2^4) + 2\beta_1^2\beta_2^2\sqrt{1+\delta^2}} = \sqrt{29.9425 + 12.5896\times 1.0204} = 6.5375$$

$$\mu_2 = \sqrt{(\beta_1^4 + \beta_2^4) - 2\beta_1^2\beta_2^2\sqrt{1+\delta^2}} = \sqrt{29.9425 - 12.5896\times 1.0204} = 4.1408,$$

Then, formulas (2.62) yield:

$$\gamma_1 = \sqrt{\frac{\sqrt{\mu_1^2 + \mu_2^2} + \mu_2}{2}} = \sqrt{\frac{7.7418 + 4.1408}{2}} = 2.4371,$$

$$\gamma_2 = \sqrt{\frac{\sqrt{\mu_1^2 + \mu_2^2} - \mu_2}{2}} = \sqrt{\frac{7.7418 - 4.1408}{2}} = 1.3412.$$

The thermally induced forces acting in the midportion of the bimaterial assembly components are given by formulas (2.8):

$$T_1^0 = -T_2^0 = (\alpha_1 - \alpha_2)\frac{\Delta t}{\lambda} = -5.4 \times 10^{-6} \frac{275}{1.4021 \times 10^{-4}} = -9.0221 \text{ kg/mm}$$

The component #1 is in compression, and the component #2 is in tension. For the trimaterial assembly, formulas (2.43) yield:

$$T_1^0 = \Delta t \frac{\lambda_2(\alpha_1 - \alpha_3) + \lambda_3(\alpha_1 - \alpha_2)}{\lambda_1\lambda_2 + \lambda_2\lambda_3 + \lambda_3\lambda_1} = 275 \frac{-14.1 \times 0.1861 \times 10^{-10} - 5.4 \times 0.2031 \times 10^{-10}}{0.5111 \times 10^{-8}}$$

$$= -19.1477 \text{ kg/mm}$$

$$T_2^0 = \Delta t \frac{\lambda_1(\alpha_2 - \alpha_3) + \lambda_3(\alpha_2 - \alpha_1)}{\lambda_1\lambda_2 + \lambda_2\lambda_3 + \lambda_3\lambda_1} = 275 \frac{-1.2160 \times 9.5 \times 10^{-10} + 0.2031 \times 5.4 \times 10^{-10}}{0.5111 \times 10^{-8}}$$

$$= -57.1357 \text{ kg/mm}$$

$$T_3^0 = \Delta t \frac{\lambda_1(\alpha_3 - \alpha_2) + \lambda_2(\alpha_3 - \alpha_1)}{\lambda_1\lambda_2 + \lambda_2\lambda_3 + \lambda_3\lambda_1} = 275 \frac{1.2160 \times 9.5 \times 10^{-10} + 0.1861 \times 14.1 \times 10^{-10}}{0.5111 \times 10^{-8}}$$

$$= 76.2834 \text{ kg/mm}$$

Clearly, the sum of these forces is zero. The components #1 and #2 are in compression, and the component #3 is in tension.

The component #1 in the bimaterial assembly experiences compressive stresses of the magnitude $\sigma_1 = \frac{T_1^0}{h_1} = -\frac{9.0221}{0.5} = -18.0442 \text{ kg/mm}^2$; and the component #2 of this assembly experiences tensile stresses of the magnitude $\sigma_2 = \frac{T_2^0}{h_2} = \frac{9.0221}{1.0} = 9.0221 \text{ kg/mm}^2$;

The components #1 and #2 in the trimaterial assembly experience compressive stresses

$$\sigma_1 = \frac{T_1^0}{h_1} = -\frac{19.1477}{0.5} = -38.2954 \text{ kg/mm}^2$$

and

$$\sigma_2 = \frac{T_2^0}{h_2} = -\frac{57.1357}{1.0} = -57.1357 \text{ kg/mm}^2,$$

respectively. The component #3 in the trimaterial assembly experiences tensile stress

$$\sigma_3 = \frac{T_3^0}{h_3} = \frac{76.2834}{2.5} = 30.5134 \text{ kg/mm}^2$$

The maximum interfacial shearing stress in the bimaterial assembly, as predicted by formula (2.19), is

$$\tau_{\max} = -kT_1^0 \tanh ka = 1.5588 \times 9.02211 \times 1.0 = 14.0645 \text{ kg/mm}^2$$

The maximum interfacial shearing stresses in the trimaterial assembly, computed by formulas (2.55), are as follows:

$$\tau_1(a) = -kT_1^0 \tanh ka = 1.6648 \times 20.0196 = 33.3286 \text{ kg/mm}^2$$

and

$$\tau_2(a) = -kT_3^0 \tanh ka = -1.6648 \times 76.2747 = -126.9821 \text{ kg/mm}^2$$

These stresses are considerably higher than the maximum interfacial stress in a bimaterial assembly, especially at the interface between the substrate and the heat sink. This stress is almost by an order of magnitude higher than the maximum interfacial shearing stress in a bimaterial assembly comprised of the chip and the ceramic substrate.

Formula (2.33) for the factor of the peeling stress at the end of a bimaterial assembly yields:

$$A = \frac{\varsigma^4}{1+\varsigma^4} k^2 T_1^0 \frac{h_{12}}{2} = -\frac{2.5811^4}{1+2.5811^4} 1.5588^2 \times 9.02211 \frac{0.5197}{2} = -5.5710 \text{ kg/mm}^2$$

The product $\beta a = 2.8452 \times 10 = 28.452$ is significant, so that formula (2.45) can be used to evaluate the maximum value of the peeling stress in the bimaterial assembly:

$$p(a) = A[\varsigma(\sqrt{2} - \varsigma) - 1] = -5.5718[2.5811(\sqrt{2} - 2.5811) - 1] = 22.3500 \text{ kg/mm}^2$$

Thus, the calculated peeling stress in a bimaterial assembly is by about 60% higher than the predicted maximum interfacial shearing stress.

For the factors A_1 and A_2 of the peeling stresses acting at the interfaces of a trimaterial assembly, formulas (2.63), with

$$k = 1.6648 \text{ mm}^{-1}, \ \zeta_1 = 1.9637, \ \zeta_2 = 0.9220, \ \xi_1 = 0.8750, \ \xi_2 = 1.9611,$$
$$h_{12} = 0.5197 \text{ mm},$$

$$h_{22} = 0.5197 \text{ mm}, \ h_2 = 1.0 \text{ mm}, \ T_1^0 = -19.1477 \text{ kg/mm}, \ T_2^0 = -57.1357 \text{ kg/mm},$$
$$T_3^0 = 75.2834 \text{ kg/mm},$$

result in the following values of these factors:

$$A_1 = -k^2 \frac{\left[\varsigma_1^4(1+\varsigma_2^4)\frac{h_{12}}{2} + \xi_1^4\xi_2^4\frac{h_2}{2}\right]T_1^0 + \xi_1^4\left[\varsigma_2^4\frac{h_{23}}{2} + (1+\varsigma_2^4)\frac{h_2}{2}\right]T_3^0}{(1+\varsigma_1^4)(1+\varsigma_2^4) - \xi_1^4\xi_2^4}$$

$$= 25.2388 \text{ kg/mm}^2$$

$$A_2 = -k^2 \frac{\xi_2^4\left[\varsigma_1^4\frac{h_{12}}{2} + (1+\varsigma_1^4)\frac{h_2}{2}\right]T_1^0 + \left[\varsigma_2^4(1+\varsigma_1^4)\frac{h_{23}}{2} + \xi_1^4\xi_2^4\frac{h_2}{2}\right]T_3^0}{(1+\varsigma_1^4)(1+\varsigma_2^4) - \xi_1^4\xi_2^4}$$

$$= 438.6403 \text{ kg/mm}^2$$

Then, with $\eta = 0.9366$, $\mu_1 = 6.5375$, $\mu_2 = 4.1408$, $\gamma_1 = 2.4371$, and $\gamma_2 = 1.3412$, we obtain the following values of the maximum peeling stresses at the interfaces #1 and #2: $p_1(a) = -13.2102$ kg/mm^2; and $p_2(a) = -229.5843$ kg/mm^2. While the maximum peeling stress $p_1(a)$ at the die-substrate interface is only about 40% of the maximum interfacial shearing stress at the interface #1 between the die and the substrate, the maximum peeling stress $p_2(a)$ at the interface #2 between the substrate and the heat sink is by a factor of 1.8 higher than the maximum shearing stress at this interface. Comparing the calculated maximum peeling stresses in the trimaterial and bimaterial assemblies, we conclude that the maximum peeling stress at the interface #1 between the die and the substrate in the trimaterial assembly is only about 60% of the peeling stress in the bimaterial assembly. This means that adding a robust heat sink to the assembly resulted in an appreciable stress relief at the die-substrate interface. The "bad news," however, is that the maximum peeling stress at the interface #2 between the substrate and the heat sink is by an order of magnitude higher than the maximum peeling stress in a bimaterial assembly.

2.8 BIMATERIAL ASSEMBLY SUBJECTED TO THERMAL STRESS: PROPENSITY TO DELAMINATION ASSESSED USING THE INTERFACIAL COMPLIANCE MODEL

2.8.1 Background/Incentive

One way to make a design for reliability (DfR) decision for a bonded assembly [39–42], as far as its interfacial fracture toughness is concerned, is by comparing the anticipated failure criterion with a critical load factor. Based on this criterion, the adhesive and the cohesive strength of the bonding material could be judged upon, and the adequate bonding material and its thickness could be selected. There are numerous proposed theories and predictive models to understand the physics behind the material failure in bonded joints (see, for example, [43–45]). The majority of models use, in one way or another, fracture mechanics concepts [46–56], and the most popular models proceed from the strain energy release rate (SERR) [57–58]. In the analysis that follows, we suggest using the interfacial compliance model [59, 60] and the probabilistic concept [61–67]; simple and physically meaningful engineering models for the assessment of the SERR in shear for a bonded assembly subjected to the change in temperature. Both the actual and the critical SERR values are random variables, and the loading (such as temperature cycling) is a step-wise process that can be described best by the extreme value distribution (EVD) model, in which both the level and the number of loadings are important. Although the analysis is carried out for the case of thermal loading and is geared in application to electronic materials and assemblies, it can be used also, with some modifications, for mechanical loading (such as the one in shear-off testing) and in numerous applied science problems beyond the electronics materials field.

2.8.2 Strain Energy Release Rate (SERR) Computed Using the Interfacial Compliance Approach

Consider a bonded bimaterial assembly manufactured at an elevated temperature and subsequently cooled down to a low (room or testing) temperature. The thermally induced stresses that arise because of the dissimilar adherend materials can possibly result in the interfacial cracking (delamination) of the assembly. The SERR is defined, as is known, as the energy dissipated during fracture (crack propagation) per unit surface length of a newly created fracture (see, for example, [39–41]):

$$G_e = -\frac{\partial U}{\partial a}. \tag{2.75}$$

Here, U is the potential strain energy available for crack growth and a is the crack length. The SERR-based failure criterion states that a crack will grow when the available (actual) SERR, G_e, exceeds its critical value (threshold) G_c:

$$G_e \geq G_c \tag{2.76}$$

The critical fracture energy is considered to be a material property and should be evaluated experimentally based on the specially designed and conducted failure-oriented accelerated tests (FOATs) (see, for example, [65]). The adequate geometry of the FOAT test specimen is certainly important.

The interfacial thermally induced shearing stress in a bonded assembly can be evaluated, in an approximate analysis, based on the interfacial compliance approach [59, 60], by the formula:

$$\tau(x) = -kT\frac{\sinh kx}{\cosh kl}, \tag{2.77}$$

where

$$T = (\alpha_1 - \alpha_2)\frac{\Delta t}{\lambda} = \frac{\Delta\alpha\Delta t}{\lambda} \tag{2.78}$$

are the thermally induced forces acting in the adherend cross sections,

$$\lambda = \lambda_1 + \lambda_2 = \frac{1}{E_1^0 h_1} + \frac{1}{E_2^0 h_2} \tag{2.79}$$

is the total longitudinal (axial) compliance of the assembly, α_1 and α_2 are the coefficients of thermal expansion (CTE) of the adherend materials, $E_1^0 = \frac{1-\nu_1}{E_1}$ and $E_2^0 = \frac{1-\nu_2}{E_2}$ are the effective Young's moduli of the materials, E_1 and E_2 are their actual Young's moduli, ν_1 and ν_2 are the Poisson's ratios, h_1 and h_2 are the thicknesses of the adherends, Δt is the change in temperature,

$$k = \sqrt{\frac{\lambda}{\kappa}} \tag{2.80}$$

is the parameter of the interfacial shearing stress,

$$\kappa = \kappa_0 + \kappa_1 + \kappa_2 \tag{2.81}$$

is the total longitudinal interfacial compliance of the assembly,

$$\kappa_0 = \frac{h_0}{G_0} \tag{2.82}$$

is the interfacial compliance of the bonding layer, h_0 is its thickness, G_0 is its shear modulus,

$$\kappa_1 = \frac{h_1}{3G_1} \quad \kappa_2 = \frac{h_2}{3G_2} \tag{2.83}$$

are the interfacial compliances of the adherends [59], and G_1 and G_2 are the shear moduli of the adherend materials. The interfacial shearing stress $\tau(x)$ is related to the forces $T(x)$ acting in the adherend cross sections as

$$T(x) = \int_{-l}^{x} \tau(\xi) d\xi, \tag{2.84}$$

where l is half the assembly length. The origin of the coordinate x is in the mid-cross section of the assembly.

Introducing (2.77) into (2.84) we find:

$$T(x) = T\left(1 - \frac{\cosh kx}{\cosh kl}\right) \tag{2.85}$$

This formula meets the zero boundary condition $T(l) = 0$ at the assembly ends, where the maximum shearing stress

$$\tau(x) = -kT \tanh kl \tag{2.86}$$

takes place.

The elastic strain energy (work), needed to deform a unit volume of the bonding material, associated with the distortion of its form and caused by the induced shearing stresses in it, can be evaluated by the formula (see, for example, [60])

$$U = \frac{3\tau_0^2}{4G_0}, \tag{2.87}$$

where

$$\tau_0 = \frac{2}{3}\sqrt{\tau_1^2 + \tau_2^2 + \tau_3^2} \tag{2.88}$$

is the octahedral shearing stress, and τ_1, τ_2, and τ_3 are the principal shearing stresses. Assuming $\tau_3 = 0$, and $\tau_1 = \tau_2 = \tau(x)$, we have

$$\tau_0 = \frac{2}{3}\sqrt{\tau_1^2 + \tau_2^2 + \tau_3^2} = \frac{2\sqrt{2}}{3}\tau(x), \tag{2.89}$$

and formula (2.87), when applied to an elementary length (segment) dx of the bond, yields

$$dU(x) = \frac{2\tau^2(x)}{3G_0} d(x). \tag{2.90}$$

The thermally induced interfacial stresses and strains are antisymmetric with respect to the mid-cross section of the assembly. The strain energy (per unit assembly width) contained in each half of the assembly length can be evaluated, using formula (2.77) and assuming that the stress is uniform over the bond thickness h_0, as

$$U = \frac{2k^2T^2h_0}{3G_0 \cosh^2 kl} \int_0^l \sinh^2 kx dx = \frac{kh_0T^2}{3G_0} \chi_0(kl), \tag{2.91}$$

where the function

$$\chi_0(kl) = \frac{\sinh 2kl - 2kl}{\cosh 2kl + 1} \tag{2.92}$$

TABLE 2.1
Function Reflecting the Effect of the Assembly Length on the Strain Energy

kl	0	0.5	1.0	1.010	2.0	3.0	4.0	5.0	∞
$\chi_0(kl)$	0	0.0689	0.3416	0.3480	0.8227	0.9655	0.9940	0.9990	1.0000
$\tanh kl$	0	0.4621	0.7616	0.7658	0.9640	0.9950	0.9994	0.9995	1.0000
$\chi_1(kl)$	0	0.1817	0.3198	0.3199	0.1362	0.0292	0.00536	0.00091	0

reflects the effect of the assembly length on the strain energy. This function changes from zero (very short assemblies and/or assemblies with very compliant bonds) to one (long assemblies and/or assemblies with stiff bonds). The function $\chi_0(kl)$ is tabulated in the second line of Table 2.1. As evident from the calculated data, the strain energy increases with an increase in the kl value, but does not practically change with an increase in the kl product, if this product reaches and exceeds the $kl = 5.0$ level.

The hyperbolic tangent $\tanh kl$ reflects, as evident from formula (2.77), the effect of the kl product on the maximum shearing stress. Table 2.1 indicates that the "saturation" of this stress starts at about $kl = 4$. Thus, the interfacial shearing stress increases faster with an increase in the product kl than the strain energy level does.

From (2.91), we find, by differentiation

$$G_e = -\frac{\partial U}{\partial l} = -\frac{\partial U}{\partial a} = \frac{kT^2}{3G_0}\frac{d\chi_0(kl)}{dl} = \frac{2k^2h_0T^2}{3G_0}\chi_1(kl), \tag{2.93}$$

where the function

$$\chi_1(kl) = kl\frac{\sinh kl}{\cosh^3 kl} \tag{2.94}$$

reflects the effect of the assembly size on the derivative of the strain energy with respect to the change in the assembly length. For the interfacial delamination crack that propagates from the assembly end (where the stress level is the highest) inward the assembly, formula (2.93) also determines the SERR, since the crack length a can be found as the difference between the constant length of the assembly (specimen) and the variable remaining length l of the still undamaged assembly, so that $da = -dl$. The function $\chi_1(kl)$ increases from zero to its maximum value of $\chi_{1,\max} = 0.3199$ at $kl \approx 1.010$ and then decreases with the further increase in the kl value. It becomes next-to-zero for kl values exceeding $kl = 4.0$.

The maximum SERR takes place for kl values that could be found from the equation $\frac{d\chi_1(kl)}{dx} = 0$, which yields

$$\sinh 2kl + 2kl(2 - \cosh 2kl) = 0 \tag{2.95}$$

This equation is fulfilled for $kl = 1.0096$. Thus, the maximum SERR takes place for not very long assemblies, because in such assemblies, although the stress level is high, the strain energy does not change significantly with an increase in the crack (delamination) length. The maximum SERR does not occur for very small size assemblies either, because, although the SERR is appreciable, the shearing stress level is low. This conclusion is important, particularly in connection with choosing the most appropriate test specimen size (see Numerical Example #1), when the critical value of the SERR is sought based on a FOAT experiment.

2.8.3 Adequate SERR Specimen's Length

2.8.3.1 Numerical Example #1

Input data:

Component #1: Young's modulus: $E_1 = 12300\ \text{kg/mm}^2$; Poisson's ratio: $\nu_1 = 0.24$; CTE: $\alpha_1 = 2.2 \times 10^{-6} 1/°C$; Thickness: $h_1 = 0.5$ mm;
Component #2: Young's modulus: $E_2 = 2000\ \text{kg/mm}^2$; Poisson's ratio: $\nu_2 = 0.30$; CTE: $\alpha_2 = 13.2 \times 10^{-6} 1/°C$; Thickness: $h_1 = 1.5$ mm;
Bonding layer (zero component): Young's modulus: $E_0 = 2000\ \text{kg/mm}^2$; Poisson's ratio: $\nu_0 = 0.40$; CTE: $\alpha_0 = 13.2 \times 10^{-6} 1/°C$; Thickness: $h_0 = 0.05$ mm;
Change in temperature: $\Delta t = 100°C$

Computed data:

"External" thermal strain: $\Delta\alpha\Delta t = 11 \times 10^6 \times 100 = 0.0011$;
Axial compliances of the assembly components:

$$\lambda_1 = \frac{1-\nu_1}{E_1 h_1} = \frac{1-0.24}{12300 \times 0.5} = 1.2358 \times 10^{-4}\ \text{mm/kg}$$

$$\lambda_2 = \frac{1-\nu_2}{E_2 h_2} = \frac{1-0.30}{2000 \times 1.5} = 2.3333 \times 10^{-4}\ \text{mm/kg}$$

Total axial compliance of the assembly:

$$\lambda = \lambda_1 + \lambda_2 = 1.2358 \times 10^{-4} + 2.3333 \times 10^{-4} = 3.5691 \times 10^{-4}\ \text{mm/kg}$$

Shear moduli of the materials:

$$G_0 = \frac{E_0}{2(1+\nu_0)} = \frac{200}{2(1+0.4)} = 71.4\ \text{kg/mm}^2$$

$$G_1 = \frac{E_1}{2(1+\nu_1)} = \frac{12300}{2(1+0.24)} = 4959.7\ \text{kg/mm}^2$$

$$G_2 = \frac{E_2}{2(1+\nu_2)} = \frac{2000}{2(1+0.3)} = 769.2\ \text{kg/mm}^2$$

Interfacial shearing compliances:

$$\kappa_0 = \frac{h_0}{G_0} = \frac{0.05}{71.4} = 7.0028 \times 10^{-4} \ \text{mm}^3/\text{kg}$$

$$\kappa_1 = \frac{h_1}{3G_1} = \frac{0.5}{3 \times 4959.7} = 0.3360 \times 10^{-4} \ \text{mm}^3/\text{kg}$$

$$\kappa_2 = \frac{h_2}{3G_2} = \frac{1.5}{3 \times 769.2} = 6.5020 \times 10^{-4} \ \text{mm}^3/\text{kg}$$

Total interfacial compliance of the assembly:

$$\kappa = \kappa_0 + \kappa_1 + \kappa_2 = 7.0028 \times 10^{-4} + 0.3360 \times 10^{-4} + 6.5020 \times 10^{4} = 13.8408 \ \text{mm}^3/\text{kg}$$

Note that the axial compliance of the assembly is due mostly to the adherends, while the interfacial compliance is due to both the adherends and the adhesive.

Parameter of the interfacial shearing stress:

$$k = \sqrt{\frac{\lambda}{\kappa}} = \sqrt{\frac{3.5691 \times 10^{-4}}{13.8408 \times 10^{-4}}} = 0.5078 \ \text{mm}^{-1}$$

Length of the test specimen (half of the undamaged assembly length) with the highest SERR:

$$2l = 2\frac{kl}{2} = 2\frac{1.010}{0.5078} = 3.9773 \ \text{mm} \approx 4.0 \ \text{mm}$$

The actual test specimen should be a little longer.

2.8.3.2 Numerical Example #2

Shear-off testing is considered for an assembly with characteristics in Numerical Example #1. How high should the measured shear-off force be at failure so that the high enough fracture toughness is assured?

The minimum shear-off force that results in the same maximum interfacial shearing stress as the thermally induced loading can be determined by the formula [61]

$$\hat{T} = \frac{\Delta\alpha\Delta t}{\lambda} \tanh kl \tanh 2kl = \frac{\Delta\alpha\Delta t}{\lambda}\left(1 - \frac{1}{\cosh 2kl}\right)$$

With $\Delta\alpha\Delta t = (13.2 - 2.0.0012)10^{-6} \times 100 = 0{,}00110$, $\lambda = 3.5691 \times 10^{-4}$ mm/kg and a short specimen with $kl = 1.010$ this formula yields: $\hat{T} = 2.2785$ kg/mm. If a long specimen is tested, then the required minimum shear-off force at failure would be

$$\hat{T} = \frac{\Delta\alpha\Delta t}{\lambda} = \frac{0.0011}{3.5691 \times 10^{-4}} 3.0820 \ \text{kg/mm}.$$

2.8.4 Probabilistic Approach: Application of the Extreme Value Distribution

When applying the probabilistic concept [62–66] for the assessment and assurance of the adequate interfacial fracture toughness of an assembly of interest, we assume that the loading process can be approximated by the extreme value distribution (EVD) (see, for example, [67]) for the SERR and that the SERR threshold (level) can be assumed to be a regular normal process. In accordance with the Appendix A results, and assuming that the process $Z(t)$ represents the process $G_c(t)$ of the SERR, and that the process $X(t)$ represents the critical SERR level $G_c(t)$ the probability that the actual random SERR G_a remains below its critical value G_c after the action of the Nth loading cycle, can be sought in the form of an integral

$$F_\eta(\eta) = \frac{1}{\sqrt{\pi}} \int_{\frac{g_a}{\sqrt{2D_c}}}^{\infty} \exp[=(\gamma - \xi)^2 - Ne^{-\delta\xi^2}]d\xi, \tag{2.96}$$

where

$$\gamma = \eta + \frac{\bar{g}_a - \bar{g}_c}{\sqrt{2D_c}}, \quad \eta = \eta(w) = \frac{w}{\sqrt{2D_c}} = \frac{\bar{g}_c - \bar{g}_a}{\sqrt{2D_c}}, \quad \delta = \frac{D_c}{D_a}. \tag{2.97}$$

Here $\bar{g}_a$ and $\bar{g}_c$ are the mean values of the actual and the critical SERR levels, respectively, and D_a and D_c are variances of the random stationary processes $G_a(t)$ and $G_c(t)$. The process $G_a(t)$ is characterized by its EVD, which depends on the number N of loadings. Equation (2.96) determines the probability that the random difference $W(t) = G_c(t) - G_a^*(t)$ of the random critical SERR value G_c and the extreme value G_a^* of the random actual G_a level remains below a certain threshold $w = \eta\sqrt{2D_c}$. To apply the integral (2.93), one has to first calculate the safety factors $\frac{\bar{g}_a - \bar{g}_c}{\sqrt{2D_c}}$ and $\frac{\bar{g}_a}{\sqrt{2D_c}}$, which is the lower limit of the integral (2.93), the variance ratio $\delta = \frac{D_c}{D_a}$, and assumes a certain η level of the safety factor based on the difference $W(t)$.

2.8.5 Probabilistic Approach: Numerical Example

Input data:

Mean value of the actual SERR process G_a: $\bar{g}_a = 0.0229$ kg/mm^3,
Mean value of the critical SERR process G_c: $\bar{g}_c = 0.0100$ kg/mm^3,
Variances of the above processes: $D_a = 1.3110 \times 10^{-4}$ kg^2/mm^6, $D_c = 0.0400 \times 10^{-4}$ kg^2/mm^6
Number of loadings $N = 10$

Computed data:

Safety factor

$$\frac{\bar{g}_a - \bar{g}_c}{\sqrt{2D_c}} = \frac{0.0229 - 0.0100}{\sqrt{2 \times 0.0400 \times 10^{-4}}} = 4.5608$$

Safety factor [lower limit of integration in the integral (2.96)

$$\frac{\bar{g}_a}{\sqrt{2D_c}} = \frac{0.0229}{\sqrt{2 \times 0.0400 \times 10^{-4}}} = 8.0964$$

Variance ratio

$$\delta = \frac{D_c}{D_a} = \frac{0.0400 \times 10^{-4}}{1.3110 \times 10^{-4}} = 0.0305$$

Formula (2.93) yields

$$F_\eta(\eta) = \frac{1}{\sqrt{\pi}} \int_{8.0964}^{\infty} \exp[-(\eta + 4.5608 - \xi)^2 - 10\exp(-0.0305\xi^2)]d\xi.$$

The calculated values of the integral (2.93) for different dimensionless SERR values are shown in Table 2.2. Clearly, the probability that the random difference between the critical and the actual SERR is below a certain level increases with an increase in this level. An example of the calculation procedure is shown in Appendix B for $\eta = 7$.

Thus, the application of the interfacial compliance approach and the probabilistic DfR concept is suggested for the assessment of the adhesive and cohesive strength of the bonding material in a bimaterial assembly subjected to the temperature change. The SERR is used as a suitable criterion of the level of the fracture toughness of the bond. The shearing mode of failure is considered. As far as the PDfR approach is concerned, it has been accounted for the random nature of both the actual and the critical SERR values. The results of the analysis can be used also beyond the field of electronic materials.

TABLE 2.2
Calculated Values of the Integral $F_\eta(\eta)$

η	0	3	4	5	6	7	8
$F_\eta(\eta)$	0	0.0750	0.2926	0.5299	0.7046	0.8384	0.9194

APPENDIX A CONVOLUTION OF EXTREME VALUE DISTRIBUTION WITH A NORMALLY DISTRIBUTED VARIABLE

The objective of the analysis that follows is to obtain a convolution of the EVD

$$F_{z^*}(z^*) = \exp\left[-N\exp\left(-\frac{(z^*-\bar{z})^2}{2D_z}\right)\right] \tag{A-1}$$

for a stationary normal random process $Z(t)$, which is the basic distribution for this EVD (see, for example, [27]), with a stationary normally distributed random variable $X(t)$, whose probability density distribution function is

$$f_x(x) = \frac{1}{\sqrt{2\pi D_x}}\exp\left[-\left(\frac{(x-\bar{x})^2}{2D_x}\right)\right]. \tag{A-2}$$

In these formulas, the basic random process $Z(t)$ is a homogeneous (the probability that the given level z^* is exceeded depends only on the duration of the time interval and is independent of the initial moment of time) and ordinary (none of the events $Z \succ z^*$ can possibly occur simultaneously with another similar event) stationary normal random process; $Z^*(t)$ are the extreme values of the process $Z(t)$; N is the number of oscillations during the time interval between two adjacent upward crossings $Z \succ z^*$ of the level z^* by the process $Z(t)$ (in such a situation the flow of the events $Z \succ z^*$ is a Poisson's process), $\bar{z}$ is the mean value of the process $Z(t)$, $\bar{x}$ is the mean value of the process $X(t)$ and D_z and D_x are variances of the processes $Z(t)$ and $X(t)$. The events $Z \succ z^*$ are assumed to be statistically independent. The number N in formula (A-1) is supposed to be not very small.

The probability density and the probability distribution functions of the random difference $W(t) = X(t) - Z^*(t)$ are as follows:

$$f_w(w) = \int_{-\infty}^{\infty} f_x(x) f_{z^*}(x-w)\,dx \tag{A-3}$$

$$F_w(w) = \int_{-\infty}^{\infty} f_x(x)dx \int_{0}^{x-w} f_{z^*}(z^*)dz^* = \int_{-\infty}^{\infty} f_x(x) F_{z^*}(x-w)\,dx \tag{A-4}$$

In these formulas, the limits of integration for the variable $X(t)$ are defined by the range, within which the function $f_x(x)$ is positive. With the distributions (A-1) and (A-2), we have:

$$F_w(w) = \frac{1}{\sqrt{2\pi D_x}}\int_{-\infty}^{\infty} \exp\left[-\left(\frac{(x-\bar{x})^2}{2D_x}\right)\right]\exp\left[-N\exp\left(-\frac{(w-x+\bar{z})^2}{2D_x}\right)\right]dx =$$

$$= \frac{1}{\sqrt{\pi}}\int_{\gamma_z}^{\infty} \exp[-(\gamma-\xi)^2 - N\exp(-\delta\xi^2)]\,d\xi \tag{A-5}$$

where a new variable $\xi = \frac{w - \bar{x} + \bar{z}}{\sqrt{2D_x}}$ of integration is introduced and notation

$$\gamma = \eta + \gamma_z - \gamma_x, \quad \eta = \eta(w) = \frac{w}{\sqrt{2D_x}}, \quad \gamma_x = \frac{\bar{x}}{\sqrt{2D_x}}, \quad \gamma_z = \frac{\bar{z}}{\sqrt{2D_x}}, \quad \delta = \frac{D_x}{D_z}, \quad \text{(A-6)}$$

is used. The $\gamma_x = \frac{\bar{x}}{\sqrt{2D_x}}$ ratio is the safety factor for the process $X(t)$, and the safety factor $\gamma_z = \frac{\bar{z}}{\sqrt{2D_x}}$ for the process $Z(t)$ can be found as $\gamma_z = \frac{\bar{z}}{\sqrt{2D_x}} = \gamma_z = \sqrt{\delta}$. The integral (A-5) determines the probability that the difference $W = X(t) - Z^*(t)$ is below the $w = \eta\sqrt{2D_x}$ value. When $N \to \infty$, $F_w(w) \to 0$: in a long run the process $Z^*(t)$ will always exceed the $X(t)$ values. When $\bar{z} \to \infty$, then $\gamma_z \to \infty$, $\gamma \to \infty$, and $F_w(w) \to 0$: when the mean value of the process $Z(t)$ is significant, the process $Z^*(t)$ will always exceed the $X(t)$ values.

When the variance D_z of the process $Z(t)$ is significantly greater than the variance D_x of the process $X(t)$, so that the variance ratio $\delta = \frac{D_x}{D_z}$ can be put equal to zero, then the integral (A-5) can be simplified:

$$F_w(w) = \frac{1}{2}e^{-N}[1 + \Phi(\gamma - \gamma_z)] = \frac{1}{2}e^{-N}\left[1 + \Phi\left(\frac{w - \bar{x}}{\sqrt{2D_x}}\right)\right] \quad \text{(A-7)}$$

where

$$\Phi(\alpha) = \frac{2}{\sqrt{\pi}}\int_0^{\alpha} e^{-t^2}dt \quad \text{(A-8)}$$

is the tabulated Laplace function.

APPENDIX B A NUMERICAL INTEGRATION EXAMPLE

This example is given in Table B-1 for the case $\eta = 7$. The integrand is as follows:

$$f(\xi) = \exp[-(11.5608 - \xi)^2 - 10\exp(-0.0305\xi^2)]$$

and the corrected sum $\Sigma_{corrected}$ is computed as the sum Σ minus half of the sum of the extreme ordinates.

$$F = \frac{1}{\sqrt{\pi}}\Delta\xi \sum_{corrected} = 0.5642 \times 0.25 \times 5.9442 = 0.8384$$

Thus, the probability that the difference between the critical value of the SERR and its actual value will be found below the (rather high) level of the probability that the difference between the critical value of the SERR and its actual value will occur below the (rather high) level of $w = \eta\sqrt{2D_x} = 7.0\sqrt{2 \times 0.0400 \times 10^{-4}} = 0.01980$ kg/mm^3 is as high as 0.8384.

TABLE B.1
Numerical Integration

ξ	$(11.5608 - \xi)^2$	$f(\xi)$
8.5964	8.7877	0.0001
8.8454	7.3680	0.0003
9.0964	6.0733	0.0010
9.3464	4.9036	0.0037
9.5964	3.8589	0.0115
9.8464	2.9392	0.0315
10.0964	2.1445	0.0750
10.3464	1.4748	0.1562
10.5964	0.9301	0.2849
10.8464	0.5104	0.4553
11.0964	0.2157	0.6379
11.8464	0.0460	0.8316
11.5969	0.0013	0.8464
11.8464	0.0816	0.8025
12.0964	0.2869	0.6689
12.3464	0.6172	0.4902
12.5964	1.0725	0.3161
12.8464	1.6528	0.1794
13.0964	2.3581	0.0897
13.3464	3.1884	0.0395
13.5964	4.1437	0.0153
13.8464	5.2240	0.0052
14.0964	6.4293	0.0016
14.3464	7.7596	0.0004
14.5964	9.2149	0.0001
$\Delta\xi = 0.25$	Σ	5.9443
	$\Sigma_{corrected}$	5.9442

REFERENCES

1. S.P. Timoshenko, "Analysis of Bi-Metal Thermostats," *Journal of the Optical Society of America*, vol. 11, 1925.
2. B.J. Aleck, "Thermal Stresses in a Rectangular Plate Clamped Along an Edge," *ASME Journal of Applied Mechanics*, vol. 16, 1949.
3. G.A. Lang, et al., "Thermal Fatigue in Silicon Power Devices," *IEEE Transactions on Electron Devices*, vol. 17, 1970.
4. G.H. Olsen, M. Ettenberg, "Calculated Stresses in Multilayered Heteroepitaxial Structures," *Journal of Applied Physics*, vol. 48, No. 6, 1977.
5. J. Vilms, D. Kerps, "Simple Stress Formula for Multilayered Thin Films on a Thick Substrate," *Journal of Applied Physics*, vol. 53, No. 3, 1982.
6. F.-V. Chang, "Thermal Contact Stresses of Bi-Metal Strip Thermostat," *Applied Mathematics and Mechanics*, vol. 4, No. 3, 1983.

7. E. Suhir, "Calculated Thermally Induced Stresses in Adhesively Bonded and Soldered Assemblies," *Proceedings of the International Symposium on Microelectronics*, ISHM, Atlanta, Georgia, Oct. 1986.
8. E. Suhir, "Stresses in Bi-Metal Thermostats," *ASME Journal of Applied Mechanics*, vol. 53, No. 3, Sept. 1986.
9. E. Suhir, "Die Attachment Design and Its Influence on the Thermally Induced Stresses in the Die and the Attachment," *Proceedings of the 37th Electrical Components Conference*, IEEE, Boston, Massachusetts, May 1987.
10. E. Suhir, "Thermal Stress Failures in Microelectronic Components - Review and Extension," in A. Bar-Cohen, A.D. Kraus, (eds.), *Advances in Thermal Modeling of Electronic Components and Systems*, Hemisphere, 1988.
11. E. Suhir, "Interfacial Stresses in Bi-Metal Thermostats," *ASME Journal of Applied Mechanics*, vol. 56, No. 3, Sept. 1989.
12. J.W. Eischen, C. Chung, J.H. Kim, "Realistic Modeling of the Edge Effect Stresses in Bimaterial Elements," *ASME Journal of Electronic Packaging*, vol. 112, No. 1, 1990.
13. J.C. Glaser, "Thermal Stresses in Compliantly Joined Materials," *ASME Journal of Electronic Packaging*, vol. 112, No. 1, 1990.
14. A.Y. Kuo, "Thermal Stress at the Edge of a Bi-Metallic Thermostat," *ASME Journal of Applied Mechanics*, vol. 57, 1990.
15. A.O. Cifuentes, "Elastoplastic Analysis of Bimaterial Beams Subjected to Thermal Loads," *ASME Journal of Electronic Packaging*, vol. 113, No. 4, 1991.
16. J.H. Lau, "Thermoelastic Solutions for a Semi-Infinite Substrate with an Electronic Device," *ASME Journal of Electronic Packaging*, vol. 114, No. 3, 1992.
17. J.H. Lau (ed.), *Thermal Stress and Strain in Microelectronics Packaging*, Van-Nostrand Reinhold, 1993.
18. E. Suhir, *Structural Analysis and Microelectronics and Fiber-Optics*, Van-Nostrand, 1997.
19. E. Suhir, "Thermal Stress Failures in Microelectronics and Photonics: Prediction and Prevention," *Future Circuits International*, No. 5, 1999.
20. E. Suhir, "Thermal Stress Modeling in Micro- and Opto-Electronics: Review and Extension," Invited Presentation, ASME Symposium dedicated to Dr. Richard Chu, IBM, Washington, DC, Nov. 2003.
21. E. Suhir, "Modeling of Thermal Stress in Microelectronic and Photonic Structures: Role, Attributes, Challenges and Brief Review," Special Issue, *ASME Journal of Electronic Packaging*, vol. 125, No. 2, June 2003.
22. E. Suhir, "Thermal Stress in Electronics and Photonics: Prediction and Prevention," Keynote presentation, Therminic, Budapest, Hungary, Sept. 2012.
23. E. Suhir, "An Approximate Analysis of Stresses in Multilayer Elastic Thin Films," *ASME Journal of Applied Mechanics*, vol. 55, No. 3, 1988.
24. E. Suhir, "Approximate Evaluation of the Elastic Interfacial Stresses in Thin Films with Application to High-Tc Superconducting Ceramics," *International Journal of Solids and Structures*, vol. 27, No. 8, 1991.
25. P.M. Hall, et al., "Strains in Aluminum-Adhesive-Ceramic Trilayers," *ASME Journal of Electronic Packaging*, vol. 112, No. 4, 1990.
26. E. Suhir, "Analysis of Interfacial Thermal Stresses in a Tri-Material Assembly," *Journal of Applied Physics*, vol. 89, No. 7, 2001.
27. E. Suhir, A. Bensoussan, L. Bechou, "Aerospace Electronic Packaging: Thermal Stress in Bi- and Tri-Material Assemblies," *2014 IEEE Aerospace Conference*, Big Sky, Montana, March 2014.
28. E. Suhir, J.J. Vuillamin, Jr, "Effects of the CTE and Young's Modulus Lateral Gradients on the Bowing of an Optical Fiber: Analytical and Finite Element Modeling," *Optical Engineering*, vol. 39, No. 12, 2000.

29. E. Suhir, "Can the Curvature of an Optical Glass Fiber be Different from the Curvature of Its Coating?" *International Journal of Solids and Structures*, vol. 30, No. 17, 1993.
30. E. Suhir, "Effect of Initial Curvature on Low Temperature Microbending in Optical Fibers," *IEEE/OSA Journal of Lightwave Technology*, vol. 6, No. 8, 1988.
31. E. Suhir, "Spring Constant in the Buckling of Dual-Coated Optical Fibers," *IEEE/OSA Journal of Lightwave Technology*, vol. 6, No. 7, 1988.
32. E. Suhir, B. Poborets, "Solder Glass Attachment in Cerdip/Cerquad Packages: Thermally Induced Stresses and Mechanical Reliability," *ASME Journal of Electronic Packaging*, vol. 112, No. 2, 1990.
33. E. Suhir, "Thermal Stress Modeling in Microelectronics and Photonics Packaging, and the Application of the Probabilistic Approach: Review and Extension," *IMAPS International Journal of Microcircuits and Electronic Packaging*, vol. 23, No. 2, 2000 (invited paper).
34. E. Suhir, J. Nicolics, "Power Core (PC) Embedding a Plurality of IC Devices and Sandwiched Between Two Dissimilar Insulated Metal Substrates (IMS'): Predicted Thermal Stresses," *Journal of Materials Science: Materials in Electronics*, vol. 27, No. 7, 2016.
35. E. Suhir, "Bi-Material Assembly Subjected to Thermal Stress: Propensity to Delamination Assessed Using Interfacial Compliance Model," *Journal of Materials Science: Materials in Electronics*, vol. 27, No. 7, 2016.
36. V. Mishkevich, E. Suhir, "Simplified Approach to the Evaluation of Thermally Induced Stresses in Bi-Material Structures," in E. Suhir, (ed.), *Structural Analysis in Microelectronics and Fiber Optics*, ASME Press, 1993.
37. E. Suhir, "Analytical Thermal Stress Modeling in Electronics and Photonics Engineering: Application of the Concept of Interfacial Compliance," *Special Issue to commemorate the 90th birthday of Richard B. Hetnarski and 40 years of the Journal of Thermal Stresses*, Feb. 2019.
38. S.P. Timoshenko, *Theory of Elastic Stability*, McGraw-Hill, 1961.
39. P. Grant, "Analysis of Adhesive Stresses in Bonded Joints," in *Symposium: Joining in Fibre Reinforced Plastics, Imperial College*, I.P.C. Science and Technology Press, 1978, p. 41.
40. R.D. Adams, W.C. Wake, *Structural Adhesive Joints in Engineering,* Elsevier Applied Science Publishers, 1984.
41. G. Fernlund, et al., "Fracture Load Predictions for Adhesive Joints," *Composites Science and Technology*, vol. 51, 1994.
42. M.N. Charalambides, et al., "Strength Prediction of Bonded Joints," *83rd Meeting of the AGARD SMP—Bolted/Bonded Joints in Polymeric Composites*, Florence, Italy, 1997.
43. C.P. Buckley, "Material Failure," Lecture Notes, University of Oxford, 2005.
44. W. Callister, *Materials Science and Engineering: An Introduction*, John Wiley and Sons, 1994.
45. J. Shackelford, *Introduction to Materials Science for Engineers*, Macmillan Publishing Company, 1985.
46. T.L. Anderson, *Fracture Mechanics: Fundamentals and Applications*, CRC Press, 1995.
47. W.S. Johnson, et al., "Applications of Fracture Mechanics to the Durability of Bonded Composite Joints," FAA Final Report DOT/FAA/AR-97/56, 1998.
48. B. Lawn, *Fracture of Brittle Solids–Second Edition*, Cambridge University Press, 1993.
49. B. Farahmand, G. Bockrath, J. Glassco, *Fatigue and Fracture Mechanics of High-Risk Parts*, Chapman & Hall, 1997.

50. W.S. Johnson, S. Mall, "A Fracture Mechanics Approach for Designing Adhesively Bonded Joints," NASA Tech Memo 85694, September, 1983.
51. W.S. Johnson, et al., "Applications of Fracture Mechanics to the Durability of Bonded Composite Joints," FAA Final Report DOT/FAA/AR-97/56, 1998.
52. G. Bao, Z. Suo, "Remarks on Crack-Bridging Concepts," *Applied Mechanics Review*, vol. 45, 1992.
53. A. Portela, *Dual Element Analysis of Crack Growth*, Computational Mechanics Publications, 1993.
54. V.Z. Parton, *Fracture Mechanics: From Theory to Practice*, Gordon and Breach Science Publishers, 1992.
55. S. Rolfe, J. Barson, *Fracture and Fatigue Control in Structures*, Prentice Hall, Inc., 1977.
56. D. Broek, *Elementary Engineering Fracture Mechanics*, Kluwer Academic Publishers Group, 1982.
57. http://en.wikipedia.org/wiki/Strain_energy_release_rate
58. P.J. Minguet, T.K. O'Brien, "Analysis of Skin/Stringer Bond Failure Using a Strain Energy Release Rate Approach," *Proceedings of the Tenth International Conference on Composite Materials (ICCM-X)*, Vancouver, British Columbia, Canada, August 1995.
59. E. Suhir, "Stresses in Bi-Metal Thermostats," *ASME Journal of Applied Mechanics*, vol. 53, No. 3, Sept. 1986.
60. E. Suhir, *Structural Analysis of Microelectronic and Fiber Optic Systems*, Van-Nostrand, 1991.
61. E. Suhir, J. Morris, L. Wang, S. Yi, "Could the Dynamic Strength of a Bonding Material in an Electronic Device Be Assessed from Static Shear-Off Test Data?" *Journal of Materials Science: Materials in Electronics*, vol. 27, No. 7, 2016.
62. E. Suhir, R. Mahajan, "Are Current Qualification Practices Adequate?" Circuit Assembly, April 2011.
63. E. Suhir, "Assuring Aerospace Electronics and Photonics Reliability: What Could and Should Be Done Differently," *2013 IEEE Aerospace Conference*, Big Sky, Montana, March 2013.
64. E. Suhir, "Predicted Reliability of Aerospace Electronics: Application of Two Advanced Probabilistic Techniques," *2013 IEEE Aerospace Conference*, Big Sky, Montana, March 2013.
65. E. Suhir, "Probabilistic Design for Reliability," *ChipScale Reviews*, vol. 14, No. 6, 2010.
66. E. Suhir, R. Mahajan, A.E. Lucero, L. Bechou, "Probabilistic Design-for-Reliability (PDfR) Concept and Novel Approach to Qualification of Aerospace Electronic Products," *IEEE Aerospace Conference*, Big Sky, Montana, March 2012.
67. E. Suhir, *Applied Probability for Engineers and Scientists*, McGraw-Hill, 1997.
68. E. Suhir, "Analytical Thermal Stress Modeling in Electronic and Photonic Systems," *ASME Applied Mechanics Reviews*, vol. 62, No. 4, 2009.

3 Thermal Stress in Assemblies with Identical Adherends

"The practical value of mathematics is, in effect, a possibility to obtain, with its help, results simpler and faster."

—Andrey N. Kolmogorov, Russian mathematician

3.1 INTRODUCTION

Stresses in adhesively bonded or soldered joints were addressed in numerous publications, including the field of electronics and photonics (see, for example, [1–32]). There is an obvious incentive to employ thermally matched materials in electronic and photonic assemblies. This is true even from the thermally induced stresses point of view, but assemblies with identical silicon adherends are attractive, first of all, from the standpoint of their electrical performance and compact size. In this chapter, we address, as suitable and useful examples and using analytical modeling, several practically important thermal stress problems associated with the employment of assemblies with identical adherends.

3.2 BELL LABS SI-ON-SI MULTI-CHIP FLIP-CHIP PACKAGING TECHNOLOGY

The Bell Labs Si-on-Si multi-chip flip-chip packaging technology [32–35], developed about 30 years ago, is a good example of such a design (Figures 3.1 and 3.2). The design could be viewed as a "giant chip" on a miniaturized board. The technology focused on low cost, high performance, compact size, and high reliability. The bottleneck of the design is reliability and, first of all, the performance of the solder Si-to-Si interconnects. Extensive testing was carried out during the development of the pioneering Bell Labs Si-on-Si design, with an objective to assess the lifetime of solder joint interconnections. A solder bump was modeled, when analytical modeling was carried out, as a short elastic cylinder clamed at its flat ends (Figure 3.3). The results are shown in Figures 3.4 and 3.5. The calculated data indicated that the highest stresses and strains occur in the axial direction, while the loadings are applied in horizontal planes. Figure 3.5 indicates also that there is an incentive for employing solder joints with elevated ratios of their stand-offs (heights) to diameter. It is noteworthy that many years later, this finding was implemented in the column-grid array technology as an attractive substitute for

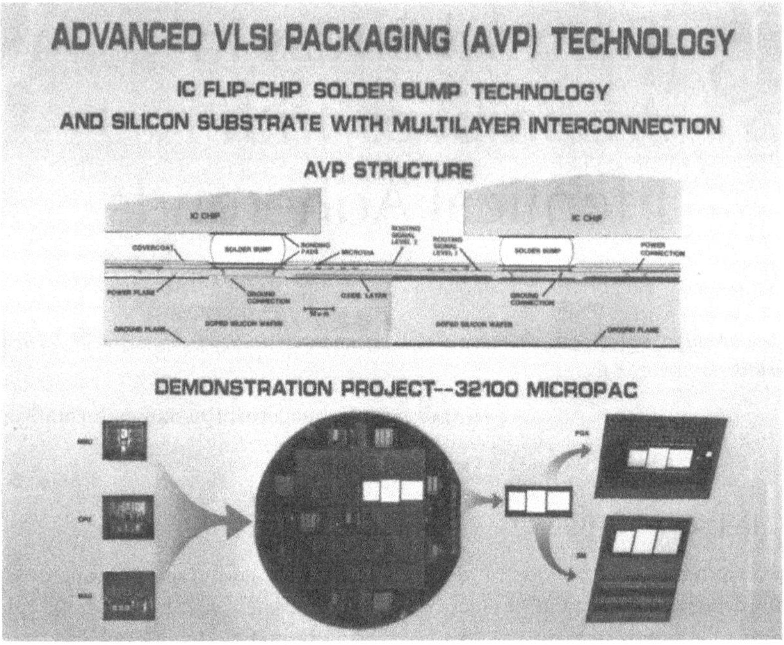

FIGURE 3.1 Bell Labs Si-on-Si technology: multichip flip-chip package design.

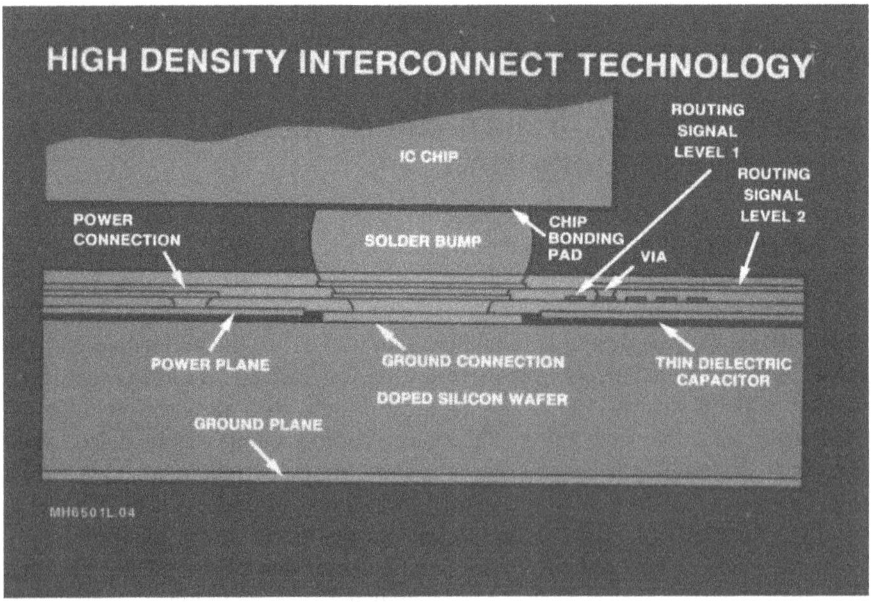

FIGURE 3.2 Solder bump in Si-on-Si technology: during temperature cycling tests, it is the mismatch between the solder and Si materials that determines solder joint fatigue strength.

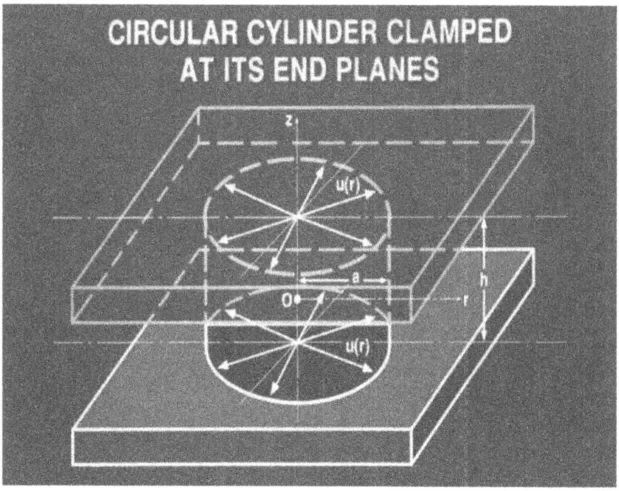

FIGURE 3.3 Si-on-Si technology: solder bump modeled as a short cylinder clamped at the ends and subjected to shearing radial tensile forces applied to its flat ends. (From E. Suhir, "Axisymmetric Elastic Deformations of a Finite Circular Cylinder with Application to Low Temperature Strains and Stresses in Solder Joints," ASME Journal of Applied Mechanics, vol. 56, No. 2, 1989.)

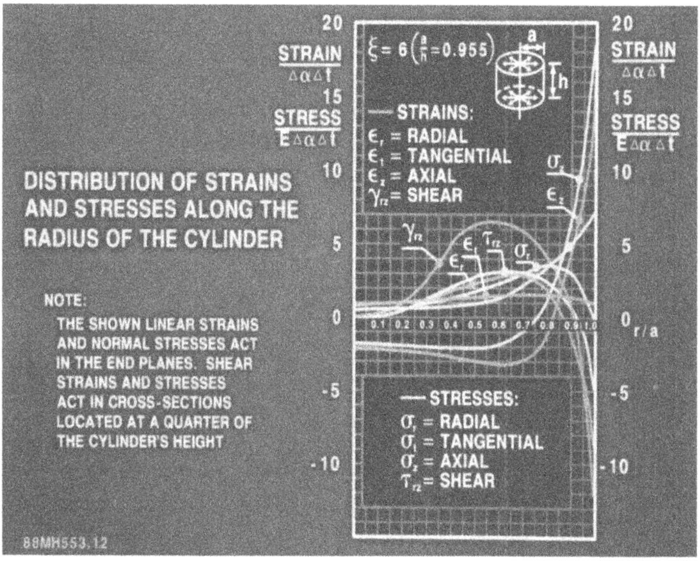

FIGURE 3.4 Calculated stresses and strains in a short cylinder whose plane ends are subjected to radial tension caused by the thermal contraction mismatch of the solder material with Si. (From E. Suhir, "Axisymmetric Elastic Deformations of a Finite Circular Cylinder with Application to Low Temperature Strains and Stresses in Solder Joints," ASME Journal of Applied Mechanics, vol. 56, No. 2, 1989.)

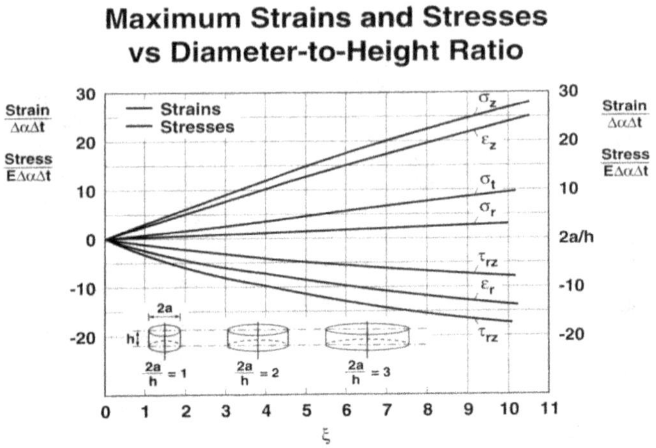

FIGURE 3.5 Maximum stresses and strains in the short cylinder shown in Figure 3.3. (From E. Suhir, "Axisymmetric Elastic Deformations of a Finite Circular Cylinder with Application to Low Temperature Strains and Stresses in Solder Joints," ASME Journal of Applied Mechanics, vol. 56, No. 2, 1989.)

ball-grid-array technology. The analytical data was confirmed by finite element analysis (FEA). Failure-oriented testing was conducted, and its results are shown in Figure 3.6. Such testing was carried out until half of the tested population failed. The experimental bathtub curve is shown in Figure 3.7. It is noteworthy that its wear-out portion occupies about half of the material's lifetime. This circumstance

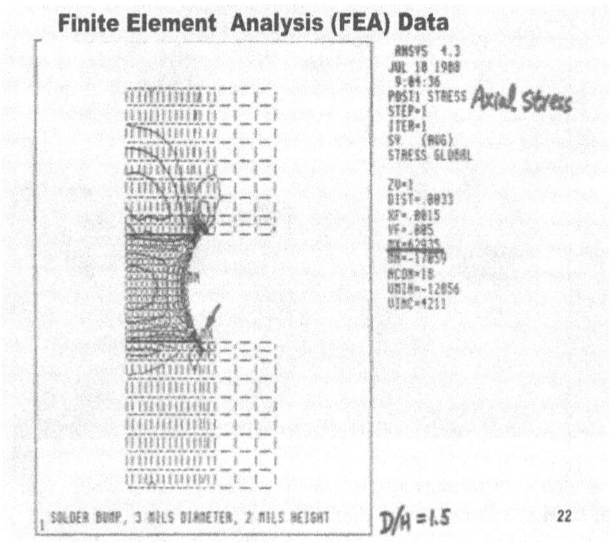

FIGURE 3.6 FEA-predicted configuration of the solder bump at low temperature conditions.

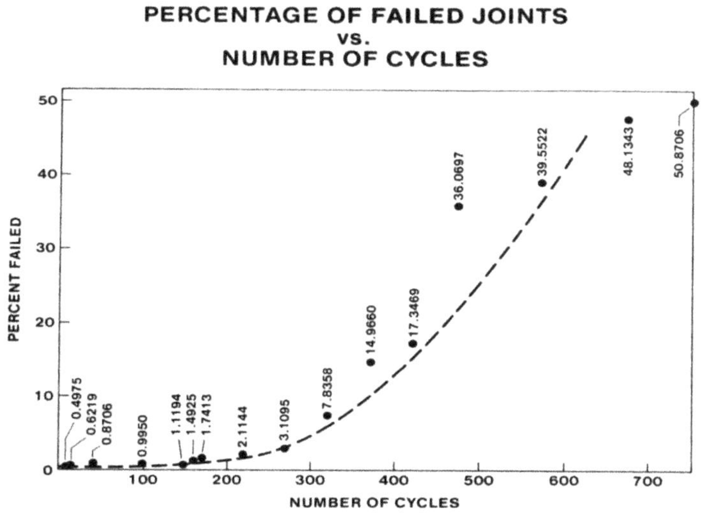

FIGURE 3.7 Percentage of failed joints versus number of cycles. (From E. Suhir, "Axisymmetric Elastic Deformations of a Finite Circular Cylinder with Application to Low Temperature Strains and Stresses in Solder Joints," ASME Journal of Applied Mechanics, vol. 56, No. 2, 1989.)

should be considered when designing solder joint interconnections; their lifetime should not be limited just by their steady-state portion.

3.3 SIMPLEST ELONGATED ASSEMBLY WITH IDENTICAL ADHERENDS

Such an assembly can be viewed as a special case of the assembly addressed in Chapter 2. Let us consider an assembly comprised of two identical elongated rectangular adherends and a low-modulus adhesive or solder bond. Based on equation (2.5), we conclude that the effective coefficient of thermal expansion (CTE) α_e of the assembly is not different of the CTE α_1 of the bonded components (components #1) of the assembly, whose effective Young's modulus and thickness are substantially greater than Young's modulus and thickness of the bonding material. Then, using formula (2.6), we conclude that the force (per unit assembly width) acting in the cross sections of the adherend materials in the midportion of the assembly is practically zero, and the force acting in the cross sections of the bonding layer (zero component) in its midportion is

$$T_0^0 = \frac{E_0}{1-\nu_0} h_0 (\alpha_0 - \alpha_1)\Delta t \tag{3.1}$$

Here, E_0 is Young's modulus of the bonding material, ν_0 is its Poisson's ratio, h_0 is its thickness, α_0 is its CTE, α_1 is the CTE of the bonded components material, and Δt is the change in temperature from the manufacturing to the low (testing, operation) temperature. The corresponding normal stress is

$$\sigma_0 = \frac{T_0^0}{h_0} = \frac{E_0}{1-\nu_0}(\alpha_0 - \alpha_1)\Delta t. \tag{3.2}$$

The axial compliance λ of the assembly is due to the adherends only and, for two identical adherends, is expressed, using formula (2.9), as

$$\lambda = 2\lambda_1 = \frac{2}{E_1^0 h_1} = 2\frac{1-\nu_1}{E_1 h_1}. \tag{3.3}$$

The interfacial compliance of the assembly can be determined, using equation (2.16), as follows:

$$\kappa = \kappa_0 + 2\kappa_1 \approx 2\kappa_1 = 2\frac{h_1}{3G_1} = \frac{4}{3}h_1\frac{1+\nu_1}{E_1}. \tag{3.4}$$

Then, the interfacial shearing stress factor is, in accordance with formula (2.15),

$$k = \sqrt{\frac{\lambda}{\kappa}} = \frac{1}{h_1}\sqrt{\frac{3}{2}\frac{1-\nu_1}{1+\nu_1}}. \tag{3.5}$$

The maximum interfacial shearing stress can be found as

$$\tau_{max} = -kT_0^0 = -kh_0\sigma_0. \tag{3.6}$$

Therefore, the assembly of interest is an elongated one and is characterized by the following parameters:

Component	CTE, α, 1/°C	Young's Modulus, *E*, kg/mm².	Poisson's Ratio, ν	Thickness, h, mm
0	7.4E-6	12,500	0.38	0.25
#1	2.4E-6	36,000	0.33	2.00

Temperature change is $\Delta t = 275°C$. Then we have:

$$\sigma_0 = \frac{E_0}{1-\nu_0}(\alpha_0 - \alpha_1)\Delta t = \frac{12{,}500}{0.62}5.0\times10^{-6}\times275 = 27.722\,\text{kg/mm}^2$$

$$k = \sqrt{\frac{\lambda}{\kappa}} = \frac{1}{h_1}\sqrt{\frac{3}{2}\frac{1-\nu_1}{1+\nu_1}} = \frac{1}{2}\sqrt{\frac{3}{2}0.5038} = 0.3726\,\text{mm}^{-1}$$

$$\tau_{max} = -kT_0^0 = -kh_0\sigma_0 = -0.3726\times0.25\times27.722 = -2.5823\,\text{kg/mm}^2$$

3.4 ASSEMBLY WITH IDENTICAL ADHERENDS SUBJECTED TO DIFFERENT TEMPERATURES: THERMAL STRESSES IN A MULTILEG THERMOELECTRIC MODULE DESIGN

3.4.1 Motivation

There is an obvious incentive to determine what could possibly be done to reduce the thermal stresses in a thermoelectric module (TEM)/power-generator design. These designs (Figures 3.8–3.10) can be modeled as assemblies comprised of two identical components subjected to different temperatures. The bonding system is a plurality of identical column-like supports (legs) spaced at equal distances from each other. It has been shown [11–13] that compliant bonds could bring down the thermally induced interfacial stresses in them, so could thinner (dimension in the horizontal direction) and longer (dimension in the vertical direction) TEM legs result in a appreciable stress relief, and, perhaps, since the device's size is always important, such a relief could be achieved even if shorter legs are employed, as long as they are thin and the spacing between them is significant? It is imperative, of course, that if thin TEM legs are employed for lower stresses, there is still enough interfacial real estate, so that the adhesive strength of the TEM assembly is not compromised. On the other hand, owing to a lower stress level in an assembly with thin legs, assurance of its interfacial strength might be less of a challenge than for a conventional assembly with stiff, thick, and closely positioned legs.

TOP
DO NOT AFFIX OVERLAYS ALONG THIS SURFACE

CALCULATED FAILURE RATE

NUMBER OF CYCLES ΔN	PERCENTAGE OF JOINTS FAILED PREVIOUSLY	PERCENTAGE OF JOINTS FAILED DURING GIVEN INTERVAL	FAILURE RATE $\lambda = \frac{(3)}{[100-(2)]\,\Delta N}$, 1/CYCLE
(1)	(2)	(3)	(4)
0 – 50	0	0.8706	1.7412×10^{-4}
50 – 100	0.8706	0.1244	0.2510×10^{-4}
100 – 150	0.9950	0.1244	0.2513×10^{-4}
150 – 200	1.1194	0.6219	1.2579×10^{-4}
200 – 250	1.7413	0.3731	0.7594×10^{-4}
250 – 300	2.1144	0.9951	2.0332×10^{-4}
300 – 350	3.1095	4.7263	9.7559×10^{-4}
350 – 400	7.8358	7.1302	15.4728×10^{-4}
400 – 450	14.9660	2.3809	5.6000×10^{-4}
450 – 500	17.3469	18.7228	45.3043×10^{-4}
500 – 600	36.0697	3.4825	5.4473×10^{-4}
600 – 700	39.5522	8.5821	14.1974×10^{-4}
700 – 750	48.1343	2.4363	10.5514×10^{-4}

FIGURE 3.8 Calculated failure rate during accelerated testing of solder joint interconnections.

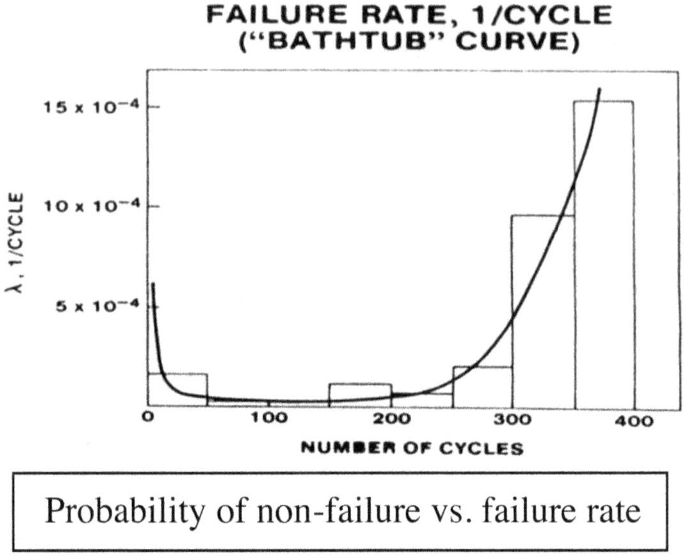

FIGURE 3.9 Experimental bathtub curve and probability of nonfailure versus failure rate and the number of cycles. (From E. Suhir, "Axisymmetric Elastic Deformations of a Finite Circular Cylinder with Application to Low Temperature Strains and Stresses in Solder Joints," ASME Journal of Applied Mechanics, vol. 56, No. 2, 1989.)

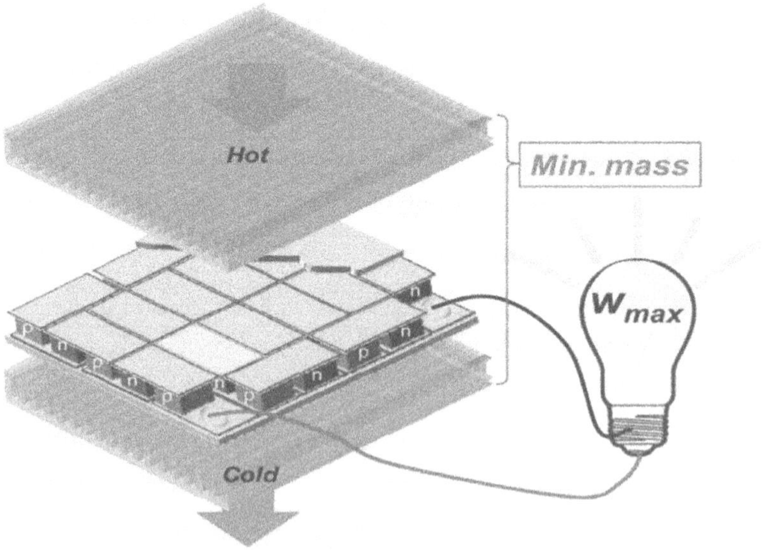

FIGURE 3.10 Thermoelectric module (TEM)/power generator. (From K. Yazawa, G. Solbrekken, A. Bar-Cohen, Thermoelectric-Powered Convective Cooling of Microprocessors, IEEE Transactions on Advanced Packaging, vol. 28, No. 2, 2005.)

3.4.2 Background

The thermoelectric cooling technology uses the Peltier effect to create heat flux between two assembly components. A cooler based on the Peltier effect (Peltier device or a solid-state refrigerator or a thermoelectric cooler or module) transfers heat from the "hot plate" to the "cold plate," thereby consuming energy.

Although the device can be used either for heating or for cooling (refrigeration), the main application in practice is cooling, and the major advantages of a Peltier cooler (compared to, say, a vapor-compression refrigerator) are its lack of moving parts or circulating liquid, as well as its small size and flexible shape. But could it also have high power efficiency, low cost, and high reliability? During the last decade, TEMs have received increased attention of the research and engineering communities.

Various aspects of the TEM technology has been addressed by a number of investigators (see, for example, [36–50]). Although the majority of the research work has been naturally focused on the thermoelectric properties of the TEM materials and modules (See Beck coefficient, electrical resistivity, thermal conductivity) and their functional (thermoelectric) performance, some investigators have pointed out the importance of the TEM short- and long-term mechanical (physical) robustness and addressed, using primarily FEA, the mechanical behavior of the TEM materials and structures. An analytical approach has been used first for the evaluation of thermal stresses in a simplified two-leg TEM design [50]. The emphasis was on the assessment of the effect of the size of the bonded regions on the maximum stress. It has been shown that the employment of thinner and longer legs could indeed result in a substantial stress relief, thereby leading to a more mechanically robust TEM. In the analysis that follows, the taken approach is generalized for the case of a TEM with any number of legs. The analysis is carried out considering that flexible (long and thin) legs provide a certain level of the mechanical compliance between the ceramic components, but do not experience axial thermal loading.

3.4.3 Basic Equation

Consider an assembly comprised of two identical components, #1 and #2, subjected to different temperatures and supported by a plurality of identical supports (legs). In Figure 3.11, an arbitrary segment of a multileg assembly is shown. There might be many similar segments to the left and to the right of the one depicted in this figure.

The longitudinal interfacial displacements of the bonded components ("adherends") can be sought in an approximate form as follows:

$$u_1(x) = -\alpha\Delta t_1 x + \lambda_1 \int_0^x T(\xi)d\xi - \kappa_1\tau(x),$$

$$u_2(x) = -\alpha\Delta t_2 x - \lambda_1 \int_0^x T(\xi)d\xi + \kappa_1\tau(x) \qquad (3.7)$$

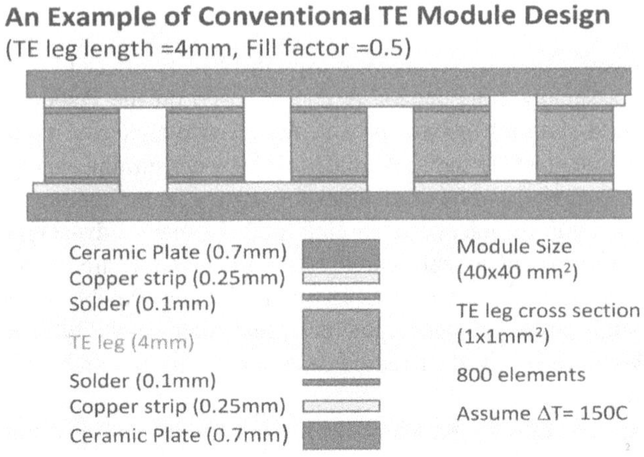

FIGURE 3.11 An example of a conventional TEM design.

Here, α is the CTE of the component material, Δt_1 and Δt_2 are the changes in the component temperatures (from the manufacturing to the operation temperature), $\lambda_1 = \frac{1-\nu_1}{E_1 h_1}$ is the axial compliance of each of the two identical components, E_1 and ν_1 are the elastic constants of their material, h_1 is component thickness, $T(x) = \int_0^x \tau(\xi)d\xi$ is the induced force acting in the cross sections of assembly components in the given cross section, $\tau(x)$ is the interfacial shearing stress, and κ_1 is the longitudinal interfacial compliance of the component. The origin of the coordinate x is in the mid-cross section of the bonded region. The first terms in formulas (3.7) are unrestricted thermal displacements. The second terms are due to the elastic displacements caused by the thermally induced forces $T(x)$, assuming that these forces are uniformly distributed over the given cross section. The third terms are, in effect, corrections to this assumption. They consider that the displacements of the point located at the interface, where the thermal forces are applied, are larger than the displacements of the inner points of the cross section. These terms are based on an assumption that the corrections in question can be evaluated as products of the (so far unknown) distributed interfacial shearing stress acting in this cross section and the (position-independent) interfacial compliance of the given assembly component. The total interfacial compliance of the assembly is due to both the components and the bond and can be evaluated using the recommendations in [11, 12]. For small $\frac{h}{l}$ values (say, below 0.25) the interfacial compliance of the assembly components loaded on one of their long sides can be evaluated as $\kappa = \frac{h}{3G}$, where $G = \frac{E}{2(1+\nu)}$ is the shear modulus of the material. For the assembly components (e.g., such as the bonding layer) loaded on both sides in an antisymmetric fashion, the factor 3 in the denominator of the previous formula for the interfacial compliance can be omitted. In another extreme case, for large $\frac{h}{l}$ values (say, above 0.5), the expression $\kappa = \frac{3-\nu}{4\pi}\frac{l}{G}$ can be obtained for the interfacial compliance [18]. Calculations carried out for typical electronic and photonic assemblies indicate that the compliances calculated for the $\frac{h}{l}$ ratios of 0.25 and 0.5 might not be very much different, and therefore, if the actual $\frac{h}{l}$ ratio falls between the above values, the compliance of the material of interest can be computed by using linear interpolation between its compliance computed for 0.25 and 0.5 aspect ratios. The condition

$$u_1(x) = u_2(x) + \kappa_0 \tau(x) \tag{3.8}$$

of the compatibility of the interfacial displacements (3.7) results in the following equation for the interfacial shearing stress $\tau(x)$:

$$\kappa\tau(x) - 2\lambda_1 \int_0^x T(\xi)d\xi = \alpha\Delta t x, \tag{3.9}$$

where the total compliance $\kappa = \kappa_0 + 2\kappa_1$ of the interface can be calculated as the sum of the interfacial compliance $\kappa_0 = \frac{h_0}{G_0}$ of a buffering (bonding) layer, if any, and the two bonded materials. Here, h_0 is the thickness of the buffering layer and G_0 is the shear modulus of its material. From (3.9), we find by differentiation:

$$\kappa\tau'(x) - 2\lambda_1 T(x) = \alpha\Delta t. \tag{3.10}$$

For the ith support (leg) in Figures 3.3 and 3.4, this equation could be written as

$$\kappa\tau_i'(x) - 2\lambda_1 T_i(x) = \alpha\Delta t. \tag{3.11}$$

The solution to this equation can be sought in the form:

$$T_i(x) = B + C_i \sinh kx + D_i \cosh kx. \tag{3.12}$$

The constant B is the particular solution to the inhomogeneous equation (3.11), the second and the third terms (containing hyperbolic functions) provide the general solution to the corresponding homogeneous equation, C_i and D_i are constants of integration, and k is thus far unknown parameter of the interfacial shearing stress. Since the interfacial shearing stress $\tau(x)$ can be evaluated as the derivative of the distributed axial force $T(x)$ i.e., $\tau(x) = T'(x)$, equation (3.12) yields

$$\tau_i'(x) = T_i''(x) = k^2(C_i \sinh kx + D_i \cosh kx). \tag{3.13}$$

Introducing the expressions (3.12) and (3.13) into the equation (3.11), we conclude that this equation is fulfilled, if the relationships $k = \sqrt{\frac{2\lambda_1}{\kappa}}$, $B = -\frac{\alpha\Delta t}{2\lambda_1}$ take place.

Equation (3.12) must satisfy the boundary conditions (Figure 3.11) $T_i(-l) = \hat{T}_{i-1}$, $T_i(l) = \hat{T}_i$. These conditions indicate that the distributed force $T(x)$ in the bonded ith region changes from $\hat{T}_{i-1}$ to $\hat{T}_i$, when moving from the left end $x = -l$ of this region to its right end $x = l$. These boundary conditions, solution (3.12), and the formula for the B value yield

$$C_i = \frac{\hat{T}_i - \hat{T}_{i-1}}{2\sinh kl}, \quad D_i = \frac{1}{\cosh kl}\left(\frac{\alpha\Delta t}{2\lambda_1} + \frac{\hat{T}_{i-1} + \hat{T}_i}{2}\right). \tag{3.14}$$

Then, solution (3.12) results in the following expression for the distributed force acting within the ith bonded region:

$$T_i(x) = -\frac{\alpha\Delta t}{2\lambda_1}\left(1 - \frac{\cosh kx}{\cosh kl}\right) + \frac{\hat{T}_i + \hat{T}_{i-1}}{2}\frac{\sinh kx}{\sinh kl} + \frac{\hat{T}_{i-1} + \hat{T}_i}{2}\frac{\cosh kx}{\cosh kl}, \quad -l \le x \le l \tag{3.15}$$

The distributed interfacial shearing stress can be found by differentiation:

$$\tau_i(x) = T'(x) = k\left[\frac{\alpha\Delta t}{2\lambda}\frac{\sinh kx}{\sinh kl} + \frac{\hat{T}_i - \hat{T}_{i-1}}{2}\frac{\cosh kx}{\cosh kl} + \frac{\hat{T}_{i-1} + \hat{T}_i}{2}\frac{\sinh kx}{\cosh kl}\right], \quad -l \le x \le l. \tag{3.16}$$

The first terms in the right parts of formulas (3.15) and (3.16) are due to the local thermal expansion (contraction) mismatch of the component materials, and the second and the third terms are caused by their global mismatch. It is the global mismatch that results in the forces applied to the bonded regions from the unbonded regions of the assembly.

3.4.4 Theorem of Three Axial Forces

From (3.16), we obtain the following expression for the interfacial shearing stress acting in the $i + 1$ leg:

$$\tau_{i+1}(x) = k\left[\frac{\alpha\Delta t}{2\lambda_1}\frac{\sinh kx}{\sinh kl} + \frac{\hat{T}_{i+1} - \hat{T}_i}{2}\frac{\cosh kx}{\cosh kl} + \frac{\hat{T}_i + \hat{T}_{i+1}}{2}\frac{\sinh kx}{\cosh kl}\right], \quad -l \le x \le l \tag{3.17}$$

The following condition of the compatibility for the longitudinal displacements at the boundary of the bonded and unbonded regions should be fulfilled:

$$\kappa[\tau_{i+1}(-l) - \tau_i(l)] = 2\lambda_1\hat{T}_i(2L - 4l). \tag{3.18}$$

The left part of this condition is the difference in the longitudinal interfacial displacements of the assembly components at the bonded region at its boundary with the unbonded portion. The right part is the difference in the longitudinal displacements of the assembly components at the unbonded portion at its boundary with the bonded region. Introducing the expressions (3.16) and (3.17) into the condition (3.18), the following equation for the forces $\hat{T}_{i-1}$, $\hat{T}_i$, and $\hat{T}_{i+1}$ can be obtained:

$$(t-c)\hat{T}_{i-1}+2(t+c+s)\hat{T}_i+(t-c)\hat{T}_{i+1}=2t\frac{\alpha\Delta t}{\lambda_1}, \tag{3.19}$$

where

$$t=\tanh kl, \quad c=\coth kl, \quad s=4kl\left(\frac{L}{2l}-1\right). \tag{3.20}$$

The recurrent equation (3.19) has the form of (and is based on the same reasoning as) the well-known theorem of three (bending) moments in the theory of continuous beams lying on separate simple supports (see, for example, [18]) and could be called therefore "theorem of three axial forces." A similar equation has been obtained earlier for the "piece-wise continuous" inhomogeneous bonding layer in a bimaterial assembly [36], that is, for an assembly with no unbonded regions. After the forces $\hat{T}_{i-1}$, $\hat{T}_i$, and $\hat{T}_{i+1}$ are determined, the shearing stress can be found by formulas (3.16) and (3.17). Equation (3.19) can be actually used in the same fashion as the "theorem of three moments" is used in structural analysis of long beams lying on simple supports, namely, by moving, say, from left to right, forming this equation for three sequential axial forces, solving it, and then evaluate the interfacial shearing stresses. Examine several special cases to make sure that the previously obtained formulas for these cases can be obtained from the more general "theorem of three axial forces."

3.4.5 Special Cases

3.4.5.1 Homogeneously Bonded Assembly

In this case, one should put $\frac{L}{2l}=1, \hat{T}_{i-1}=\hat{T}_{i+1}=0, \hat{T}_i=\hat{T}$, and equation (3.19) yields [12]

$$\hat{T}=\frac{\alpha\Delta t}{2\lambda_1}\left(1-\frac{1}{\cosh 2kl}\right). \tag{3.21}$$

If, in addition, the kl product is significant, such as the assembly is long and/or stiff, then $\hat{T}=\hat{T}_\infty=\frac{\alpha\Delta t}{2\lambda_1}$. The maximum interfacial shearing stress can be found in this case as

$$\tau_{\max}=k\hat{T}_\infty\tanh kl=k\frac{\alpha\Delta t}{2\lambda_1}\tanh kl. \tag{3.22}$$

For a long and/or stiff assembly, tanh kl can be put equal to one and therefore

$$\tau_{\max}=k\hat{T}_\infty=k\frac{\alpha\Delta t}{2\lambda_1}.$$

3.4.5.2 Assembly Bonded at the Ends (Two-Legged TEM)

By putting $\hat{T}_{i-1} = \hat{T}_{i+1} = 0$, $\hat{T}_i = \hat{T}$, we obtain from equation (3.19) [23]

$$\hat{T} = \frac{\alpha\Delta t}{\lambda_1} \frac{\tanh kl}{\tanh kl + \coth kl + 4kl\left(\frac{L}{2l} - 1\right)} = \frac{\alpha\Delta t}{2\lambda_1} \frac{\tanh kl}{2kl\left(\frac{L}{2l} - 1\right) + \coth 2kl} \tag{3.23}$$

For the interfacial shearing stress, formulas (3.16) and (3.17) yield

$$\tau(x) = k\frac{\alpha\Delta t}{2\lambda_1}\frac{\sinh kx}{\cosh kl} + k\hat{T}\frac{\cosh k(l-x)}{\sinh 2kl}$$

$$= k\frac{\alpha\Delta t}{2\lambda_1}\left[\frac{\sinh kx}{\cosh kl} + \frac{\tanh kl}{2kl\left(\frac{L}{2l} - 1\right)\sinh 2kl + \cosh 2kl}\cosh k(l-x)\right] \tag{3.24}$$

The maximum shearing stress can be found as: $\tau_{\max} = \tau(l) = \tau_\infty \chi(kl)$, where $\tau_\infty = k\frac{\alpha\Delta t}{2\lambda_1}$ is the interfacial shearing stress at the ends of an infinitely large assembly (or a two-legged assembly with very short-and-thick legs), and the function

$$\chi(kl) = \tanh kl\left[1 + \left(2kl\left(\frac{L}{2l} - 1\right)\sinh 2kl + \cosh 2kl\right)^{-1}\right] \tag{3.25}$$

considers the effects of the finite values of the product kl and the ratio $\frac{L}{2l}$. The function $\chi(kl)$ is tabulated in Table 3.1.

As evident from the computed data, the function $\chi(kl)$ can be put equal to 1 for kl values exceeding 4.

TABLE 3.1
Function $\chi(kl)$ that Determines the Effect of the Finite Size of the Assembly and the Bonded Regions in it on the Maximum Shearing Stress

kl	0	1	2	3	4	5
$L/2l$	x	x	x	x	x	x
1	0	0.9640	0.9994	0.9999	1.0000	1.0000
2	0	0.8278	0.9730	0.9957	0.9995	1.0000
3	0	0.8022	0.9698	0.9954	0.9994	1.0000
4	0	0.7909	0.9686	0.9953	0.9994	1.0000
5	0	0.7845	0.9680	0.9952	0.9994	1.0000
6	0	0.7804	0.9676	0.9952	0.9994	1.0000
∞	0	0.7616	0.9659	0.9950	0.9994	1.0000

TABLE 3.2
Function $\chi_m(kl)$ that Determines the Effect of the Finite Size of the Assembly and the Bonded Regions in it on the Maximum Shearing Stress

Kl	0	0.1	0.5	1	2	3	4	5	∞
L/2*l*	x	x	x	x	x	x	x	x	x
1	1.0000	1.0000	1.0000	1.0000	1.0000	1.0000	1.0000	1.0000	1.0000
2	0.3333	0.3326	0.3161	0.2758	0.1942	0.1422	0.1110	0.0909	0
3	0.2000	0.1995	0.1877	0.1599	0.1075	0.0766	0.0588	0.0476	0
4	0.1429	0.1425	0.1335	0.1126	0.0744	0.0524	0.0400	0.0323	0
5	0.1111	0.1108	0.1036	0.0869	0.0568	0.0398	0.0303	0.0244	0
6	0.0909	0.0906	0.0846	0.0708	0.0460	0.0321	0.0244	0.0196	0
∞	0	0	0	0	0	0	0	0	0

3.4.5.3 Midportion of a Long Multilegged Assembly

By putting in equation (3.18) $\hat{T}_{i-1} = \hat{T}_i = \hat{T}_{i+1} = \hat{T}$, we obtain: $\hat{T} = \frac{\alpha\Delta t}{2\lambda_1}\chi_m(kl)$, where the factor

$$\chi_m(kl) = \left[1 + 2kl\left(\frac{L}{2l} - 1\right)\coth kl\right]^{-1} \tag{3.26}$$

is tabulated in Table 3.2. As evident from the calculated data, this factor can be very small, when the ratio $L/2l$ is significant, such as when the leg spacing is large, no matter how large or small the product kl is. This behavior is quite different of the behavior of the two-legged assembly. Clearly, when the force $\hat{T}$ in the midportion of the assembly is small, the maximum interfacial shearing stress at the assembly end will be low as well.

The maximum interfacial shearing stress at the end of the assembly can be computed as $\tau_{max} = k\hat{T}\tanh kl$, where $\hat{T}$ the axial force is expressed by formula (3.23) and the hyperbolic tangent $\tanh kl$ considers the effect of the finite size of the bonded area on the maximum interfacial shearing stress.

3.4.6 TEM Designs in Figures 3.11 and 3.12

Let us apply equation (3.19) to the five-legged structures (designs) shown in Figures 3.11 (conventional design with thick-and-long legs) and 3.12 (novel UCSC design with thin-and-short legs). Going in this equation from left to right, putting sequentially $i = 1,2,3,4$ and considering the symmetry of the structures in question with respect to their mid-cross sections, we obtain the following two equations for the forces $\hat{T}_1$ and $\hat{T}_2$ acting in the spans between the two peripheral legs (on each side of the assembly) and between the second and the third legs, respectively (Figure 3.13):

$$2(t+c+s)\hat{T}_1 + (t-c)\hat{T}_2 = 2t\frac{\alpha\Delta t}{\lambda_1}, \quad (t-c)\hat{T}_1 + (3t+c+s)\hat{T}_2 = 2t\frac{\alpha\Delta t}{\lambda_1}. \tag{3.27}$$

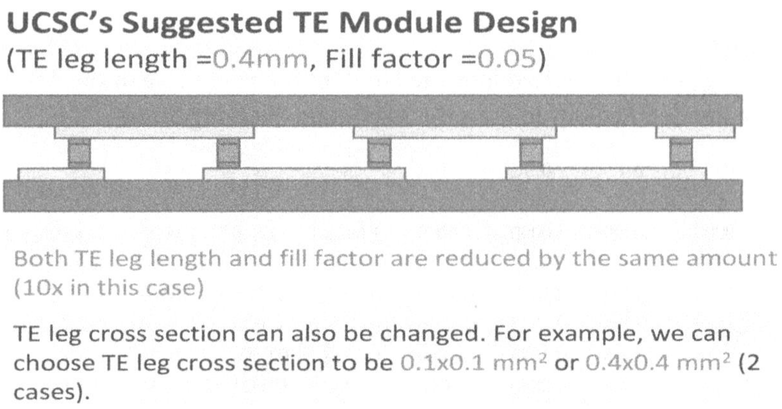
UCSC's Suggested TE Module Design
(TE leg length =0.4mm, Fill factor =0.05)
Both TE leg length and fill factor are reduced by the same amount (10x in this case)
TE leg cross section can also be changed. For example, we can choose TE leg cross section to be 0.1x0.1 mm² or 0.4x0.4 mm² (2 cases).

FIGURE 3.12 UCSC TEM design with thinner legs (in our analyses, the upper plate is component #1 and the lower plate is component #2).

Since the right parts of these equations are the same, the forces $\hat{T}_1$ and $\hat{T}_2$ have to be related as

$$(t+3c+2s)\hat{T}_1-(2t+2c+s)\hat{T}_2=0. \tag{3.28}$$

From one of equations (3.27) and equation (3.28) we find:

$$\hat{T}_1=2t\frac{\alpha\Delta t}{\lambda_1 D}(2t+2c+s),\quad \hat{T}_2=2t\frac{\alpha\Delta t}{\lambda_1 D}(t+3c+2s), \tag{3.29}$$

where

$$D=5t^2+c^2+10+2s^2+8ts+4cs \tag{3.30}$$

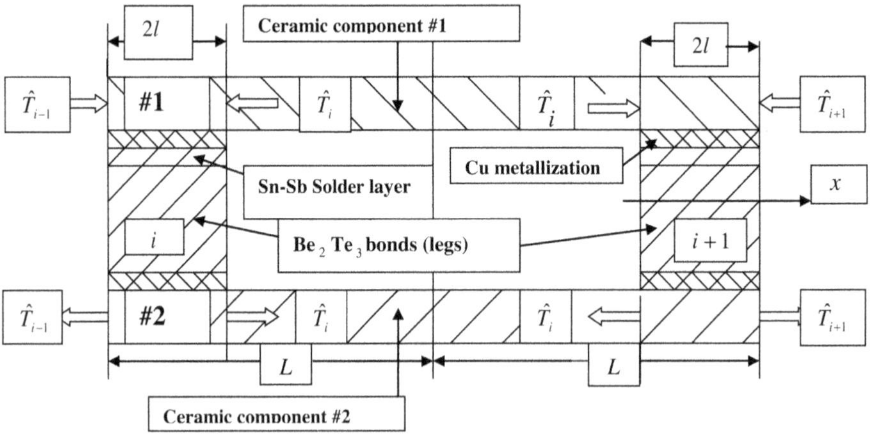
2l
Ceramic component #1
2l
$\hat{T}_{i-1}$
#1
$\hat{T}_i$
$\hat{T}_i$
$\hat{T}_{i+1}$
Cu metallization
Sn-Sb Solder layer
x
i
Be 2 Te 3 bonds (legs)
i + 1
$\hat{T}_{i-1}$
#2
$\hat{T}_i$
$\hat{T}_i$
$\hat{T}_{i+1}$
L
L
Ceramic component #2

FIGURE 3.13 A segment of a TEM structure.

is the determinant of the system of equations (3.27). The forces $\hat{T}_1$ and $\hat{T}_2$ will be equal if the condition $t - c = s$ takes place. This condition is fulfilled for a long assembly with a continuous bond, when $\tanh kl = \coth kl = 1$ and $\frac{L}{2l} = 1$.

Equation (3.29) indicates that the induced forces $\hat{T}_1$ and $\hat{T}_2$ will be small and, hence, the induced thermal stresses will be low as well, if the $t = \tanh kl$ value is low (short bonded areas with highly compliant interfaces) and/or if the determinant D is significant. The latter condition takes place, as one could see from equation (3.30), if the span $s = 4kl\left(\frac{L}{2l} - 1\right)$ is large. With low $t = \tanh kl$ values, such as with low kl values, this relationship will take place when the $\frac{L}{2l}$ ratio is significantly larger than unity.

Consider now a numerical example for the designs in Figures 3.11 and 3.12.

Input data:

Material	Thickness (height), mm	Length, mm	CTE, 1/°C	Young's Modulus, GPa (kg/mm.sq)	Poisson's Ratio	Yield Stress, GPa (kg/mm.sq)	Ultimate Stress, GPa (kg/mm.sq)
Ceramic Component	0.70	–	6.5E-6	380 (38775)	0.26	–	–
Copper strip (metallization) on ceramic	0.25	4.00	17.0E-6	115 (11735)	0.31	70.0 (7143)	250.0 (25510)
Sn–Sb solder layer	0.10	4.00	27.0E-6	44.5 (4540)	0.33	26 (2653)	41 (4184)
Be_2Te_3bond (leg)	4.00	4.00	16.8E-6	47.0 (4795)	0.40	–	–

The difference in temperature between the hot and the cold components is $\Delta t = 130°C$.

Calculated data:

- Axial compliance of one ceramic component:

$$\lambda_1 = \frac{1 - \nu_1}{E_1 h_1} = \frac{1 - 0.26}{38775 \times 0.7} = 2.7264 \times 10^{-5} \text{ mm/kg}$$

- Interfacial compliances for an infinitely long assembly:
  - Ceramic component:

$$\kappa_1 = \frac{h_1}{3G_1} = \frac{0.7}{3 \times 15387} = 1.5164 \times 10^{-5} \text{ mm}^3/\text{kg}$$

  - Copper layer:

$$\kappa_c = \frac{h_c}{3G_c} = \frac{0.25}{4479} = 5.5882 \times 10^{-5} \text{ mm}^3/\text{kg}$$

– Solder layer:

$$\kappa_s = \frac{h_s}{3G_s} = \frac{0.10}{1707} = 5.8582 \times 10^{-5}\ \text{mm}^3/\text{kg}$$

– Be_2Te_3bond (leg):

$$\kappa_l = \frac{h_l}{G_l} = \frac{0.5}{1713} = 29.1886 \times 10^{-5}\ \text{mm}^3/\text{kg}$$

Total:

$$\kappa = 2(\kappa_1 + \kappa_c + \kappa_s) + \kappa_l = 2(1.5164 \times 10^{-5} + 5.5882 \times 10^{-5}) + 29.1886 \times 10^{-5}$$

$$= 55.1142 \times 10^{-5}\ \text{mm}^3/\text{kg}$$

- Parameter of the interfacial shearing stress:

$$k = \sqrt{\frac{2\lambda_1}{\kappa}} = \sqrt{\frac{2 \times 2.7264 \times 10^{-5}}{55.1142 \times 10^{-5}}} = 0.3145\,\text{mm}^{-1}$$

- Maximum interfacial shearing stress:

$$\tau_\infty = k\frac{\alpha \Delta t}{2\lambda_1} = 0.3145\frac{6.5 \times 10^{-6} \times 130}{2 \times 2.7264 \times 10^{-5}} = 4.8737\,\text{kg/mm}^2$$

This stress is well below the yield stress in shear for the solder material that can be assessed as

$$\tau_Y = \frac{2653}{\sqrt{3}} = 1532\,\text{kg/mm}^2$$

Let, for example, the length of the assembly is $2L = 20$ and the thickness (dimension in the horizontal direction) of each leg is $2l = 1.0$ mm, so that the $\frac{L}{2l}$ ratio is $\frac{L}{2l} = 10$. Other input data are the same as in the table. The axial compliances λ_1 of each of the ceramic components, and the interfacial compliances κ_1 and κ_s of the ceramic components and of the solder layers are not different of those previously calculated. The other computed data are as follows:

- Interfacial compliance of one of the copper layers:

$$\kappa_c = \frac{3 - \nu_c}{4\pi}\frac{l}{G_c} = \frac{3 - 0.31}{12.5616}\frac{0.5}{4479} = 2.3897 \times 10^{-5}\ \text{mm}^3/\text{kg}$$

- Interfacial compliance of the leg:

$$\kappa_l = \frac{3-\nu_l}{4\pi}\frac{l}{G_l} = \frac{3-0.4}{12.5616}\frac{0.5}{1713} = 6.0392\times10^{-5}\,\text{mm}^3/\text{kg}$$

- Total interfacial compliance:

$$\kappa = 2(\kappa_1+\kappa_c+\kappa_s)+\kappa_l = 2(1.5164\times10^{-5}+2.3897\times10^{-5}+5.8582\times10^{-5})+$$

$$6.0392\times10^{-5} = 25.5678\times10^{-5}\,\text{mm}^3/\text{kg}$$

- Parameter of the interfacial shearing stress:

$$k = \sqrt{\frac{2\lambda_1}{\kappa}} = \sqrt{\frac{2\times2.7264\times10^{-5}}{25.5678\times10^{-5}}} = 0.4618\,\text{mm}^{-1} = 461.8098\,\text{m}^{-1},$$

The product $kl = 0.4771\times0.5 = 0.2309$

- Maximum interfacial shearing stress in an infinitely long assembly:

$$\tau_\infty = k\frac{\alpha\Delta t}{2\lambda_1} = 0.4618\frac{6.5\times10^{-6}\times130}{2\times2.7264\times10^{-5}} = 7.1563\,\text{kg/mm}^2$$

- Factor that considers the role of the kl value and the length ratio $\frac{L}{2l}$:

$$\chi(kl) = \tanh kl\left[1+\frac{1}{2kl\left(\frac{L}{2l}-1\right)\sinh 2kl+\cosh 2kl}\right]$$

$$= 0.2269\left(1+\frac{1}{0.4618\times9.0\times0.4784+1.1085}\right) = 0.3002$$

- Maximum shearing stress

$$\tau_{\max} = \tau(l) = \tau_\infty\chi(kl) = 7.1563\times0.3002 = 2.1481\,\text{kg/mm}^2$$

This stress is only 44.0% of the stress in a long assembly with a homogeneous bond.

The midportion of a long multilegged assembly can be addressed by putting $\hat{T}_{i-1} = \hat{T}_i = \hat{T}_{i+1} = \hat{T}$ in equation (3.19). Then, we obtain:

$$\hat{T} = \frac{\alpha\Delta t}{2\lambda_1}\chi_m(kl), \tag{3.31}$$

where the factor

$$\chi_m(kl) = \left[1+2kl\left(\frac{L}{2l}-1\right)\coth kl\right]^{-1} \tag{3.32}$$

TABLE 3.3
Function $\chi_m(kl)$ that Determines the Effect of the Finite Size of the Assembly and the Bonded Regions in it on the Maximum Shearing Stress

kl	0	0.1	0.5	1	2	3	4	5	∞
$L/2l$	x	x	x	x	x	x	x	x	x
1	1.0000	1.0000	1.0000	1.0000	1.0000	1.0000	1.0000	1.0000	1.0000
2	0.3333	0.3326	0.3161	0.2758	0.1942	0.1422	0.1110	0.0909	0
3	0.2000	0.1995	0.1877	0.1599	0.1075	0.0766	0.0588	0.0476	0
4	0.1429	0.1425	0.1335	0.1126	0.0744	0.0524	0.0400	0.0323	0
5	0.1111	0.1108	0.1036	0.0869	0.0568	0.0398	0.0303	0.0244	0
6	0.0909	0.0906	0.0846	0.0708	0.0460	0.0321	0.0244	0.0196	0
∞	0	0	0	0	0	0	0	0	0

is tabulated in Table 3.3. As evident from the calculated data, this factor can be very small, when the ratio $L/2l$ is significant, such as when the leg spacing is large, no matter how large or small the product kl might be. This behavior is quite different of the behavior of the two-legged assembly (see Table 3.1 data). Clearly, when the force $\hat{T}$ in the midportion of the assembly is small, the maximum interfacial shearing stress at the assembly end will be small as well.

The maximum interfacial shearing stress at the end of the assembly can be computed as

$$\tau_{\max} = k\hat{T}\tanh kl, \tag{3.33}$$

where $\hat{T}$ the axial force is expressed by formula (3.31) and the hyperbolic tangent tanh kl considers the effect of the finite size of the bonded area on the maximum interfacial shearing stress. With the input data from the previous examples, formula (3.32) yields

$$\chi_m(kl) = \left[1 + 2kl\left(\frac{L}{2l} - 1\right)\coth kl\right]^{-1} = \frac{1}{1 + 2\times 0.1941\times 9.0\times 5.2151} = 0.2225$$

The induced forces in the midportion of the assembly are

$$\hat{T} = \frac{\alpha\Delta t}{2\lambda_1}\chi_m(kl) = \frac{\alpha\Delta t}{2\lambda_1} = \frac{6.5\times 10^{-6}\times 130}{2\times 2.7264\times 10^{-5}}\times 0.22253 = 3.4485\,\text{kg/mm}$$

The maximum interfacial stress is

$$\tau_{\max} = k\hat{T}\tanh kl = 0.3882\times 3.4485\times 0.19175 = 0.2567\,\text{kg/mm}^2$$

This stress is significantly lower than the stress in a two-legged assembly: it is only 5.3% of the stress in a long assembly with a homogeneous bond and 12.0% of the maximum stress in a two-legged assembly.

3.4.7 TEM Designs in Figures 3.11 and 3.12

Consider now five-legged structures (designs) shown in Figures 3.11 (conventional design with thick-and-long legs) and 3.12 (novel UCSC design with thin-and-short legs). Going in this equation from left to right, putting sequentially i = 1,2,3,4 and considering the symmetry of the structures in question with respect to their mid-cross sections, we obtain the following two equations for the forces $\hat{T}_1$ and $\hat{T}_2$ acting in the spans between the two peripheral legs (on each side of the assembly) and between the second and the third legs, respectively:

$$2(t+c+s)\hat{T}_1 + (t-c)\hat{T}_2 = 2t\frac{\alpha\Delta t}{\lambda_1}, \quad (t-c)\hat{T}_1 + (3t+c+s)\hat{T}_2 = 2t\frac{\alpha\Delta t}{\lambda_1}. \tag{3.34}$$

Since the right parts of these equations are the same, the forces $\hat{T}_1$ and $\hat{T}_2$ should be related as

$$(t+3c+2s)\hat{T}_1 - (2t+2c+s)\hat{T}_2 = 0. \tag{3.35}$$

From one of equations (3.34) and equation (3.35) we find:

$$\hat{T}_1 = 2t\frac{\alpha\Delta t}{\lambda_1 D}(2t+2c+s), \quad \hat{T}_2 = 2t\frac{\alpha\Delta t}{\lambda_1 D}(t+3c+2s), \tag{3.36}$$

where

$$D = 5t^2 + c^2 + 10 + 2s^2 + 8ts + 4cs \tag{3.37}$$

is the determinant of the system of equations (3.34). The forces $\hat{T}_1$ and $\hat{T}_2$ will be equal if the following condition takes place:

$$t - c = s. \tag{3.38}$$

This condition is fulfilled for a long assembly with a continuous bond, when tanh kl = coth kl = 1 and $\frac{L}{2l} = 1$. Solution (3.36) indicates that the induced forces $\hat{T}_1$ and $\hat{T}_2$ will be small and, hence, the induced thermal stresses will be low as well, if the t = tanh kl value is low (short bonded areas with compliant interfaces) and/or if the determinant D is significant. The latter condition takes place, as one could see from (3.37), if the span $s = 4kl\left(\frac{L}{2l} - 1\right)$ is large. With low t = tanh kl values, i.e., with low kl values, this will take place, when the $\frac{L}{2l}$ ratio is significantly larger than unity. This is indeed the case in the novel UCSC Figure 3.12 design as compared to the conventional design in Figure 3.11.

Let us obtain the actual numerical assessments of the induced stresses in these designs, assuming the same materials characteristics, but different geometries of the two structures.

Let, for example, half the distance between the outer sides of the two adjacent legs: $L = 1.25$ mm; leg thickness: $2l = 1.0$ mm; leg height: $h = 4.0$ mm. Materials characteristics and the geometrical characteristics of other structural elements are the same as already chosen. The axial compliances λ_1 and the interfacial compliances κ_1 of the ceramic components, and the interfacial compliances κ_c and κ_s of the copper and the solder layers are not different of those already calculated. The other computed data are as follows:

- Interfacial compliance of the $Be_2\,Te_3$ bond (leg):

$$\kappa_l = \frac{3-\nu_l}{4\pi}\frac{l}{G} = \frac{3-0.4}{12.5664}\frac{0.5}{1712.5} = 6.0409\times10^{-5}\ \text{mm}^3/\text{kg}$$

- Total interfacial compliance:

$$\begin{aligned}\kappa &= 2(\kappa_1+\kappa_c+\kappa_s)+\kappa_l \\ &= 2(1.5164\times10^{-5}+5.5882\times10^{-5}+5.8582\times10^{-5})+6.0409\times10^{-5} \\ &= 31.9665\times10^{-5}\,\text{mm}^3/\text{kg}\end{aligned}$$

- Parameter of the interfacial shearing stress:

$$k = \sqrt{\frac{2\lambda_1}{\kappa}} = \sqrt{\frac{2\times2.7264\times10^{-5}}{31.9665\times10^{-5}}} = 0.4130\,\text{mm}^{-1}$$

The product $kl = 0.4130\times0.5 = 0.2065$

- With

$$L = 1.25\,mm, 2l = 1.0\,mm, L/2l = 1.25, t = \tanh kl = 0.2036, c = \coth kl = 4.9114,$$

$$s = 4kl\left(\frac{L}{2l}-1\right) = 0.8260\times0.25 = 0.2065,$$

we have:

$$D = 5t^2+c^2+10+2s^2+8ts+4cs = 0.2073+24.1218+10+0.0853+0.3363+4.0568 = 38.8045$$

$$2t\frac{\alpha\Delta t}{\lambda_1 D} = 0.4072\times\frac{6.5\times10^{-6}\times130}{2.7264\times10^{-5}\times38.8045} = 0.3252$$

- Then

$$\hat{T}_1 = 2t\frac{\alpha\Delta t}{\lambda_1 D}(2t+2c+s) = 0.3252\times16.4365 = 3.3939\,\text{kg/mm}$$

$$\hat{T}_2 = 2t\frac{\alpha\Delta t}{\lambda_1 D}(t+3c+2s) = 0.3252\times15.3508 = 4.9921\,\text{kg/mm}$$

- The maximum shearing stress at the end of the assembly is

$$\tau(-l) = -k\left(\frac{\alpha\Delta t}{2\lambda_1}\tanh kl + \frac{\hat{T}_1}{\sinh 2kl}\right) = -0.4130\left(\frac{6.5\times10^{-6}\times130}{2\times2.7264\times10^{-5}}\times0.2036+\frac{3.3939}{0.4185}\right)=$$

$$= -1.3031 - 3.3493 = -4.6524\,\text{kg/mm}^2$$

Because of rather stiff legs and relatively small spacing between them, this value is only by 4.5% lower than the maximum shearing stress in an assembly with a homogeneous bond.

Let, for example, for the design in Figure 3.12, half the distance between the outer sides of the two adjacent legs L = 1.25 mm; leg thickness: $2l$ = 0.4 mm; leg height: h = 0.1 mm. Materials characteristics and the geometrical characteristics of other structural elements are the same as in the table for the input data. Axial and interfacial compliances of the ceramic component, and the interfacial compliances of the copper and the solder layers, are not different of those previously calculated. The other computed data are as follows:

- Interfacial compliance of the Be_2Te_3bond (leg):

$$\kappa_l = \frac{3-\nu_l}{4\pi}\frac{l}{G} = \frac{3-0.4}{12.5664}\frac{0.2}{1712.5} = 2.4164\times10^{-5}\,\text{mm}^3/\text{kg}$$

Total interfacial compliance:

$$\kappa = 2(\kappa_1 + \kappa_c + \kappa_s) + \kappa_l$$

- Parameter of the interfacial shearing stress

$$= 2(1.5164\times10^{-5} + 5.5882\times10^{-5} + 5.8582\times10^{-5}) + 2.4164\times10^{-5}$$
$$= 28.3420\times10^{-5}\,\text{mm}^3/\text{kg}$$

$$\kappa = \sqrt{\frac{2\lambda_1}{\kappa}} = \sqrt{\frac{2\times2.7264\times10^{-5}}{28.3420\times10^{-5}}} = 0.4386\,\text{mm}^{-1}$$

The product $kl = 0.4386\times0.2 = 0.08772$
With

$$L = 1.25\,\text{mm}, 2l = 0.4\,\text{mm}, L/2l = 3.125, t = \tanh kl = 0.08750, c = \coth kl = 11.4292,$$

$$s = 4kl\left(\frac{L}{2l} - 1\right) = 0.3500\times2.125 = 0.7438,$$

we have:

$$D = 5t^2 + c^2 + 10 + 2s^2 + 8ts + 4cs =$$
$$0.0383 + 130.6266 + 10 + 1.1065 + 0.5207 + 34.0042 = 176.2963$$

and

$$2t\frac{\alpha\Delta t}{\lambda_1 D}=0.1750\times\frac{6.5\times10^{-6}\times130}{2.7264\times10^{-5}\times176.2963}=0.03076$$

Then:

$$\hat{T}_1=2t\frac{\alpha\Delta t}{\lambda_1 D}(2t+2c+s)=0.03076\times(0.1750+22.8584+0.7438)=0.7314\,\text{kg/mm}$$

$$\hat{T}_2=2t\frac{\alpha\Delta t}{\lambda_1 D}(t+3c+2s)=0.03076\times(0.0875+34.2876+1.4876)=1.1031\,\text{kg/mm}$$

The maximum shearing stress at the end of the assembly is

$$\tau(-l)=-k\left(\frac{\alpha\Delta t}{2\lambda_1}\tanh kl+\frac{\hat{T}_1}{\sinh 2kl}\right)=-0.4386\left(\frac{6.5\times10^{-6}\times130}{2\times2.7264\times10^{-5}}\times0.0875+\frac{0.7314}{0.1763}\right)$$

$$=-0.5947-1.8196=-2.4143\,\text{kg/mm}^2$$

Because of the thin and, hence, compliant legs and relatively large spacing between them, this value is only 51.9% of the maximum shearing stress in the conventional TEM design and is 49.5% of the maximum shearing stress in an assembly with a continuous bond. In addition, this novel design is significantly thinner than the conventional TEM design.

3.4.8 Design in Figure 3.12 for a High-Temperature Power Generation TEM

In the calculations carried out in this section, we address a TEM design that could be supposedly used in high-temperature power generation. Accordingly, we consider a Si or SiGe thermoelectric material with tentative properties $\alpha=2.6\times10^{-6}$ 1/°C, $E=150$ GPa $=15306$ kg/mm^2, $\nu=0.28$, and Gold100 solder with $\alpha=14.2\times10^{-6}$ 1/°C, $E=78.5$ GPa $=8010$ kg/mm^2, $\nu=0.42$. The tensile yield stress of gold is about 200 GPa = 20408 kg/mm^2, and its ultimate stress is about 220 GPa = 22449 kg/mm^2.

In the case of shear loading, one could tentatively assume

$$\tau_Y=\frac{20408}{\sqrt{3}}=11783\,\text{kg/mm}^2,$$

and

$$\tau_u=\frac{22449}{\sqrt{3}}=12961\,\text{kg/mm}^2.$$

We consider that the hot component's temperature can be as high as 800–1000°C and the cold component's temperature could be about 300–500°C, so that the difference in temperature could be as high as $\Delta t = 700$°C. Since the melting point of the gold solder is as high as 1063°C, these operation temperatures seem to be acceptable. The geometries of all the materials are the same as in the example carried out for the Figure 3.11 design. The axial compliances λ_1 and the interfacial compliances κ_1 of the ceramic components, and the interfacial compliance κ_c of the copper are not different of those already calculated. The other computed data are as follows:

- Solder layer:

$$\kappa_s = \frac{h_s}{G_s} = \frac{0.10}{2820.5} = 3.5455 \times 10^{-5}\ \text{mm}^3/\text{kg}$$

- Si (or SiGe) bond (leg)

$$\kappa_l = \frac{3-\nu_l}{4\pi}\frac{l}{G} = \frac{3-0.27}{12.5664}\frac{0.2}{6026.0} = 0.72103 \times 10^{-5}\ \text{mm}^3/\text{kg}$$

Total interfacial compliance:

$$\kappa = 2(\kappa_1 + \kappa_c + \kappa_s) + \kappa_l$$

$$= 2(1.5164 \times 10^{-5} + 5.5882 \times 10^{-5} + 3.5455 \times 10^{-5}) + 0.72103 \times 10^{-5}$$

$$= 22.0212 \times 10^{-5}\ \text{mm}^3/\text{kg}$$

Parameter of the interfacial shearing stress:

$$\kappa = \sqrt{\frac{2\lambda_1}{\kappa}} = \sqrt{\frac{2 \times 2.7264 \times 10^{-5}}{22.0212 \times 10^{-5}}} = 0.497\ \text{mm}^{-1}$$

The product $kl = 0.4976 \times 0.2 = 0.09952$

With

$$L = 1.25\ \text{mm}, 2l = 0.4\ \text{mm}, L/2l = 3.125, t = \tanh kl = 0.08750, c = \coth kl = 10.0425,$$

$$s = 4kl\left(\frac{L}{2l} - 1\right) = 0.3981 \times 2.125 = 0.8459,$$

we have:

$$D = 5t^2 + c^2 + 10 + 2s^2 + 8ts + 4cs = 0.0496 + 100.8518 + 10 + 1.4311 + 0.6739 + 33.9798 = 146.9862$$

and

$$2t\frac{\alpha\Delta t}{\lambda_1 D} = 1.1992 \times \frac{6.5 \times 10^{-6} \times 700}{2.7264 \times 10^{-5} \times 146.9862} = 1.3616$$

Then:

$$\hat{T}_1 = 2t\frac{\alpha\Delta t}{\lambda_1 D}(2t+2c+s) = 1.3616\times(0.1992+20.0850+0.8459) = 28.7707\,\text{kg/mm}$$

$$\hat{T}_2 = 2t\frac{\alpha\Delta t}{\lambda_1 D}(t+3c+2s) = 1.3616\times(0.09958+30.1275+1.6918) = 43.4607\,\text{kg/mm}$$

The maximum shearing stress at the end of the assembly is:

$$\tau(-l) = -k\left(\frac{\alpha\Delta t}{2\lambda_1}\tanh kl + \frac{\hat{T}_1}{\sinh 2kl}\right)$$

$$= -0.4976\left(\frac{6.5\times10^{-6}\times700}{2\times2.7264\times10^{-5}}\times0.09958 + \frac{28.7707}{0.2004}\right)$$

$$= -4.1347 - 71.4386 = -75.5733\,\text{kg/mm}^2$$

Although this stress exceeds significantly, by a factor of 31.3, the stress in the assembly in Figure 3.4 with the Be_2 Te_3 as the thermoelectric material (leg) and Sn–Sb solder, it is still well below the shearing yield stress of 11783 kg/mm^2 in the gold as solder material, so that the robustness of the design will not be compromised. It is noteworthy that, since the melting point of the gold solder is as high as 1063°C, the difference in temperature from the fabrication temperature, when the thermal stress is next-to-zero, and the low testing temperature of about, say, –37°C, when the thermal stresses are the highest, is 1100°C. In this case, the predicted maximum stress of 75.5733 kg/mm^2 should be increased by the factor of $\frac{1100}{700} = 1.571$, and the predicted low temperature maximum shearing stress will become 118.8 kg/mm^2 but even this stress is only about 10% of the shearing yield stress for the gold solder. Although this result looks promising, a significant additional effort, both theoretical and experimental should be undertaken to evaluate the long-term brittle and fatigue strength of the considered module.

3.4.9 Ultrathin and Long (Beam-Like) Legs

In the aforementioned analyses, we assumed that the legs are sufficiently thick and short, so that no bending deformations should be considered, and only shearing stresses occur in the legs. In the analysis carried out in this section, we address a situation when the legs are thin (horizontal dimension) and long (vertical dimension) enough, so that they become beam-like configured and experience bending deformations. These deformations might be accompanied, however, by lateral (in the horizontal direction) shear.

Accordingly, we consider a short clamped-clamped beam, whose ends are offset, and its elastic curve is expressed as follows:

$$w(z) = \Delta + \frac{Nh^3}{12EI}\left(2 - 3\frac{z^2}{h^2}\right) \tag{3.39}$$

Here, Δ is the ends offset (since the beam is thin, all its points at the clamped cross sections are assumed to have the same ends offset), h is its height (vertical dimension), EI is its flexural rigidity, and N is the lateral force (acting in the horizontal direction and the same for all the beam's cross sections). This force is related to the bending moments M_0 at the beam's cross sections by the formula: $N = \frac{2M_0}{h}$. The origin of the vertical (in the direction of the leg's length and the TEM thickness) coordinate z is at the lower clamped end of the beam. The distributed bending moment and the lateral force are

$$M(z) = M_0 - Nz = \frac{6EI\Delta}{h^2}\left(1 - \frac{2z}{h}\right), \quad N = \frac{12EI\Delta}{h^3}, \tag{3.40}$$

and the strain energy due to bending is

$$U_b = \int_0^h \frac{M^2(z)dz}{2EI} = \frac{6EI\Delta^2}{h^3}. \tag{3.41}$$

Assuming that the lateral shearing stress $\tau(x)$ is distributed over the leg's cross section in a parabolic fashion

$$\tau(x) = C(l^2 - x^2) \tag{3.42}$$

we obtain the following formula for the strain energy due to the lateral shear (per unit leg's height h and its unit width b):

$$\frac{U_s'}{b} = \int_{-l}^{l} \frac{\tau^2(x)dx}{2G} = \frac{8}{15}C^2\frac{l^5}{G}, \tag{3.43}$$

where G is the shear modulus of the material. The lateral force per unit leg's width can be found as

$$\frac{N}{b} = \int_{-l}^{l} \tau(x)dx = \frac{4}{3}Cl^3. \tag{3.44}$$

Solving this equation for the constant C and introducing the obtained expression into (3.43), we have:

$$U_s' = \frac{3}{10}\frac{N^2}{Gbl}, \tag{3.45}$$

so that the total strain energy due to lateral shear is

$$U_s = \frac{3}{10}\frac{N^2 h}{Gbl}. \tag{3.46}$$

The total strain energy due to both bending and lateral shear is

$$U_t = U_b + U_s = U_b(1+\eta_s), \tag{3.47}$$

where the parameter

$$\eta_s = \frac{U_s}{U_b} = \frac{1}{20}\frac{N^2h^4}{GblEI\Delta^2} = \frac{48}{5}(1+\nu)\frac{l^2}{h^2} \tag{3.48}$$

considers the effect of the lateral shear on the total strain energy. Assuming, for instance, that the beam model is applicable if the η_s value does not exceed 0.75, and putting $\nu = 1/3$, we conclude that these model could be applied for $\frac{h}{l}$ values exceeding 4.0 For considerably lower length-to-thickness ratios, the model of the type addressed in the previous sections (and in [13, 14]) should be applied.

The result expressed by formula (3.48) can be interpreted as that the lateral force N_s with consideration of the effect of the lateral shear can be found from the lateral force N due to bending only as follows:

$$N_s = (1+\eta_s)N. \tag{3.49}$$

From (3.42) and (3.44), we obtain:

$$\tau(x) = C(l^2 - x^2) = \frac{3}{4}\frac{N_s}{bl^3}(l^2 - x^2) = \frac{3}{4}\frac{(1+\eta_s)}{bl^3}\frac{12EI\Delta}{h^3}(l^2 - x^2)$$

$$= 6E(1+\eta_s)\frac{\Delta l^2}{h^3}\left(1 - \frac{x^2}{l^2}\right). \tag{3.50}$$

The maximum lateral shearing stress value is therefore

$$\tau_{\max} = \tau(0) = 6E(1+\eta_s)\frac{\Delta l^2}{h^3}. \tag{3.51}$$

The maximum normal stress due to bending is

$$\tau_{\max} = 6E\frac{\Delta l}{h^2}. \tag{3.52}$$

Let, for instance, a two-leg $2L = 20.0$ mm long TEM is considered and two $2l = 0.2$ mm thick and $h = 2.0$ mm long $Be_2\,Te_3$ legs are employed. Assuming that these legs are thin enough, so that they do not resist the horizontal displacements of the ceramics, one can assess the relative displacement of the component ends as follows:

$$\Delta = \alpha\Delta t L = 6.5\times10^{-6}\times130\times10 = 0.008450\,\text{mm}$$

Assuming also that the entire leg is made of the Be_2Te_3 material, formula (3.48) yields

$$\eta_s = \frac{48}{5}(1+\nu)\frac{l^2}{h^2} = \frac{48}{5}(1+0.40)\frac{0.1^2}{2.0^2} = 0.0336,$$

and the maximum lateral shearing stress predicted by formula (3.49) and the maximum normal bending stress predicted by formula (3.52) are

$$\tau_{\max} = 6E(1+\eta_s)\frac{\Delta l^2}{h^3} = 6\times 4795\times 1.0336\times\frac{0.00845\times 0.01}{2.0^3} = 0.3141\,\text{kg/mm}^2$$

$$\sigma_{\max} = 6E\frac{\Delta l}{h^2} = 6\times 4795\times\frac{0.00845\times 0.1}{2.0^2} = 6.0777\,\text{kg/mm}^2$$

Based on these data, we conclude that thin-and-long legs should be considered for lower stresses, but not to such an extent that appreciable bending deformations become possible.

Thus, a practically useful and physically meaningful analytical (mathematical) thermal stress model has been developed for an assembly comprised of identical components subjected to different temperature. The case of an assembly with an inhomogeneous bond is considered. The model is applied to the case of a multilegged Bismuth-Telluride-Alloy–based TEM design. We assessed, based on the developed model, the effect of the size and compliance of the bonded regions on the maximum interfacial shearing stress responsible for the physical robustness of the module.

We have determined that the employment of thinner and shorter legs could result in considerably lower interfacial thermal stresses, and, at the same time, lead to a significantly thinner module. We have found that although thin-and-long legs should be considered for lower stresses, but not to such an extent that appreciable bending deformations become possible.

The future work should include, but will not be limited to, the FEA analyses aimed at the confirmation of the developed analytical model, as well as experimental evaluations of the stress-at-failure for the TEM designs of interest. Particularly, shear-off testing of the suitable experimental specimens should be carried out. Although this result looks promising, a significant additional effort, both theoretical and experimental, should be undertaken to evaluate the long-term brittle and fatigue strength of the module.

3.5 PREDICTED THERMAL STRESS IN A CIRCULAR BONDED ASSEMBLY WITH IDENTICAL ADHERENDS

3.5.1 MOTIVATION

The objective of the analysis that follows is to develop an easy-to-use, simple, and physically meaningful analytical ("mathematical") stress model for the evaluation of the thermally induced stresses in a circular adhesively bonded assembly with identical adherends. The assembly is fabricated at an elevated temperature and is subsequently cooled down to a low (e.g., room) temperature. Thermal stresses arise because of the different coefficients of thermal expansion (contraction) of the dissimilar materials of the two adherends with the adhesive layer. The developed model can be helpful in stress–strain analyses and physical design of electronic and photonic assemblies of the type in question.

3.5.2 Assumptions

The following assumptions are used in our study.

- The adherends can be treated as circular plates experiencing small deflections.
- The engineering theory of bending of plates (see, for example, [51]) can be used to predict their mechanical behavior.
- The "peeling" stresses (i.e., the interfacial normal stresses acting in the through-thickness direction of the assembly) do not affect the interfacial shearing stresses and therefore do not have to be accounted for when evaluating the shearing stresses.
- The "peeling" stresses can be determined based on the evaluated shearing stresses.
- The interfacial compliance of the assembly in its plane is due to the joint interfacial compliance

$$\kappa_0 = \frac{2h_0}{3G_0} = \frac{4}{3}(1+\nu_0)\frac{h_0}{E_0} \tag{3.53}$$

of the two adherends and the compliance

$$\kappa_1 = \frac{h_1}{6G_1} = \frac{1}{3}(1+\nu_1)\frac{h_1}{E_1}, \tag{3.54}$$

of the bonding layer [6]. In these formulas, h_0 is the thickness of one of the adherends, h_1 is the thickness of the bonding layer, $G_0 = \frac{E_0}{2(1+\nu_0)}$ is the shear modulus of the adherend material, $G_1 = \frac{E_1}{2(1+\nu_1)}$ is the shear modulus of the bonding material, E_0 is the Young's modulus of the adherend material, E_1 is the Young's modulus of the bonding material, and ν_0 and ν_1 are the Poisson's ratios of the adherend and the bonding materials.
- The interfacial radial displacements $u_1(r)$ of the bonding material in the radial direction can be evaluated as the sum of the stress-free radial displacements $\alpha_1 \Delta tr$; the radial displacements $u(r)$ due to the thermally induced forces caused by the thermal contraction mismatch of the bonding material and the material of the adherends; and the displacements $\kappa_1\tau_0(r)$ of the interfacial point at the given radius r with respect to the displacements $u(r)$ of the inner points of the cross section:

$$u_1(r) = -\alpha_1 \Delta tr + u(r) + \kappa_1\tau_0(r). \tag{3.55}$$

In this formula, $u_1(r)$ are the total interfacial radial displacements of the bonding material, α_1 is the coefficient of thermal expansion (contraction) of the bonding material, Δt is the change in temperature, r is the current radius, $u(r)$ are the (stress-dependent) radial displacements of the bonding layer, $\tau_0(r)$ is the interfacial shearing stress in the given cross section, and κ_1 is the interfacial compliance of the bonding layer defined by formula (3.54).

- The displacements $u(r)$ in formula (3.55) can be evaluated based on the Hooke's law, and are considered the same for all the points of the given circumferential cross section.
- The third term in (3.55) considers the deviation of the given cross section from planarity: the interfacial radial displacements are somewhat larger than the displacements of the inner points of the cross section.
- The interfacial radial displacements, $u_0(r)$, in the adherends can be evaluated as

$$u_0(r) = -\alpha_0 \Delta t r - \kappa_0 \tau_0(r), \tag{3.56}$$

where α_0 is the coefficient of thermal expansion (contraction) of the adherend material, $\tau_0(r)$ is the interfacial shearing stress, and κ_0 is the interfacial compliance of the adherend. This compliance can be computed by formula (3.53).

- The first term in the right part of formula (3.56) is the stress-free thermal contraction of the adherends. The second term is due to the interfacial shearing stress. This term reflects an assumption that the interfacial radial displacements of the adherends, caused by their interaction with the bonding layer, are proportional to the interfacial shearing stress in the given cross section and are not affected by the stresses and strains in the adjacent cross sections.
- Formula (3.56) also reflects an assumption that the radial displacements of the inner portions of the given cross section of the adherend are not affected by the displacements of a substantially thinner and low modulus bonding layer.

3.5.3 Basic Equation

Formulas (3.55) and (3.56) for the interfacial radial displacements in the adherends and in the bond, and the condition $u_0(r) = u_1(r)$ of the compatibility of these displacements result in the following formula for the radial displacements in the bonding layer:

$$u(r) = \Delta\alpha \Delta t r - \kappa \tau_0(r). \tag{3.57}$$

Here, $\Delta\alpha = \alpha_1 - \alpha_0$ is the difference in the CTE (contraction) of the bonding material and the material of the adherends, and $\kappa = \kappa_0 + \kappa_1$ is the total interfacial compliance of the assembly. Expression (3.57) and the Cauchy formulas [54]

$$\varepsilon_r = u'(r), \quad \varepsilon_\theta = \frac{u(r)}{r}, \tag{3.58}$$

for the normal radial, ε_r, strains and the normal circumferential (tangential), ε_θ, strains yield:

$$\varepsilon_r = \Delta\alpha\Delta t - \kappa\tau_0'(r), \quad \varepsilon_\theta = \Delta\alpha\Delta t - \kappa\frac{\tau_0(r)}{r}. \tag{3.59}$$

The corresponding radial, σ_r, and circumferential, σ_θ, normal stresses in the bonding layer can be evaluated, using Hooke's law equations [54]

$$\sigma_r = \frac{E_1}{1-\nu_1^2}(\varepsilon_r + \nu_1\varepsilon_\theta), \quad \sigma_\theta = \frac{E_1}{1-\nu_1^2}(\varepsilon_\theta + \nu_1\varepsilon_r), \tag{3.60}$$

as follows:

$$\left.\begin{aligned} \sigma_r &= \frac{E_1}{1-\nu_1^2}\left\{(1+\nu_1)\Delta\alpha\Delta t - \kappa\left[\tau_0'(r) + \nu_1\frac{\tau_0(r)}{r}\right]\right\} \\ \sigma_\theta &= \frac{E_1}{1-\nu_1^2}\left\{(1+\nu_1)\Delta\alpha\Delta t - \kappa\left[\frac{\tau_0(r)}{r} + \nu_1\tau_0'(r)\right]\right\} \end{aligned}\right\} \tag{3.61}$$

Since the thickness, h_1, of the bonding layer is small, and the interfacial shearing stress should be symmetric with respect to the horizontal midplane of the assembly, the gradient $\frac{\partial\tau_{rz}}{\partial z}$ of the interfacial shearing stress, τ_{rz}, in the through-thickness direction, z, can be represented using the following approximate formula:

$$\frac{\partial\tau_{rz}}{\partial z} \cong \frac{2\tau_0(r)}{h_1}. \tag{3.62}$$

Introducing formulas (3.61) and (3.62) into the equilibrium equation [54]

$$\frac{\partial\sigma_r}{\partial r} + \frac{\partial\sigma_{rz}}{\partial z} + \frac{\sigma_r - \sigma_\theta}{r} = 0, \tag{3.63}$$

we obtain the following basic differential equation for the shearing stress function, $\tau_0(r)$:

$$\tau_0''(r) + \frac{\tau_0'(r)}{r} - \frac{\tau_0'(r)}{r^2} - k^2\tau_0(r) = 0, \tag{3.64}$$

where

$$k = \sqrt{\frac{2\lambda_1}{\kappa}} \tag{3.65}$$

is the parameter of the interfacial shearing stress, and

$$\lambda_1 = \frac{1-\nu_1^2}{E_1 h_1} \tag{3.66}$$

is the radial compliance of the bonding layer.

3.5.4 Solution to the Basic Equation

Equation (3.64) has the following solution:

$$\tau_0(r) = C_0 r I_0(kr) + C_1 a I_1(kr), \tag{3.67}$$

where C_0 and C_1 are the constants of integration, a is the assembly radius, k is the parameter of the interfacial shearing stress, and $I_0(kr)$ and $I_1(kr)$ are the modified Bessel functions of the first kind of zero and first order, respectively (see, for examples, [55, 56]). These functions obey the following rules of differentiation:

$$I_0'(kr) = kI_1(kr), \quad I_1'(kr) = kI_0(kr) - \frac{I_1(kr)}{r}. \tag{3.68}$$

Introducing the sought solution (3.67), with consideration of formula (3.68), into the basic equation (3.64), we find that the following equation should be fulfilled for any radius, r:

$$C_0 k[I_1(kr) - krI_0(kr)] = 0. \tag{3.69}$$

Hence, one should put $C_0 = 0$, and solution (3.67) can be simplified:

$$\tau_0(r) = C_1 a I_1(kr). \tag{3.70}$$

Note that the modified Bessel function $I_1(kr)$ plays the role of the hyperbolic sine in the solution obtained earlier for an elongated rectangular plate [12].

Substituting solution (3.70) into the first equation (3.61), we obtain the following expression for the radial stress:

$$\sigma_r = \frac{E_1}{1-\nu_1^2}\left\{(1+\nu_1)\Delta\alpha\Delta t - \kappa a C_1\left[kI_0(kr) + \nu_1 \frac{I_1(kr)}{r}\right]\right\}. \tag{3.71}$$

Since the edge $r = a$ of the assembly is stress-free, the condition

$$\sigma_r(a) = 0 \tag{3.72}$$

should be fulfilled. Then we obtain:

$$C_1 = \frac{\Delta\alpha\Delta t}{\kappa} \frac{1+\nu}{kaI_0(ka)-(1-\nu_1)I_1(ka)}, \tag{3.73}$$

and solution (3.70) results in the following formula for the interfacial shearing stress:

$$\tau_0(r) = \tau_{\max} \frac{I_1(kr)}{I_1(ka)}, \tag{3.74}$$

where the maximum value

$$\tau_{\max} = \tau_0(a) = \frac{(1+\nu_1)aI_1(kr)}{kaI_0(ka)-(1-\nu_1)I_1(ka)} \frac{\Delta\alpha\Delta t}{\kappa} \tag{3.75}$$

of this stress takes place at the end $r = a$ of the assembly.

3.5.5 Large and/or Stiff Assemblies

With a small bond thickness, h_1, the radial in-plane compliance, λ_1, defined by formula (3.66), is large, and so is the parameter, k, of the interfacial shearing stress, expressed by formula (3.65). For large arguments z, the modified Bessel function of the order n can be evaluated by the approximate formula [55]:

$$I_n(z) = \frac{e^z}{\sqrt{2\pi z}}. \tag{3.76}$$

Then, with

$$I_0(ka) \cong I_1(ka) \cong \frac{e^{ka}}{\sqrt{2\pi ka}}, \quad I_0(kr) \cong I_1(kr) \cong \frac{e^{kr}}{\sqrt{2\pi kr}}, \tag{3.77}$$

we obtain solution (3.73) in the form:

$$\tau_0(r) = \tau_{\max}\sqrt{\frac{a}{r}}e^{-k(a-r)}, \tag{3.78}$$

where

$$\tau_{\max} = \frac{(1+\nu_1)a}{ka-(1-\nu_1)} \frac{\Delta\alpha\Delta t}{\kappa} \approx (1+\nu_1)\frac{\Delta\alpha\Delta t}{k\kappa} \tag{3.79}$$

is the maximum interfacial shearing stress. As evident from formula (3.78), the interfacial shearing stress, $\tau_0(r)$, concentrates around a narrow peripheral ring, and is next to zero for the inner radii of the bonding layer ($r \ll a$). Note that for large and/or stiff enough assemblies, the maximum interfacial shearing stress is assembly size independent.

3.5.6 Normal Stresses in the Bonding Layer

Introducing formula (3.74) for the interfacial shearing stress into formulas (3.61), we obtain the following expressions for the normal radial, σ_r, and the normal circumferential, σ_θ, stresses in the bonding layer:

$$\sigma_r = \sigma_1\left[1-\frac{kaI_0(kr)-(1-\nu_1)a\dfrac{I_1(kr)}{r}}{kaI_0(ka)-(1-\nu_1)I_1(ka)}\right],$$

$$\sigma_\theta = \sigma_1\left[1-\frac{\nu_1 kaI_0(kr)+(1-\nu_1)a\dfrac{I_1(kr)}{r}}{kaI_0(ka)-(1-\nu_1)I_1(ka)}\right] \qquad (3.80)$$

where

$$\sigma_1 = \frac{E_1}{1-\nu_1}\Delta\alpha\Delta t \qquad (3.81)$$

is the normal stress in the midportion of the bonding layer. Formula (3.80) defines the stress in a thin film fabricated on a thick substrate formed by the two adherends. The expressions in the brackets in formula (3.80) are, in effect, factors, which consider the role of the finite radius, a, of the assembly. These factors indicate the change in the normal stresses in the bonding layer, when the current radius r changes from zero to the radius a of the assembly.

In the case of a large size (large a values) and/or stiff (large k values) assemblies, formulas (3.80) can be simplified, considering the relationships (3.77), as follows:

$$\sigma_r = \sigma_1\left[1-\sqrt{\frac{a}{r}}e^{-k(a-r)}\right], \quad \sigma_\theta = \sigma_1\left[1-\nu_1\sqrt{\frac{a}{r}}e^{-k(a-r)}\right]. \qquad (3.82)$$

Formulas (3.82) indicate that the normal stresses, σ_r and σ_θ, in the bonding layer are uniformly distributed over the inner portion of the assembly ($r \ll a$), and can be predicted on the basis of formula (3.81), that is, assuming that the adherends play the role of a thick substrate, and the bonding layer plays the role of a thin film fabricated on this substarte. The normal radial stress, σ_r, is zero at the bond edge, $r = a$. The normal circumferential stress, σ_θ, at the edge ($r = a$) can be found as $\sigma_\theta = E_1\Delta\alpha\Delta t$. This stress is by the factor of $\frac{1}{1-\nu_1}$ lower than the stress, σ_1, in the inner portion of the film.

3.5.7 Bow

The assembly as a whole does not experience any bowing. Each of the adherends, however, can bow with respect to the horizontal midplane of the assembly. We seek the angles of rotation, $w'(r)$, of the adherend cross sections in the form:

$$w'(r) = A_1 r + A_2 a I_1(kr), \qquad (3.83)$$

where A_1 and A_2 are thus far unknown constants. Then we obtain, by differentiation:

$$w''(r) = A_1 + A_2 a\left[kI_0(kr) - \frac{I_1(kr)}{r}\right], \tag{3.84}$$

and the radial bending moment, M_r, acting on the adherend, can be evaluated as [51]

$$M_r = -D_0\left[w''(r) + \nu_0\frac{w'(r)}{r}\right] = -D_0\left\{(1+\nu_0)A_1 + A_2 a\left[kI_0(kr) - (1-\nu_0)\frac{I_1(kr)}{r}\right]\right\}, \tag{3.85}$$

where

$$D_0 = \frac{E_0 h_0^3}{12(1-\nu_0^2)} \tag{3.86}$$

is the flexural rigidity of one of the adherends treated as a thin plate. On the other hand, the radial bending moment, M_r, can be determined, based on the first formula in (3.79) for the radial normal stress, σ_r, as follows:

$$M_r = \sigma_r h_1\frac{h_0}{2} = \sigma_1\frac{h_0 h_1}{2}\left[1 - a\frac{kI_0(kr) - (1-\nu_1)\dfrac{I_1(kr)}{r}}{kaI_0(ka) - (1-\nu_1)I_1(ka)}\right]. \tag{3.87}$$

In an approximate analysis, one can assume that Poisson's ratio, ν_0, of the adherend material in the expression in the brackets in formula (3.85) can be substituted with Poisson's ratio, ν_1, of the material of the film. Then, comparing the expressions (3.84) and (3.85), we conclude that the constants A_1 and A_2 can be evaluated by the formulas:

$$A_1 = -6\frac{E_1^*}{E_0^*}\frac{h_1}{h_2}\Delta\alpha\Delta t, \quad A_2 = 6\frac{E_1^*}{E_0^*}\frac{h_1}{h_0^2}\frac{(1+\nu_0)\Delta\alpha\Delta t}{kaI_0(ka) - (1-\nu_1)I_1(ka)}, \tag{3.88}$$

where

$$E_0^* = \frac{E_0}{1-\nu_0}, \quad E_1^* = \frac{E_1}{1-\nu_1} \tag{3.89}$$

are the effective Young's moduli of the adherend and the bonding materials, respectively.

Introducing formulas (3.88) into expression (3.84), we obtain the following formula for the adherend curvature:

$$K(r) = w''(r) = -6\frac{E_1^*}{E_0^*}\frac{h_1}{h_0^2}\left[1 - (1+\nu_0)\frac{kaI_0(kr) - a\dfrac{I_1(kr)}{r}}{kaI_0(ka) - (1-\nu_1)I_1(ka)}\right]\Delta\alpha\Delta t. \tag{3.90}$$

The rotation angles, expressed formula (3.82), are:

$$w'(r) = -6\frac{E_1^*}{E_0^*}\frac{h_1 a}{h_0^2}\left[\frac{r}{a} - (1+\nu_0)\frac{I_1(kr)}{kaI_0(ka) - (1-\nu_1)I_1(ka)}\right]\Delta\alpha\Delta t. \qquad (3.91)$$

The deflection function, $w(r)$, can be found by integration:

$$w(r) = -3\frac{E_1^*}{E_0^*}\frac{a^2}{h_0^2}h_1\left[\frac{r^2}{a^2} - 2(1+\nu_0)\frac{I_0(kr)}{ka(kaI_0(ka) - (1-\nu_1)I_1(ka))}\right]\Delta\alpha\Delta t + A_0, \quad (3.92)$$

where the constant of integration, A_0, is the displacement of the adherend as a non-deformable rigid body. Since we are interested in the elastic displacements only, this constant can be chosen in an arbitrary fashion. Choosing it, for instance, in such a way that $w(a) = 0$, one can obtain formula (3.92) in the form:

$$w(r) = 3\frac{E_1^*}{E_0^*}\frac{a^2}{h_0^2}h_1\left[1 - \frac{r^2}{a^2} - 2(1+\nu_0)\frac{I_0(ka) - I_0(kr)}{ka[kaI_0(ka) - (1-\nu_1)I_1(ka)]}\right]\Delta\alpha\Delta t. \qquad (3.93)$$

The maximum deflection takes place at the center of the substrate ($r = 0$) and is

$$w_{max} = f\left[1 - 2(1+\nu_0)\frac{I_0(ka) - 1}{ka[kaI_0(ka) - (1-\nu_1)I_1(ka)]}\right], \qquad (3.94)$$

where

$$f = 3\frac{E_1^*}{E_0^*}\frac{a^2}{h_0^2}h_1\Delta\alpha\Delta t \qquad (3.95)$$

is the maximum deflection in the case of a large and/or stiff structure. The term in the brackets in formula (3.94) reflects the effect of the finite size of the assembly on the maximum deflection.

In the case of a large and/or a stiff assembly, formula (3.93) yields

$$w(r) = f\left(1 - \frac{r^2}{a^2}\right), \qquad (3.96)$$

and the induced radial curvature can be evaluated as

$$K(r) = w''(r) = -2\frac{f}{a^2} = -6\frac{E_1^*}{E_0^*}\frac{h_1}{h_0^2}\Delta\alpha\Delta t. \qquad (3.97)$$

This formula indicates that the thermally induced curvatures of the adherends are proportional to the ratio, E_1^*/E_0^*, of the effective Young's moduli of the adhesive and the adherend materials; to the thickness, h_1, of the bonding layer; and to the thermal mismatch strain $\Delta\alpha\Delta t$ between the materials of the adhesive and the adherends; and is inversely proportional to the thickness, h_0, of the adherend squared. For thick enough adherends this curvature is next-to-zero.

3.5.8 Bending Stresses in the Adherends

The radial, M_r, and the circumferential, M_θ, bending moments in the adherends can be evaluated, using formula (3.90) for the curvature and formula (3.91) for the angles of rotation, as follows:

$$\left.\begin{aligned} M_r &= -D_0\left(w'' + \nu_0 \frac{w'}{r}\right) = E_1^* \frac{h_0 h_1}{2}\left[1 - \frac{kaI_0(kr) + (1+\nu_0)a\frac{I_1(kr)}{r}}{kaI_0(ka) - (1-\nu_1)I_1(ka)}\right]\Delta\alpha\Delta t \\ M_\theta &= -D_0\left(\nu_0 w'' + \frac{w'}{r}\right) = E_1^* \frac{h_0 h_1}{2}\left[1 - \frac{kaI_0(kr) + (1-\nu_0)a\frac{I_1(kr)}{r}}{kaI_0(ka) - (1-\nu_1)I_1(ka)}\right]\Delta\alpha\Delta t \end{aligned}\right\}. \tag{3.98}$$

The corresponding maximum bending stresses in the adherends are

$$\left.\begin{aligned} \sigma_r &= \frac{6M_r}{h_0^2} = 3\sigma_1 \frac{h_1}{h_0}\left[1 - \frac{kaI_0(kr) + (1+\nu_0)a\frac{I_1(kr)}{r}}{kaI_0(ka) - (1-\nu_1)I_1(ka)}\right] \\ \sigma_\theta &= \frac{6M_\theta}{h_0^2} = 3\sigma_1 \frac{h_1}{h_0}\left[1 - \frac{\nu_0 kaI_0(kr) + (1-\nu_0)a\frac{I_1(kr)}{r}}{kaI_0(ka) - (1-\nu_1)I_1(ka)}\right] \end{aligned}\right\}, \tag{3.99}$$

where the stress σ_1 is expressed by formula (3.81). For sufficiently large and/or stiff assemblies,

$$\sigma_r = 3\sigma_1 \frac{h_1}{h_0}\left[1 - \sqrt{\frac{a}{r}}e^{-k(a-r)}\right], \quad \sigma_\theta = 3\sigma_1 \frac{h_1}{h_0}\left[1 - \nu_0\sqrt{\frac{a}{r}}e^{-k(a-r)}\right]. \tag{3.100}$$

Comparing these formulas with formula (3.82) for the normal stresses in the bonding material, we conclude that, in an approximate analysis, the normal bending stresses in the adherends can be assumed to be proportional to the corresponding in-plane normal stresses in the bonding layer (this layer does not experience, of course, any bending), at the same current radius, r, and can be obtained by multiplying these stresses by the reduction factor of $3\frac{h_1}{h_0}$.

3.5.9 Peeling Stress

Since the bonding layer does not experience bending, the total radial bending moment acting in its cross sections must be zero. The lateral load, $q_0(r)$, acting on the bond, is due to the interfacial "peeling" stress, $p_0(x)$, and the interfacial shearing stress, $\tau_0(r)$, and can be found as

$$q_0(r) = p_0(r) - \frac{h_1}{2}\tau_0'(r). \tag{3.101}$$

The total bending moment will be zero, if this load is zero, such as if

$$p_0(r) = \frac{h_1}{2}\tau_0'(r) = p_{\max}\frac{kaI_0(kr) - a\dfrac{I_1(kr)}{r}}{kaI_0(ka) - I_1(ka)}, \tag{3.102}$$

where

$$p_{\max} = \frac{3}{4}\frac{1+\nu_1}{1+\nu_0}\frac{h_1}{h_0}E_0\frac{kaI_0(ka) - I_1(ka)}{kaI_0(ka) - (1-\nu_1)I_1(ka)}\Delta\alpha\Delta t \tag{3.103}$$

is the "peeling" stress at the edge $r = a$. Considering formula (3.75) for the maximum shearing stress, $\tau_{\max}$, one can write formula (3.102) as follows:

$$p_{\max} = \frac{h_1}{2a}\left[ka\frac{I_0(ka)}{I_1(ka)} - 1\right]\tau_{\max}. \tag{3.104}$$

In the case of sufficiently large and/or stiff assemblies, formulas (3.102), (3.103) and (3.104) yield:

$$p_0(r) = p_{\max}\sqrt{\frac{a}{r}}e^{-k(a-r)}, \quad p_{\max} = \frac{kh_1}{2}\tau_{\max}. \tag{3.105}$$

The second formula in (3.104) indicates that for thin enough bonding layers, the maximum peeling stress can be very low, even if the factor of the interfacial shearing stress is significant.

3.5.10 Numerical Example

Input Data

Component	Adherends	Adhesive
Young's modulus	$E_0 = 7384$ kg/mm^2	$E_1 = 500$ kg/mm^2
Poisson's ratios	$\nu_0 = 0.25$	$\nu_1 = 0.45$
CTE	$\alpha_0 = 0.5 \times 10^{-6}$ 1/°C	$\alpha_1 = 60.5 \times 10^{-6}$ 1/°C
Thickness	$h_0 = 1.5$ mm	$h_1 = 0.05$ mm
Assembly radius $a = 50.8$ mm; Temperature Change $\Delta t = 40$°C		

Calculated Data

Axial compliance of the bonding layer: $\lambda_1 = \frac{1-\nu_1^2}{E_1 h_1} = \frac{1-0.45^2}{500 \times 0.05} = 0.0319\,\text{mm/kg}$

Interfacial compliance of the two adherends:

$$\kappa_0 = \frac{4}{3}(1+\nu_0)\frac{h_0}{E_0} = \frac{4}{3} \times 1.25 \times 2.0314 \times 10^{-4} = 3.3857 \times 10^{-4}\,\text{mm}^3/\text{kg}$$

Interfacial compliance of the bonding layer:

$$\kappa_1 = \frac{1}{6}(1+\nu_1)\frac{h_1}{E_1} = \frac{1}{6} \times 1.45 \times 1.000 \times 10^{-4} = 0.2417 \times 10^{-4}\,\text{mm}^3/\text{kg}$$

Total $\kappa = \kappa_0 + \kappa_1 = 3.3857 \times 10^{-4} + 0.2417 \times 10^{-4} = 3.6274 \times 10^{-4}\,\text{mm}^3/\text{kg}$

Parameter of the interfacial shearing stress: $k = \sqrt{\frac{2\lambda_1}{\kappa}} = \sqrt{\frac{0.0638}{3.6274 \times 10^{-4}}} = 13.2621\,\text{mm}^{-1}$

Parameter $ka = 13.2621 \times 50.8 = 673.7152$ is significant, and therefore simplified formulas for large and stiff assemblies can be used to evaluate stresses.

Maximum interfacial shearing stress:

$$\tau_{max} \approx (1+\nu_1)\frac{\Delta\alpha\Delta t}{k\kappa} = 1.45\frac{0.0024}{13.2621 \times 3.6274 \times 10^{-4}} = 0.7234\,\text{kg/mm}^2$$

Maximum longitudinal displacement (at the meniscus at the assembly edge)

$$u_{max} = \kappa\tau_{max} = 3.6274 \times 10^{-4} \times 0.7234 = 2.6240 \times 10^{-4}\,\text{mm} = 0.2624\,\mu\text{m}$$

Normal tensile stress in the mid-portion of the assembly

$$\sigma_1 = \frac{E_1}{1-\nu_1}\Delta\alpha\Delta t = \frac{500}{0.55} \times 0.0024 = 2.1818\,\text{kg/mm}^2$$

Effective Young's moduli

$$E_0^* = \frac{E_0}{1-\nu_0} = \frac{7384}{0.75} = 9845.3\ \text{kg/mm}^2, \quad E_1^* = \frac{E_1}{1-\nu_1} = \frac{500}{0.55} = 909.1\,\text{kg/mm}^2$$

Maximum deflections of the adherends:

$$f = 3\frac{E_1^*}{E_0^*}\frac{a^2}{h_0^2}h_1\Delta\alpha\Delta t = 3 \times 0.0923 \times 1146.9511 \times 0.05 \times .0024 = 0.0381\,\text{mm} = 38.1\,\mu\text{m}$$

Maximum peeling stress: $p_{max} = \frac{kh_1}{2}\tau_{max} = 0.3316 \times 0.7234 = 0.2398\,\text{kg/mm}^2$

Thus, a simple, easy-to-apply and physically meaningful analytical ("mathematical") stress model is developed for the prediction of the stresses in a circular adhesively bonded assembly with identical adherends. The developed model can be helpful for stress–strain analyses and physical design of electronic and photonic assemblies of the type in question, and particularly holographic memory assemblies.

REFERENCES

1. S.P. Timoshenko, "Analysis of Bi-Metal Thermostats," *Journal of the Optical Society of America*, No. 11, 1925.
2. O. Volkersen, "Die Niekraftverteilung in Zugbeanspruchten mit Konstanten Laschenquerschritten," *Luftfahrforschung*, vol.15, 1938.
3. M. Goland, E. Reissner, "The Stresses in Cemented Joints," *ASME Journal of Applied Mechanics*, vol. 11, 1944.
4. N.A. de Bruyne, "The Strength of Glued Joints," *Aircraft Engineering*, vol. 16, 1944.
5. B.J. Aleck, "Thermal Stresses in a Rectangular Plate Clamped Along an Edge," *ASME Journal of Applied Mechanics*, vol. 16, 1949.
6. G.A. Lang, et al., "Thermal Fatigue in Silicon Power Devices," *IEEE Transactions on Electron Devices*, vol. 17, 1970.
7. R. Zeyfang, "Stresses and Strains in a Plate Bonded to a Substrate: Semiconductor Devices," *Solid State Electronics*, vol. 14, 1971.
8. B.A. Boley, J.H. Weiner, *Theory of Thermal Stresses*, Quantum Publishers, 1974.
9. F.-V. Chang, "Thermal Contact Stresses of Bi-Metal Strip Thermostat," *Applied Mathematics and Mechanics*, vol. 4, No. 3, Tsinghua Univ., Beijing, China, 1983.
10. R.D. Adams, W.C. Wake, *Structural Adhesive Joints in Engineering*, Elsevier, 1984.
11. E. Suhir, "Calculated Thermally Induced Stresses in Adhesively Bonded and Soldered Assemblies," *Procedure of the International Symposium on Microelectronics*, ISHM, Atlanta, Georgia, 1986 (best paper award).
12. E. Suhir, "Stresses in Bi-Metal Thermostats," *ASME Journal of Applied Mechanics*, vol. 53, No. 3, Sept. 1986.
13. E. Suhir, "An Approximate Analysis of Stresses in Multilayer Elastic Thin Films," *ASME Journal of Applied Mechanics*, vol. 55, No. 3, 1988.
14. E. Suhir, "Interfacial Stresses in Bi-Metal Thermostats," *ASME Journal of Applied Mechanics*, vol. 56, No. 3, September 1989.
15. J.H. Lau, "A Note on the Calculation of Thermal Stresses in Electronic Packaging by Finite-Element Method," *ASME Journal of Electronic Packaging*, vol. 111, No. 12, 1989.
16. A.Y. Kuo, "Thermal Stresses at the Edge of a Bimetallic Thermostat," *ASME Journal of Applied Mechanics*, vol. 56, 1989.
17. J.W. Eischen, C. Chung, J.H. Kim, "Realistic Modeling of the Edge Effect Stresses in Bi-material Elements," *ASME Journal of Electronic Packaging*, vol. 112, No. 1, 1990.
18. W.C. Carpenter, "A Comparison of Numerous Lap Joint Theories for Adhesively Bonded Joints," *Journal of Adhesion*, vol. 35, No. 1, 1991.
19. R. John, G. A. Hartman, J.P. Gallagher, "Crack Growth Induced by Thermal-Mechanical Loading," *Experimental Mechanics*, vol. 32, No. 2, 1992.
20. J.H. Lau (ed.), *Thermal Stress and Strain in Microelectronics Packaging*, Van-Nostrand Reinhold, 1993.
21. E. Suhir, "'Global' and 'Local' Thermal Mismatch Stresses in an Elongated Bi-Material Assembly Bonded at the Ends," in E. Suhir, (ed.), *Structural Analysis in Microelectronic and Fiber-Optic Systems*, Symposium Proceedings, ASME Press, 1995.

22. E. Suhir, "Thermal Stress Failures in Microelectronics and Photonics: Prediction and Prevention," *Future Circuits International*, vol. 5, 1999.
23. E. Suhir, "Thermal Stress in a Bi-Material Assembly Adhesively Bonded at the Ends," *Journal of Applied Physics*, vol. 89, No.1, 2001.
24. E. Suhir, "Analysis of Interfacial Thermal Stresses in a Tri-Material Assembly," *Journal of Applied Physics*, vol. 89, No. 7, 2001.
25. J-S. Bae, S. Krishnaswamy, "Subinterfacial Cracks in Bimaterial Systems Subjected to Mechanical and Thermal Loading," *Engineering Fracture Mechanics*, vol. 68, No. 9, 2001.
26. J.-S. Hsu, et al., "Photoelastic Investigation on Thermal Stresses in Bonded Structures," *SPIE Congrès Experimental Mechanics*, vol. 4537, 2002.
27. E. Suhir, "Thermal Stress in an Adhesively Bonded Joint with a Low Modulus Adhesive Layer at the Ends," *Journal of Applied Physics*, April 2003.
28. D. Sujan, et al., "Engineering Model for Interfacial Stresses of a Heated Bimaterial Structure with Bond Material Used in Electronic Packages," *IMAPS Journal of Microelectronics and Electronic Packaging*, vol. 2, No. 2, 2005.
29. E. Suhir, "Interfacial Thermal Stresses in a Bi-Material Assembly with a Low-Yield-Stress Bonding Layer," *Modeling and Simulation in Materials Science and Engineering*, vol. 14, 2006.
30. E. Suhir, "Thermal Stress in a Bi-Material Assembly with a "Piecewise-Continuous" Bonding Layer: Theorem of Three Axial Forces," *Journal of Applied Physics, D*, vol. 42, 2009.
31. E. Suhir, "On a Paradoxical Situation Related to Bonded Joints: Could Stiffer Mid-Portions of a Compliant Attachment Result in Lower Thermal Stress?" *ASME Journal of Solid Mechanics and Materials Engineering*, vol. 3, No. 7, 2009.
32. E. Suhir, "Analytical Thermal Stress Modeling in Electronic and Photonic Systems," *ASME Applied Mechanics Reviews*, vol. 62, No. 4, 2009.
33. K.-L. Tai, "Si-on-Si MCM Technology and Its Applications," *International Journal of High Speed Electronics and Systems*, vol. 2, No. 4, 1991.
34. E. Suhir, "Mechanical Behavior and Reliability of Solder Joint Interconnections in Thermally Matched Assemblies," 42nd ECTC., IEEE, San Diego, California, May 1992.
35. E. Suhir, "Axisymmetric Elastic Deformations of a Finite Circular Cylinder with Application to Low Temperature Strains and Stresses in Solder Joints," *ASME Journal of Applied Mechanics*, vol. 56, No. 2, 1989.
36. E. Suhir, "Adhesively Bonded Assemblies with Identical Nondeformable Adherends and 'Piecewise Continuous' Adhesive Layer: Predicted Thermal Stresses and Displacements in the Adhesive," *International Journal of Solids and Structures*, vol. 37, 2000.
37. G. Min, D.M. Rowe, "A Novel Thermoelectric Converter Employing Fermi Gas/Liquid Interfaces," *Journal of Physics, D, Applied Physics*, vol. 32, L1, 1999.
38. J.W. Stevens, "Optimum Design of Small DT Thermoelectric Generation Systems," *Energy Conversation Management*, 42, 2001.
39. K. Yazawa, G. Solbrekken, A. Bar-Cohen, "Thermoelectric- Powered Convective Cooling of Microprocessors," *IEEE Transactions of Advanced Packaging*, vol. 28 No. 2, 2005.
40. E.E. Antonova, D.C. Looman, "Finite Elements for Thermoelectric Device Analysis," ICT 2005 24th International Conference on Thermoelectrics, 2005.
41. K. Fukutani, A. Shakouri, "Design of Bulk Thermoelectric Modules for Integrated Circuit Thermal Management," *IEEE CPMT Transactions*, vol. 29, No. 4, 2006.
42. P.M. Mayer, R.J. Ram, "Optimization of Heat Sink-Limited Thermoelectric Generators," *Nanoscale Microscale Thermophysical Engineering*, vol. 10, 2006.

43. L.E. Bell, "Cooling, Heating, Generating Power, and Recovering Waste Heat with Thermoelectric Systems," *Science*, vol. 321, 2008.
44. T.H. Clin, et al., "Numerical Simulation of the Thermomechanical Behavior of Extruded Bismuth Telluride Alloy Module," *Journal of Electronic Materials*, vol. 38, No. 7, 2009.
45. V. Leonov, R.J.M. Vullers, "Wearable Thermoelectric Generators for Body-Powered Devices," *Journal of Electronic Materials*, vol. 38, No. 7, 2009.
46. E. Suhir, "Thermal Stress in a Bi-Material Assembly with a "Piecewise-Continuous" Bonding Layer: Theorem of Three Axial Forces," *Journal of Applied Physics, D*, vol. 42, 2009.
47. K. Yazawa, A. Shakouri, "Cost-Efficiency Trade-off and the Design of Thermoelectric Power generators," *Environmental Science and Technology*, 2010.
48. D. Kraemer, "High-Performance Flat-Panel Solar Thermoelectric Generators with High Thermal Concentration," *Natural Materials*, vol. 10, 2011.
49. J.-L. Gao, et al., "Thermal Stress Analysis and Structure Parameter Selection for a Be_2Te_3- Based Thermoelectric Module," *Journal of Electronic Materials*, vol. 40, No. 5, 2011.
50. E. Suhir, A. Shakouri, "Assembly Bonded at the Ends: Could Thinner and Longer Legs Result in a Lower Thermal Stress in a Thermoelectric Module (TEM) Design?" *ASME Journal of Applied Mechanics*, vol. 79, No. 6, 2012.
51. S.P. Timoshenko, S. Woinowsky-Krieger, *Theory of Plates and Shells*, McGraw-Hill, 1959.
52. E. Suhir, C. Gu, C. Cao, "Predicted Thermal Stresses in a Circular Assembly with Identical Adherends and with Application to a Holographic Memory Design," *ASME Journal of Applied Mechanics*, vol. 79, No. 1, 2011.
53. E. Suhir, "Approximate Evaluation of the Interfacial Shearing Stress in Cylindrical Double-Lap Shear Joints with Application to Dual-Coated Optical Fibers," *International Journal of Solids and Structures*, vol. 31, No. 23, 1994.
54. S.P. Timoshenko, J.N. Goodier, *Theory of Elasticity*, 3rd ed., McGraw-Hill, 1970.
55. I.N. Sneddon, *Special Functions of Mathematical Physics and Chemistry*, Oliver and Boyd, 1956.
56. E. Janke, F. Emde, F. Lösch, *Tafeln Höherer Functionen*, B. G., Tenbrer Verlagsgesellschaft, 1960 (in German).

4 Inelastic Strains in Solder Joint Interconnections

"Everyone knows that we live in the era of engineering, however, he rarely realizes that literally all our engineering is based on mathematics and physics."

—Bartel Leendert van der Waerden, Dutch mathematician

4.1 BACKGROUND/MOTIVATION

Solder materials are widely used in microelectronics and optoelectronics. Tin- and tin-lead–based solders have been used in radio engineering to provide electrical connection since the early 1920s. Since the mid-1980s, tin-lead-eutectic–based solder joint interconnections (SJI) were and still are used also as a mechanical support in flip-chip designs (first level interconnections). Ball-grid array (BGA) and column-grid array (CGA) interconnections are also used to attach a package to its substrate (second level of interconnections). SJIs are the bottleneck of electronic and photonic packaging reliability. This is mostly because they are typically subjected to low-cycle fatigue conditions, such as experiencing inelastic deformations in operation conditions, not to mention accelerated testing, such as temperature cycling or drop tests. Inelastic strains in solder joints have been addressed by numerous investigators (see, for example, [1–10]).

Interfacial thermal stresses, whether elastic or inelastic, concentrate at the peripheral portions of solderly bonded assemblies. Figure 4.1 explains the physics of this phenomenon. Consider a bimaterial bonded assembly manufactured at an elevated temperature and subsequently cooled down to a low (operation or testing) temperature. The adherend (bonded) element in the midportion of an assembly component is subjected either to tension [the component with a higher coefficient of thermal expansion (CTE)] or compression (the component with a lower CTE), and the induced forces acting on this element are more or less the same on both sides of the element. The situation is different for an element located at the assembly end. There are no external forces acting on it from the outside. Because of that, the force acting on the inner side of the element has to be equilibrated by the other assembly component through the assembly interface. This circumstance leads to the interfacial shearing stress at the assembly end. But that is not all. The force acting in the component's body and the interfacial force act in different planes and form a bending moment and, since there is no external bending moment acting at the end of the component, this bending moment has to be equilibrated by the other assembly component through the interface. This circumstance causes the stress acting in the through-thickness direction of the assembly—the pealing stress. The interfacial shearing stress is antisymmetric: it is zero in the mid-cross section of the assembly

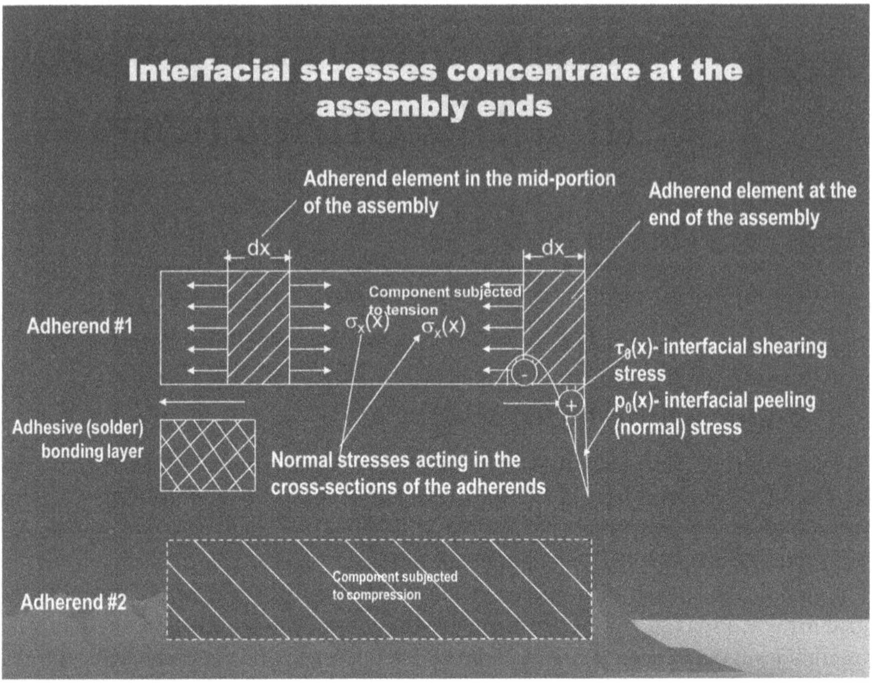

FIGURE 4.1 Thermally induced interfacial stresses concentrate at the bonded assembly ends.

and acts in the opposite directions at the assembly "halves." The peeling stress, however, has to be self-equilibrated, because there are no external forces acting on the assembly in its through-thickness direction. This means that the areas of the peeling stress diagram above and below the interface should be equal (although they are, as a rule, configured differently: the portion of the peeling stress diagram at the assembly end is "sharper" than its inner portion). The peeling stress is symmetric with respect to the mean cross section of the assembly. Since the interfacial stresses, both shearing and peeling, are the largest at the assembly ends, it is the peripheral portions of a soldered assembly that are most likely to experience plastic deformations, and this happens if the maximum thermally induced strain exceeds the yield stress of the bonding material—the solder. The low-cycle fatigue conditions, when a soldered assembly is subjected to temperature cycling during testing or in actual operation, make such a bonding material vulnerable and thereby responsible for the fatigue strength of the assembly.

As to the effect of the assembly size (Figure 4.2), it has been shown [11–14] that this effect depends not only on this size per se, but also on the parameter $k = \sqrt{\frac{\lambda}{\kappa}}$ of the interfacial shearing stress. This parameter, as this formula indicates, increases with an increase in the axial compliance λ of the adherends and decreases with an increase in its interfacial compliance κ. That is why a reliable adhesively bonded or

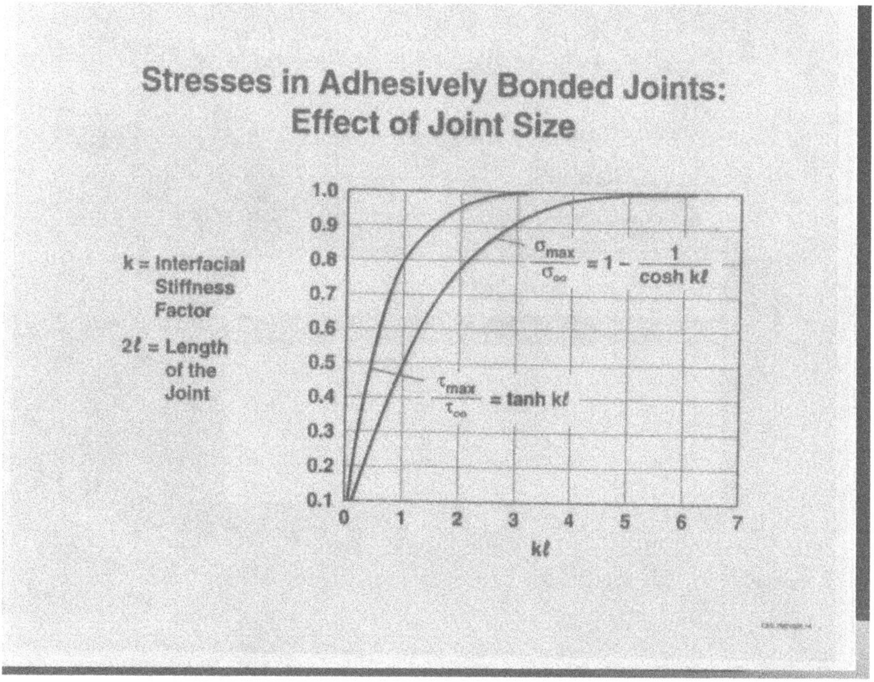

FIGURE 4.2 Effect of the assembly size on the interfacial shearing stress.

soldered assembly is the one that is characterized by stiff adherends and a compliant bond. When the product *kl* is above 3.5–4.0, the further increase in the assembly size does not affect the level and the distribution of the stresses: the stress fields at the end portions of long and/or stiff enough assemblies do not interact. Mathematically (see Figure 4.2) this circumstance manifests itself for the interfacial shearing stress through the factor of tanh *kl*, that becomes equal to one for *kl* values exceeding 3.5–4.0.

In the subsequent analysis, we develop an approximate analytical model for the assessment of interfacial stresses in a bimaterial soldered assembly with a low-yield stress of the bonding material. The analysis has been carried out initially for a photonic assembly, in which a tin-silver solder was used to attach a vulnerable GaAs photonic chip to a copper submount (Figure 4.3) [7] and was extended later to BGA and other SJI designs [8–11].

The analysis is carried out under the major assumption that the bonding material is linearly elastic at the strain level below its yield strain and is ideally plastic at the levels exceeding this strain. It is clear that the previously obtained elastic solution (see, for example, [12–14]), on one hand, and the present ideally-elastic/ideally-plastic solution, on the other, addresses the two extreme cases in the behavior of the bonding material. The more general and a more realistic situation, when the bonding material experiences elasto-plastic deformations above the yield point, is beyond the scope of the present analysis.

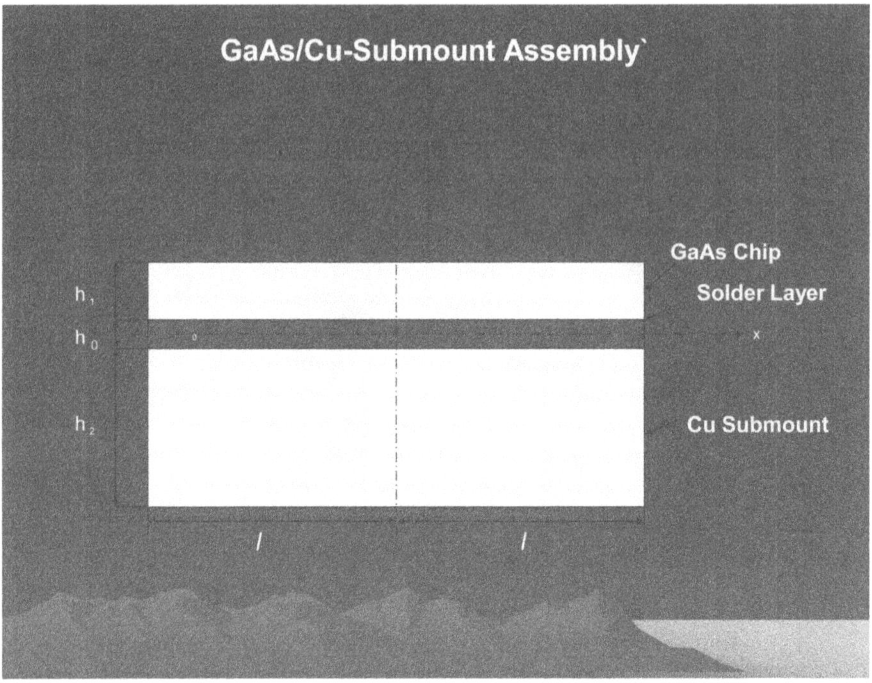

FIGURE 4.3 GaAs photonic chip mounted on a copper submount (substrate) using tin-silver solder, characterized by a low-yield strain. (From E. Suhir, "Interfacial Thermal Stresses in a Bi-Material Assembly with a Low-Yield-Stress Bonding Layer," Journal of Applied Physics-D, Modeling and Simulation in Materials Science and Engineering, vol. 14, 2006.)

4.2 ASSUMPTIONS

The following major assumptions are used in this analysis:

- Only the longitudinal cross section of the package-substrate assembly can be considered.
- The bonded components (the chip and the submount/substrate) can be treated, from the standpoint of stress/structural analysis, as elongated rectangular plates that experience linear elastic deformations.
- Approximate methods of structural analysis (strength-of-materials) and materials physics, rather than rigorous methods of elasticity and plasticity, can be used to evaluate stresses and displacements.
- At least one of the assembly components (substrate/submount) is thick and stiff enough, so that this component and the assembly as a whole do not experience bending deformations. The thinner component, the chip, might experience, however, some bending with respect to the thicker component.
- The bonding material (solder) behaves in a linearly elastic fashion, when the induced shearing strain is below its yield point, and is ideally plastic, when this strain exceeds the materials yield strain.

- The yield stress in shear, τ_Y, if unknown, can be assessed from the measured yield stress in tension, σ_Y, by the von Mises formula

$$\tau_Y = \sigma_Y / \sqrt{3} \tag{4.1}$$

- The interfacial shearing stresses can be evaluated without considering the effect of peeling; the peeling stress can be determined, if necessary, from the evaluated interfacial shearing stress.
- The peeling stress is proportional to the deflections of the thinner component of the assembly—the chip, with respect to the thicker component—the submount.

4.3 SHEARING STRESS

4.3.1 Basic Equation

Let an elongated soldered bimaterial assembly (Figure 4.3) be manufactured at an elevated temperature and subsequently cooled down to a low (room or testing) temperature. In an approximate analysis, the longitudinal interfacial displacements, $u_1(x)$ and $u_2(x)$, of the adherends (the assembly components) can be sought, within the assembly's elastic midportion, $-x_* \le x \le x_*$, ($\pm x_*$ are the abscissas of the boundaries of the elastic midportion of the assembly), in the form [12]:

$$u_1(x) = -\alpha_1 \Delta t x + \lambda_1 \int_0^x T(\xi)d\xi - \kappa_1 \tau(x), \quad u_2(x) = -\alpha_2 \Delta t x + \lambda_2 \int_0^x T(\xi)d\xi - \kappa_2 \tau(x), \tag{4.2}$$

where α_1 and α_2 are the coefficients of thermal expansion (contraction) of the materials, Δt is the change in temperature,

$$\lambda_1 = \frac{1-\nu_1}{E_1 h_1}, \quad \lambda_2 = \frac{1-\nu_2}{E_2 h_2}, \tag{4.3}$$

are the longitudinal axial compliances of the assembly components, h_1 and h_2 are the component thicknesses (in accordance with one of our assumptions, the thickness, h_2, of the thicker component—the submount—is significantly greater than the thickness, h_1, of the thinner component), E_1 and E_2 are the Young's moduli of the component materials, ν_1 and ν_2 are their Poisson's ratios,

$$\kappa_1 = \frac{h_1}{3G_1} = \frac{2}{3}(1+\nu_1)\frac{h_1}{E_1}, \quad \kappa_2 = \frac{h_2}{3G_2} = \frac{2}{3}(1+\nu_2)\frac{h_2}{E_2} \tag{4.4}$$

are the interfacial compliances of the assembly components, G_1 and G_2 are the shear moduli of the component materials, $\tau(x)$ is the interfacial shearing stress,

$$T(x) = \int_{-x_*}^{x} \tau(\zeta)d\zeta - \tau_Y l_* \tag{4.5}$$

are the thermally induced forces acting in the cross sections of the assembly components, τ_Y is the yield stress of the bonding material, and l_* is the length of the plastic zone at the ends of the assembly. The length, l_*, can be defined as $l_* = l - x_*$, where l is half the assembly length. The origin, 0, of the coordinate, x, is in the mid-cross section of the assembly.

The first terms in the right parts of the expressions (4.2) are unrestricted (stress-free) displacements. The second terms determine the displacements due to the thermally induced forces, $T(x)$, that arise in the cross sections of the assembly components, because of the thermal contraction mismatch of the dissimilar materials of the soldered components. These terms are defined based on Hooke's law assuming that all the points of the given cross section have the same longitudinal displacements, that is, the assembly's cross sections remain flat (undistorted) despite the change in the states of stress and strain. The third terms in the right part of equations (4.2) account for the inaccuracy of such an assumption and consider the fact that the interfacial displacements are somewhat larger than the displacements of the inner points of the cross sections. The structure of these additional terms reflects an assumption that the displacements, which are responsible for the distortion in the planarity of the component's cross section, are proportional to the interfacial shearing stress acting in this cross section (Figure 4.4). It is also assumed that these additional displacements are not affected by the stresses and strains in the adjacent cross sections and can be assessed as the product of the interfacial compliance of the

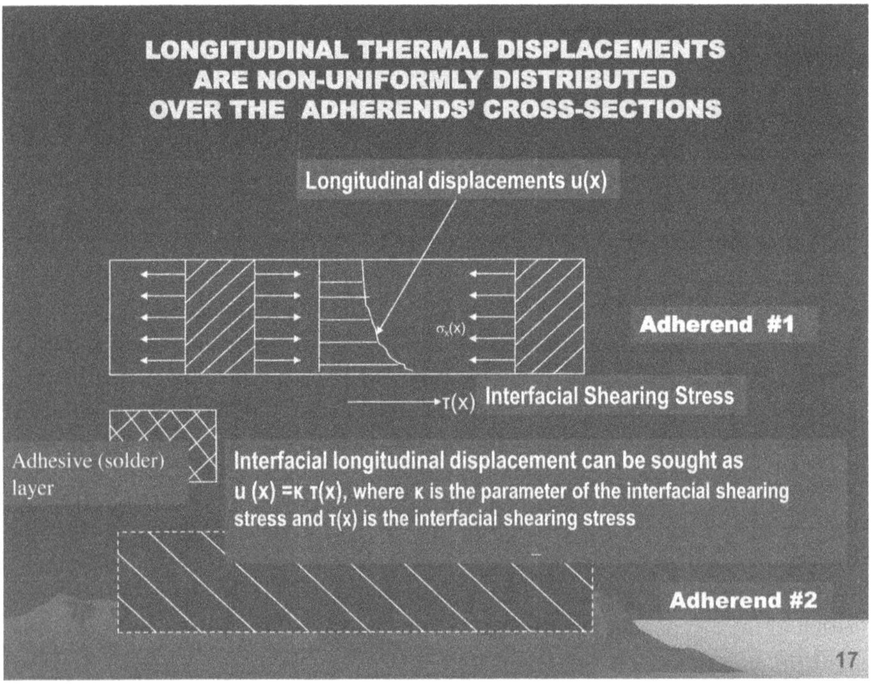

FIGURE 4.4 The longitudinal displacements of the given cross section of the adherend are somewhat larger than the displacements of the inner points of the cross section.

assembly component (which is known in advance) and the sought induced shearing stress acting in this cross section.

While the structural analysis approach is used in this book for the evaluation of stresses and displacements, the coefficients of proportionality (interfacial compliances) between the interfacial displacements and the interfacial shearing stresses are evaluated on the basis of the theory of elasticity solution (see, for example [38]). This solution was obtained using Ribiére treatment of the problems for long-and-narrow strips subjected to the distributed shearing loads applied to one or to both of their long edges.

The condition of the compatibility of the interfacial displacements, $u_1(x)$ and $u_2(x)$, can be written, considering the compliance

$$\kappa_0 = \frac{h_0}{G_0} = 2(1+\nu_0)\frac{h_0}{E_0} \tag{4.6}$$

of the bonding layer, as

$$u_1(x) = u_2(x) - \kappa_0 \tau(x) \tag{4.7}$$

Here, E_0 and ν_0 are the elastic constants of the bonding material (the solder below the yield stress), and h_0 is the thickness of the bonding layer. Introducing formulas (4.2) into the compatibility condition (4.7), we obtain the following basic integral equation for the shearing stress function, $\tau(x)$, in the elastic midportion of the assembly:

$$\kappa\tau(x) - \lambda\int_0^x T(\xi)d\xi = \Delta\alpha\Delta t x. \tag{4.8}$$

Here, $\Delta\alpha = \alpha_2 - \alpha_1$ is the thermal expansion (contraction) mismatch of the materials of the soldered components (the adherends), $\lambda = \lambda_1 + \lambda_2$ is the total longitudinal axial compliance of the assembly, and $\kappa = \kappa_0 + \kappa_1 + \kappa_2$ is its total longitudinal interfacial compliance. It is noteworthy that, in the case of a thin and/or low modulus bonding layer, only the two soldered components (the adherends) determine the axial compliance of the assembly. As to the interfacial compliance, both the soldered components (the adherends) and the bonding layer ("adhesive") contribute to the interfacial compliance: the role of a thin and low modulus bonding layer is typically comparable with the role of thick and high modulus bonded components ("adherends"), as one could see from the numerical example at the end of this analysis (see also Chapter 2).

4.3.2 Boundary Conditions

In the case when plastic strains occur in the bonding material, the following conditions must be fulfilled at the boundary, $x = x_*$, between the "inner" (linearly elastic) and the "outer" (ideally plastic) zones:

$$\tau(x_*) = \tau_Y, \quad T(x_*) = -\tau_Y l_* \tag{4.9}$$

The first condition in (4.9) indicates that the shearing stress at the boundary between the elastic and the plastic zones must be equal to the yield stress of the bonding material. The second condition follows from formula (4.5): the shearing stress, $\tau(x)$, is self-equilibrated, and therefore the integral in (4.5) is zero for $x = x_*$. Physically, this condition is due to the fact that, since the interfacial shearing stress at the peripheral portions of the assembly is constant (is equal to the yield stress, τ_Y), the thermally induced force $T(x)$ changes linearly at these portions, from $\tau_Y l_*$, at the boundary of the elastic and the plastic zones, to zero at the assembly ends. The sign "minus" in front of the second boundary condition in (4.9) indicates that the force at the boundary should be compressive (negative) for the compressed component of the assembly. In the case of a purely elastic state of strain ($l_* = 0$), the following boundary condition should be fulfilled:

$$T(l) = \int_{-l}^{l} \tau(x)dx = 0 \tag{4.10}$$

This condition reflects the fact that there are no external longitudinal forces acting at the end cross sections of the assembly components.

4.3.4 Elasto-Plastic Solution

From (4.8), we find, by differentiation with respect to the coordinate, x:

$$\kappa\tau'(x) - \lambda T(x) = \Delta\alpha\Delta t. \tag{4.11}$$

The next differentiation, considering the relationship (4.5), yields

$$\tau''(x) - k^2\tau(x) = 0, \tag{4.12}$$

where

$$k = \sqrt{\frac{\lambda}{\kappa}} \tag{4.13}$$

is the parameter of the interfacial shearing stress. Equation (4.11) has the following solution in the elastic midportion of the assembly:

$$\tau(x) = \tau_Y \frac{\sinh kx}{\sinh kx_*}. \tag{4.14}$$

It is clear that this solution satisfies the first condition in (4.9). Introducing sought solution (4.14) into formula (4.5), we conclude that the second condition in (4.9) is also satisfied. Introducing solution (4.14) into basic equation (4.8), we find that the relative length $\frac{l_*}{l}$ of the plastic zone could be determined from the following transcendental equation:

$$\frac{l_*}{l} = \frac{1}{kl}\left[\frac{\tau_{\max}^{\infty}}{\tau_Y} - \coth\left(kl\left(1-\frac{l_*}{l}\right)\right)\right], \tag{4.15}$$

where

$$\tau_{\max}^{\infty} = k\frac{\Delta\alpha\Delta t}{\lambda} \tag{4.16}$$

is the maximum elastic interfacial shearing stress at the end of an infinitely long assembly [11–14]. As evident from equation (4.15), no plastic zones could possibly occur ($l_* \le 0$), if the ratio $\frac{\tau_{\max}^{\infty}}{\tau_Y}$ of the maximum elastic shearing stress in an infinitely long assembly to the yield stress of the bonding material is equal or smaller than coth kl:

$$\frac{\tau_{\max}^{\infty}}{\tau_Y} \le \coth kl. \tag{4.17}$$

Indeed, for long (large l values) and/or stiff (large k values) assemblies, when coth kl could be considered equal to one, condition (4.17) is equivalent to the requirement that the yield stress is simply larger than the maximum elastic interfacial shearing stress. In such a situation, no plastic stresses could possibly occur. If the kl value is small, then condition (4.17) yields

$$\tau_Y \ge \frac{\Delta\alpha\Delta t}{\kappa}l. \tag{4.18}$$

Thus, no plastic deformations could possibly occur, if, in a short and/or compliant assembly, condition (4.18) is fulfilled, that is, if the yield stress τ_Y is high, the assembly compliance κ in the denominator in the right part of condition (4.18) is significant, the thermal strain $\Delta\alpha\Delta t$ in the numerator is low, and the size, l, of the assembly is small.

Equation (4.15), if solved for the lengths ratio in the parentheses, yields

$$\frac{l_*}{l} = 1 - \frac{1}{2kl}\ln\frac{\frac{\tau_{\max}^{\infty}}{\tau_Y} - kl\frac{l_*}{l} + 1}{\frac{\tau_{\max}^{\infty}}{\tau_Y} - kl\frac{l_*}{l} - 1}. \tag{4.19}$$

If the yield stress τ_Y is low and, for this reason, the stress ratio $\frac{\tau_{\max}^{\infty}}{\tau_Y}$ is significantly larger than one, then, as evident from (4.19), $l_* = l$, such as the entire interfacial zone is occupied by the plastic strains (stresses), regardless of whether the kl value is large or small. If the above stress ratio is significantly smaller than one, then equation (4.19) yields

$$\frac{l_*}{l} = 1 - \frac{1}{2kl} \ln \frac{kl\frac{l_*}{l} - 1}{kl\frac{l_*}{l} + 1}. \tag{4.20}$$

As evident from this equation, plastic deformations might still take place if the k value is large, despite the low l value.

4.3.5 Possible Numerical Procedure for Solving the Elasto-Plastic Equations

The transcendental equations (4.15) and (4.19) can be solved numerically. Let, for example, the stress ratio be $\frac{\tau_{\max}^{\infty}}{\tau_Y} = 2$, and the parameter kl is also $kl = 2$. Then, equation (4.15) yields

$$\frac{l_*}{l} = 1 - \frac{1}{2} \coth\left[2\left(1 - \frac{l_*}{l}\right)\right] \tag{4.21}$$

This equation has the following solution: $\frac{l_*}{l} = 0.4002$. This result could be obtained by simply assuming different length ratios, plotting the function in the right part of equation (4.21) and accepting, as a solution, the length ratio, which is equal to the computed value of this function in the left part of the equation. The same result can be obtained from equation (4.19). Another approach is to employ a rapidly converging iterative numerical procedure based on the well-known Newton's formula

$$x_{n+1} = x_n - \frac{f(x_n)}{f'(x_n)} \tag{4.22}$$

for solving a transcendental equation $f(x) = 0$. The formula of the (4.22) type can be obtained from equation (4.15) as follows:

$$\left(\frac{l_*}{l}\right)_{n+1} = \left(\frac{l_*}{l}\right)_n + \frac{\frac{1}{kl}\left(\frac{\tau_{\max}^{\infty}}{\tau_Y} - (z_*)_n\right) - \left(\frac{l_*}{l}\right)_n}{2(z_*)_n^2 - 1}, \tag{4.23}$$

where the notation

$$z_* = \coth\left[kl\left(1 - \frac{l_*}{l}\right)\right] \tag{4.24}$$

is used. For $\frac{\tau_{\max}^{\infty}}{\tau_Y} = 2$ and $kl = 2$, assuming, in the zero approximation, $\left(\frac{l_*}{l}\right)_0 = 0.5$, we obtain:

$$\left(\frac{l_*}{l}\right)_1 = 0.4361, \quad \left(\frac{l_*}{l}\right)_2 = 0.4101, \quad \left(\frac{l_*}{l}\right)_3 = 0.4026,$$
$$\left(\frac{l_*}{l}\right)_4 = 0.4007, \quad \left(\frac{l_*}{l}\right)_5 = 0.4003, \quad \left(\frac{l_*}{l}\right)_6 = 0.4002,$$

No iterations are necessary for sufficiently long and stiff assemblies, characterized by large kl values, since the hyperbolic cotangent will be equal to "one" for any (but large enough) value of the argument. Indeed, let kl be equal to $kl = 10$. Then, assuming that the $\frac{l_*}{l}$ ratio is significantly smaller than one, we have $z_* \approx 1$ and equation (4.23) yields

$$\left(\frac{l_*}{l}\right)_{n+1} = \frac{2}{kl}\left(\frac{\tau_{\max}^{\infty}}{\tau_Y} - 1\right) = \frac{2}{10}(2-1) = 0.2$$

4.3.6 Predicted Lengths of the Plastic Zones Based on an Elastic Solution

We proceed from equation (4.12) and seek its elastic solution in the form similar to (4.14):

$$\tau(x) = C\frac{\sinh kx}{\sinh kl}. \tag{4.25}$$

Introducing (4.25) into equation (4.12), we conclude that the following relationships must be fulfilled:

$$k = \sqrt{\frac{\lambda}{\kappa}}, \quad C = \frac{k}{\lambda}\Delta\alpha\Delta t \tanh kl. \tag{4.26}$$

The first formula in (4.26) is the same as formula (4.13). This is because formula (4.13) defines the parameter k of the elastic interfacial shearing stress. With formulas (4.26), solution (4.25) yields

$$\tau(x) = k\frac{\Delta\alpha\Delta t}{\lambda}\frac{\sinh kx}{\cosh kl} = \tau_{\max}^{\infty}\tanh kl\frac{\sinh kx}{\sinh kl} = \tau_{\max}^{\infty}\frac{\sinh kx}{\cosh kl}, \tag{4.27}$$

where $\tau_{\max}^{\infty}$ is the maximum shearing stress at the end of a long and/or stiff assembly ($kl \to \infty$). This stress is expressed by formula (4.16). Putting $\tau_Y = \tau(x_*)$ in formula

(4.27), we obtain the following formula for the relative length $\frac{l_*}{l} = 1 - \frac{x_*}{l}$ of the plastic zone:

$$\frac{l_*}{l} = 1 - \frac{1}{kl} \ln[z(1+\sqrt{1+z^{-2}})], \tag{4.28}$$

where

$$z = \frac{\tau_Y}{\tau_{max}^{\infty}} \cosh kl. \tag{4.29}$$

For $\frac{\tau_{max}^{\infty}}{\tau_Y} = 2$ and $kl = 2$, formula (4.28) yields $\frac{l_*}{l} = 0.3054$. Comparing this value with the value, $\frac{l_*}{l} = 0.4002$ obtained using the elasto-plastic solution, we conclude that in the case in question the prediction based on an elastic solution underestimates considerably (by about 24%) the length of the plastic zones. The underestimation is even greater (about 80%) in the numerical example carried out in the last section of this analysis.

4.4 PEELING STRESS

4.4.1 Basic Equation

The basic equation for the peeling stress, $p(x)$, can be obtained using the following equation of equilibrium for the thinner (more flexible) component (#1) of the assembly treated as an elongated thin plate:

$$\int_{-x_*}^{x} \int_{-x_*}^{x} p(\xi)d\xi d\xi + D_1 w''(x) = \frac{h_1}{2} T(x) = \frac{h_1}{2} \left[\int_{-x_*}^{x} \tau(\xi)d\xi - \tau_Y l_* \right]. \tag{4.30}$$

Here

$$D_1 = \frac{E_1 h_1^3}{12(1-\nu_1^2)} \tag{4.31}$$

is the flexural rigidity of the component, and $w(x)$ is its deflection function with respect to the thicker component that, in accordance with our assumption, does not experience bending deformations. Equation (4.30) indicates that the "external" bending moment experienced by component #1 and expressed by the right part in equation (4.30) and the first term in the left part should be equilibrated by the elastic bending moment, which is expressed by the second term in the left part of this equation. This term is proportional, for small deflections, to the second derivative (curvature) of the deflection function, $w(x)$.

In accordance with one of our assumptions, the peeling stress, $p(x)$, can be sought as

$$P(x) = Kw(x), \tag{4.32}$$

where K is the spring constant of the elastic foundation, provided by the bonding layer and, generally speaking, also by the assembly components themselves. Excluding the deflection function, $w(x)$, from equations (4.30) and (4.32), we obtain the following equation for the sought peeling stress function, $p(x)$:

$$\int_{-x*}^{x}\int_{-x*}^{x} p(\xi)d\xi d\xi + \frac{D_1}{K}p''(x) = \frac{h_1}{2}T(x) \tag{4.33}$$

After differentiating this equation twice with respect to the coordinate x and considering relationship (4.5), we obtain the following basic equation for the peeling stress function $p(x)$:

$$p^{IV}(x) + 4\beta^4 p(x) = 2\beta^4 h_1 \tau'(x), \tag{4.34}$$

where

$$\beta = \sqrt[4]{\frac{K}{4D_1}} \tag{4.35}$$

is the parameter of the peeling stress. Equation (4.34) has the form of the equation of bending of a beam lying on a continuous elastic foundation (see, for example, [18]) and loaded by a distributed load whose magnitude is proportional to the rate of changing (the gradient) in the interfacial shearing stress along the assembly. Since the shearing stress is constant outside the elastic region, the peeling stress is zero in this region. Note, that in the close proximity to the boundary between the elastic and plastic zones the peeling stress might change in a manner that violates its proportionality to the derivative of the shearing stress. This is due to the fact that the approximate solution (4.18) ignores the singularity at the boundary between the elastic and the inelastic zones, and therefore the behavior of the peeling stress in the proximity of this boundary might be different of what is anticipated by equation (4.34).

The peeling stress should be self-equilibrated within the elastic region and therefore the following conditions of equilibrium with respect to the bending moments and the lateral forces should be fulfilled:

$$\int_{-x*}^{x}\int_{-x*}^{x} p(\xi)d\xi d\xi = 0, \quad \int_{-x*}^{x} p(\xi)d\xi = 0. \tag{4.36}$$

From (4.33) we find, by differentiation:

$$\int_{-x_*}^{x} p(\xi)d\xi + \frac{D_1}{K} p'''(x) = \frac{h_1}{2}\tau(x). \tag{4.37}$$

The relationships (4.33) and (4.37), with consideration of equilibrium conditions (4.36), result in the following boundary conditions for the peeling stress function, *p(x)*:

$$p''(x_*) = -\frac{h_1 K l_*}{4D_1}\tau_Y = -2\beta^4 h_1 l_* \tau_Y, \quad p'''(x_*) = \frac{h_1 K}{2D_1}\tau_Y = 2\beta^4 h_1 \tau_Y. \tag{4.38}$$

The peeling stress in the zones of plastic shearing strains should be zero, because it follows from equation (4.34): the shearing stress function is equal to the yield stress in these zones, and, hence, does not change along the assembly.

4.4.2 Solution to the Basic Equation

Equation (4.34) has the form of an equation of a beam lying on a continuous elastic foundation (see, for instance, [18]). Accordingly, we seek the solution to this equation in the form:

$$p(x) = C_0 V_0(\beta x) + C_2 V_2(\beta x) = A\frac{\cosh kx}{\cosh kx_*}, \tag{4.39}$$

where the functions $V_i(\beta x)$, $i = 0,1,2,3$, are expressed as

$$\begin{gathered} V_0(\beta x) = \cosh\beta x \cos\beta x, \quad V_2(\beta x) = \sinh\beta x \sin\beta x, \\ V_{1,3}(\beta x) = \frac{1}{\sqrt{2}}(\cosh\beta x \sin\beta x \pm \sinh\beta x \cos\beta x) \end{gathered} \tag{4.40}$$

These functions have the following properties (see, for example, [18]):

$$\begin{aligned} V_0'(\beta x) &= \beta\sqrt{2V_3}(\beta x), \quad & V_1'(\beta x) &= \beta\sqrt{2V_0}(\beta x), \\ V_2'(\beta x) &= \beta\sqrt{2V_1}(\beta x), \quad & V_3'(\beta x) &= \beta\sqrt{2V_2}(\beta x) \end{aligned} \tag{4.41}$$

which make their use of convenience. As evident from expression (4.39), the peeling stress function, *p(x),* has its maximum value (zero derivative) at the origin, and, unlike the interfacial shearing stress, is symmetric with respect to the mid-cross section of the assembly. The first two terms in (4.39) provide the general solution to the homogeneous equation, which corresponds to the inhomogeneous equation (4.34), and the third term is the particular solution to this equation. Introducing this term into equation (4.34), we obtain:

$$A = \frac{kh_1\tau_Y}{2(1+\eta^4)}\coth kx_*, \tag{4.42}$$

where the ratio

$$\eta = \frac{k}{\beta\sqrt{2}} \tag{4.43}$$

characterizes the relative role of the interfacial shearing and peeling stresses, and x_* is the abscissa of the boundary of the elastic zone.

Using conditions (4.38), the following equations for the constants C_0 and C_2 of integration can be obtained:

$$\begin{aligned} V_2(\beta x_*)C_0 - V_0(\beta x_*)C_2 &= \tau_Y \frac{\beta h_1}{\sqrt{2}}\left(\sqrt{2}\beta l_* + \frac{\eta^3}{1+\eta^4}\coth kx_*\right) = \tau_Y \frac{\beta h_1}{\sqrt{2}} F_1 \\ V_1(\beta x_*)C_0 + V_3(\beta x_*)C_2 &= -\tau_Y \frac{\beta h_1}{\sqrt{2}} \frac{1}{1+\eta^4} = \tau_Y \frac{\beta h_1}{\sqrt{2}} F_2 \end{aligned} \tag{4.44}$$

where the following notation is used:

$$F_1 = \sqrt{2}\beta l_* + \frac{\eta^3}{1+\eta^4}\coth kx_*, \quad F_2 = -\frac{1}{1+\eta^4}. \tag{4.45}$$

Equations (4.44) have the following solutions:

$$C_0 = 2\tau_Y\beta h_1 \frac{V_0(\beta h_*)F_2 + V_3(\beta x_*)F_1}{\sinh 2\beta x_* + \sin 2\beta x_*}, \quad C_2 = 2\tau_Y\beta h_1 \frac{V_2(\beta h_*)F_2 - V_1(\beta x_*)F_1}{\sinh 2\beta x_* + \sin 2\beta x_*}. \tag{4.46}$$

For long enough elastic zones (which is usually the case in actual soldered assemblies), solution (4.39) can be simplified:

$$\begin{aligned} p(x) = \tau_Y\beta h_1 e^{-\beta(x_*-x)} &\left[\frac{F_1}{\sqrt{2}}(\sin[\beta(x_* - x)] - \cos[\beta(x_* - x)]) + F_2\cos[\beta(x_* - x)]\right] \\ &+ \frac{\tau_Y kh_1}{2(1+\eta^4)} \end{aligned} \tag{4.47}$$

and the hyperbolic function in the first formula in (4.45) can be put equal to one.

4.5 NUMERICAL EXAMPLE

Let a laser GaAs **chip** (Young's modulus $E_1 = 8775$ kg/mm^2; shear modulus $G_1 = 3367.3$ kg/mm^2; CTE $\alpha_1 = 6.5 \times 10^{-6}$1/°C; thickness $h_1 = 0.150$ mm; width $b_1 = 0.4$ mm; length $2l_1 = 2.4$ mm) be soldered, using 96.5%Ag3.5%Sn silver-tin **solder** (Young's modulus $E_a = 1939.0$ kg/mm^2; shear modulus $G_a = 737$ kg/mm^2; estimated yield stress in shear $\tau_Y = 1.5306$ kg/mm^2; thickness $h_a = 1\text{mil} = 0.025$ mm), onto a copper **submount/substrate** (Young's modulus $E_2 = 11224.5$ kg/mm^2; shear modulus $G_2 = 4693.9$ kg/mm^2; CTE $\alpha_2 = 17.8 \times 10^{-6}$1/°C; thickness $h_2 = 0.5$ mm; width $b_2 = 4$ mm; length $2l_2 = 4.8$ mm. In the computations that follow, we assume $l = l_1 = l_2 = 1.2$ mm; and, for the sake of simplicity and for illustrative purposes, we assume also that the chip and the substrate have the same width.

The soldering temperature is about 158°C, so that the change in temperature can be assumed to be $\Delta t = 125$°C. The Poisson's ratios of the bonded components and the bonding material are

$$\nu_1 = \frac{E_1}{2G_1} - 1 = 0.3; \quad \nu_a = \frac{E_a}{2G_a} - 1 = 0.32; \quad \nu_2 = \frac{E_2}{2G_2} - 1 = 0.2.$$

Then, formulas (4.3), (4.4), and (4.6) result in the following axial

$$\lambda_1 = \frac{1-\nu_1}{E_1 h_1} = \frac{1-0.3}{8775.5 \times 0.150} = 5.3178 \times 10^{-4} \text{ mm/kg},$$

$$\lambda_2 = \frac{1-\nu_2}{E_2 h_2} = \frac{1-0.3}{11224.5 \times 0.50} = 1.4254 \times 10^{-4} \text{ mm/kg}$$

$$\lambda = \lambda_1 + \lambda_2 = 5.3178 \times 10^{-4} + 1.4254 \times 10^{-4} = 6.7432 \times 10^{-4} \text{ mm/kg}$$

and interfacial

$$\kappa_1 = \frac{h_1}{3G_1} = \frac{0.150}{3 \times 3367.3} = 0.1485 \times 10^{-4} \text{ mm}^3\text{/kg}$$

$$\kappa_2 = \frac{h_2}{3G_2} = \frac{0.50}{3 \times 4693.9} = 0.3551 \times 10^{-4} \text{ mm}^3\text{/kg}$$

$$\kappa_a = \frac{h_a}{G_a} = \frac{0.025}{737} = 0.3321 \times 10^{-4} \text{ mm}^3\text{/kg}$$

$$\kappa = \kappa_a + \kappa_1 + \kappa_2 = 0.3321 \times 10^{-4} + 0.1485 \times 10^{-4} + 0.3551 \times 10^{-4}$$
$$= 0.8357 \times 10^{-4} \text{ mm}^3\text{/kg}$$

compliances.

The parameter k of the interfacial shearing stress, in accordance with formula (4.13), is

$$k = \sqrt{\frac{\lambda}{\kappa}} = 2.8406 \text{ mm}^{-1}$$

Hence, the product kl is $kl = 3.4087$. From (4.16), we find that the maximum elastic interfacial shearing stress, if the bonding material behaves elastically under any level of loading, is

$$\tau_{\max}^{\infty} = k\frac{\Delta\alpha\Delta t}{\lambda} = 5.9502 \text{ kg/mm}^2$$

The ratio $\tau_{\max}^{\infty} / \tau_Y$ is 3.8875, and equation (4.19) yields $\frac{l_*}{l} = 0.7334$, that is, about 26.6% of the interface's length experiences elastic deformations.

The flexural rigidity of the thinner component #1 can be found by formula (4.31) as

$$D_1 = \frac{E_1 h_1^3}{12(1-\nu_1^2)} = 2.7122 \text{ kg} \times \text{mm}.$$

In an approximate analysis, we assume that the spring constant of the elastic foundation can be put equal to the Young's modulus of the bonding material, so that $K = E_0 = 1939$ kgf/mm^2. The coordinate x_* of the plastic zone is $x_* = l - l_* = 0.3199$ mm, and therefore $kx_* = 0.9088$, and $\coth kx_* = 1.3878$. Hence, in the case in question, the elastic zone cannot be assumed to be infinitely long. Formulas (4.31), (4.38), and (4.39) yield:

$$\beta = \sqrt[4]{\frac{K}{4D_1}} = 3.6564 \text{ mm}^{-1}; \quad \beta x_* = 1.1697; \quad \eta = \frac{k}{\beta\sqrt{2}} = 0.54934;$$

$$A = \frac{kh_1\tau_Y}{2(1-\eta^4)}\coth kx_* = 0.7185 \text{ kg/mm}^2;$$

The functions $V_i(\beta x_*)$, which are expressed by formulas (4.36), have the following values:

$$V_0(\beta x_*) = V_0(1.1697) = 0.6917; \quad V_1(\beta x_*) = V_1(1.1697) = 1.5537;$$
$$V_2(\beta x_*) = V_2(1.1697) = 1.3350; \quad V_3(\beta x_*) = V_3(1.1697) = 0.7458;$$

Formulas (4.45) and (4.46) yield:

$$F_1 = 4.7651; \quad F_2 = -0.9165; \quad C_0 = 0.8401 \text{ kg/mm}^2; \quad C_2 = -2.4822 \text{ kg/mm}^2$$

Then, the peeling stress function expressed by solution (4.47) is as follows:

$$p(x) = 0.8401V_0(3.6564x) - 2.4822V_2(3.6564x) + 0.4988\cosh(2.8406x)$$

The calculated interfacial shearing and the peeling stresses are plotted in Figure 4.5. As evident from this plot, one cannot simply truncate the diagram for the shearing stress, obtained on the basis of an elastic solution, to evaluate the length of the plastic

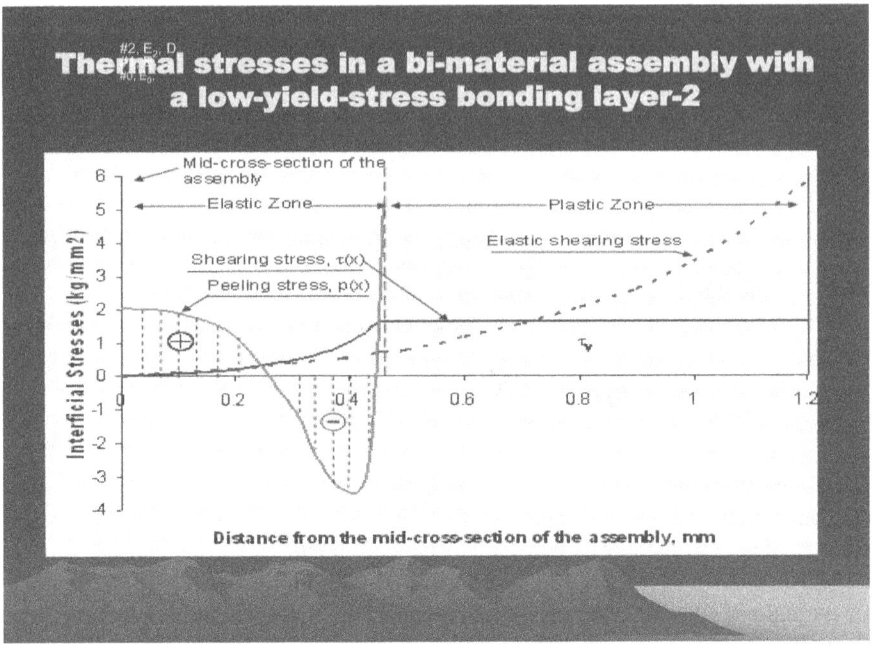

FIGURE 4.5 Calculated shearing and peeling stresses in an optical chip-submount assembly.

zone. Such an approach will result in a substantial, by about 80%, underestimation of the length of the plastic zone. Clearly, the elastic approach cannot be used to assess the peeling stress either.

4.6 THE CASE OF A BGA ASSEMBLY

4.6.1 Background/Motivation

In today's BGA packaging technologies (second level interconnections), solders are employed to attach surface mount packages to printed circuit boards (PCBs). While the reliability of SJIs in flip-chip designs has been improved dramatically owing to the introduction of epoxy encapsulants (underfills), the BGA systems remain the reliability bottleneck in the today's electronic packaging engineering. BGA SJIs experience thermal loading caused by the thermal expansion (contraction) mismatch of the dissimilar materials in the structure, when the system (on the board level) is subjected to the change in temperature, and/or because of the temperature gradients. Solder materials, even lead-free solders, are characterized by low-yield stresses, and, because of that, the package peripheral joints, where thermal loading is the largest, often exhibit inelastic deformations.

During temperature cycling, BGA SJIs operate in the so called low-cycle fatigue mode, and their durability is determined by the level of the inelastic strains in the material and the size of the peripheral zones, in which inelastic strains occur. It is these strains that are responsible for the accumulated damage and for the finite lifetime of the interconnections. The objective of the analysis that follows is to establish, by using predictive analytical modeling, the size of this zone. Then, one of the Coffin–Manson-type of equations (such as, say, Anand's model in the ANSYS software) can be applied to evaluate the fatigue lifetime of the BGA SJIs. The developed model is a modification and an extension of the models developed previously for the case of an elastic bonding layer (see preceding two chapters), as well as for an inelastic bond [7].

The following additional (in comparison with the previously considered problem for a homogeneous bonding layer) assumption is used in this analysis: the inhomogeneous ("discrete") BGA structure can be substituted by a homogeneous (continuous) bonding layer of the same thickness (height). It has been shown [11] that one can get away with employing a simpler model intended for an assembly with a homogeneous bond if the gaps between the supports (BGA balls or CGA columns) are small, so that the ratio of the pitch (the distance between the joint centers) to the joint widths is below 5, and the product kl of the parameter k of the interfacial shearing stress and half the assembly length l in the equivalent homogeneously bonded assembly is above 2.5. This is indeed the case for actual BGA and CGA systems. It is noteworthy that this finding can also be used in other areas of mechanical/structural engineering when there is an intent to simplify the calculations by replacing a model for beams on separate supports by using a model intended for a beam lying on a continuous elastic foundation.

4.6.2 Basic Equation

Consider an elongated soldered bimaterial assembly of the second level of interconnections (package to its PCB or ceramic substrate) (Figure 4.6). The assembly is manufactured at an elevated temperature and subsequently cooled down to a low temperature. The longitudinal interfacial displacements, $u_1(x)$ and $u_2(x)$, of the adherends (package and its substrate) can be sought, within the elastic midportion,

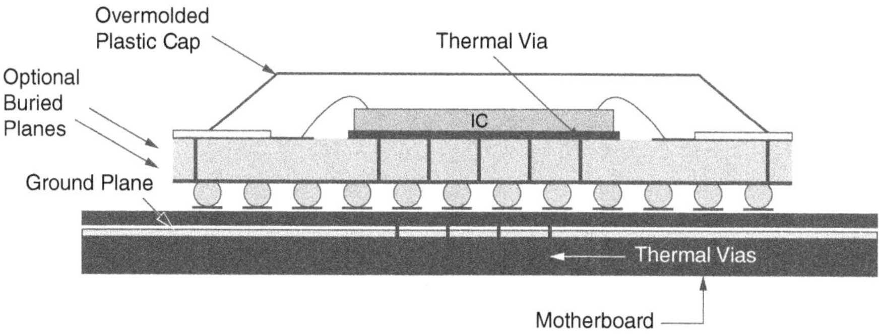

FIGURE 4.6 Ball-grid array (BGA) package.

$-x_* \leq x \leq x_*$, of the assembly, where $\pm x_*$ are the coordinates of the boundaries of the elastic midportion, in the form (4.2). All the formulas and derivations already carried out for the assembly with a continuous solder layer are applicable to the case in question as well.

The ratio $\frac{\tau_Y}{\tau_{max}^{\infty}}$ of the yield stress τ_Y of the solder material to the maximum value τ_{max}^{∞} of the induced shearing stress in a long-and-stiff assembly under an assumption of its elastic behavior, no matter how significant the loading might be expressed, based on the aforementioned analysis for the assembly with a continuous bond, is

$$\frac{\tau_Y}{\tau_{max}^{\infty}} = \frac{1}{(kl)\frac{l_*}{l} + \coth\left(kl\left(1+\frac{l_*}{l}\right)\right)} \tag{4.48}$$

Here $\frac{l_*}{l}$ is the ratio of the length of the inelastic zone to half of the assembly length and the product kl is expressed as $kl = l\sqrt{\frac{\lambda}{\kappa}}$. The calculated $\frac{l_*}{l}$ ratios based on relationship (4.48) are given in Table 4.1 and plotted in Figure 4.6.

TABLE 4.1
The Calculated Relative Lengths of the Inelastic Zone to the Half of the Assembly Length as a Function of the Product of the Factor of the Interfacial Shearing Stress and Half the Assembly Length

kl	0	0.5	1	2	3	4	5	10	∞
$\frac{l_*}{l}$	x	x	x	x	x	x	x	x	x
0	0	0.4621	0.7616	0.9640	0.9950	0.9994	1.0000	1.0000	1.0000
0.05	0	0.4374	0.7134	0.8728	0.8646	0.8326	0.7999	0.6667	0
0.10	0	0.4132	0.6684	0.7961	0.7639	0.7135	0.6667	0.5000	0
0.20	0	0.3660	0.5862	0.6734	0.6186	0.5556	0.4998	0.3333	0
0.30	0	0.3202	0.5116	0.5782	0.5180	0.4530	0.3997	0.2500	0
0.40	0	0.2753	0.4421	0.5000	0.4432	0.3822	0.3328	0.2000	0
0.50	0	0.2308	0.3754	0.4323	0.3839	0.3292	0.2846	0.1667	0
0.60	0	0.1863	0.3094	0.3696	0.3334	0.2869	0.2477	0.1429	0
0.70	0	0.1415	0.2420	0.3066	0.2860	0.2500	0.2172	0.1250	0
0.80	0	0.0958	0.1705	0.2363	0.2346	0.2125	0.1882	0.1111	0
0.90	0	0.0489	0.0915	0.1456	0.1547	0.1605	0.1501	0.1000	0
0.95	0	0.0247	0.0477	0.0838	0.1045	0.1128	0.1132	0.0952	0
0.99	0	0.0001	0.0050	0.0099	0.0192	0.0297	0.0345	0.0401	0
1.00	0	0	0	0	0	0	0	0	0

Note that the lengths l_* of the inelastic zones do not change when the kl product exceeds the $kl \approx 4$ value, i.e., when $\tanh kl$ can be put equal to one. Because of that the $\frac{l_*}{l}$ ratios decrease with an increase in the l values, and so do the $\frac{\tau_Y}{\tau_{\max}^{\infty}}$ ratios, when the kl product is larger than $kl \approx 4$. The stress ratios $\frac{\tau_Y}{\tau_{\max}^{\infty}}$ have their maxima

$$\left(\frac{\tau_Y}{\tau_{\max}^{\infty}}\right)_{\max} = \frac{1}{\frac{\frac{l_*}{l}}{1-\frac{l_*}{l}}\ln\frac{1+\sqrt{1-\frac{l_*}{l}}}{\sqrt{\frac{l_*}{l}}} + \coth\left(\ln\frac{1+\sqrt{1-\frac{l_*}{l}}}{\sqrt{\frac{l_*}{l}}}\right)} \tag{4.49}$$

for the product

$$kl = \frac{1}{1-\frac{l_*}{l}}\ln\frac{1+\sqrt{1-\frac{l_*}{l}}}{\sqrt{\frac{l_*}{l}}}. \tag{4.50}$$

of the parameter of the interfacial shearing stress and half the assembly lengths shown in Figure 4.6. As evident from expression (4.48), no plastic zones could possibly occur ($l_* \le 0$), if the condition

$$\frac{\tau_Y}{\tau_{\max}^{\infty}} \ge \tanh kl \tag{4.51}$$

takes place. Indeed, for long (large l values) and/or stiff (large k values) assemblies, when $\tanh kl$ could be considered equal to one, condition (4.48) is equivalent to the requirement that the yield stress is simply larger than the maximum elastic interfacial shearing stress. In such a situation, no plastic stresses could occur. If the kl value is small, condition (4.48) results in the relationship:

$$\tau_Y \ge \frac{\Delta\alpha\Delta t}{\kappa} l. \tag{4.52}$$

Thus, no plastic deformations could occur, if, in a short and/or compliant assembly, condition (4.52) is fulfilled, that is, if the yield stress τ_Y is high, the assembly compliance κ in the denominator in the right part of condition (4.52) is significant, the thermal strain $\Delta\alpha\Delta t$ in the numerator is low, and the size, l, of the assembly is small. Formula (4.52) indicates, particularly, that elevated interfacial compliances are highly desirable from the standpoint of avoiding inelastic stresses in the bonding material.

Equation (4.49), if solved for the length ratio in the parentheses, yields

$$\frac{l_*}{l} = 1 - \frac{1}{2kl} \ln \frac{\frac{\tau_{max}^{\infty}}{\tau_Y} - kl\frac{l_*}{l} + 1}{\frac{\tau_{max}^{\infty}}{\tau_Y} - kl\frac{l_*}{l} - 1} \tag{4.53}$$

If the yield stress τ_Y is low and, for this reason, the stress ratio $\frac{\tau_{max}^{\infty}}{\tau_Y}$ is significantly larger than one, then, as evident from (4.53), $l_* = l$, i.e., the entire interfacial zone is occupied by the plastic strains (stresses), regardless of whether the kl value is large or small. If the stress ratio $\frac{\tau_{max}^{\infty}}{\tau_Y}$ is significantly smaller than one, then equation (4.53) yields

$$\frac{l_*}{l} = 1 - \frac{1}{2kl} \ln \frac{kl\frac{l_*}{l} - 1}{kl\frac{l_*}{l} + 1} \tag{4.54}$$

As evident from this equation, plastic deformations might still take place, if, because of the low interfacial compliance, the k value is large, despite the low l value.

4.6.3 Numerical Example #1 (PCB Substrate)

Input data

Structural Element	Package	PCB	Solder (96.5%Ag3.5%Sn)
Element's number	1	2	0
Effective Young's modulus, kg/mm^2	8775.5	2321.4	1939.0
Poisson's Ratio	0.30	0.30	0.30
Shear modulus, kg/mm^2	3367.3	892.7	1958.8
CTE, 1/°C	6.5×10^{-6}	15.0×10^{-6}	x
Thickness, mm	2.0	1.5	0.2

Estimated yield stress of the solder material in shear $\tau_Y = 1.5306$ kg/mm^2
Soldering temperature 158°C; Assumed change in temperature $\Delta t = 130$°C
Half Package Length $l = 15$ mm

Computed data

Axial and interfacial compliances:

$$\lambda_1 = \frac{1-\nu_1}{E_1 h_1} = \frac{1-0.3}{8775.5\times 2.0} = 3.9884\times 10^{-5}\ \text{mm/kg};$$

$$\lambda_2 = \frac{1-\nu_2}{E_2 h_2} = \frac{1-0.3}{2321.4\times 1.5} = 20.1028\times 10^{-5}\ \text{mm/kg};$$

$$\lambda = \lambda_1 + \lambda_2 = 24.0912\times 10^{-5}\ \text{mm/kg}$$

$$\kappa_1 = \frac{h_1}{3G_1} = \frac{2.0}{3\times 3367.3} = 19.7983\times 10^{-5}\ \text{mm}^3/\text{kg},$$

$$\kappa_2 = \frac{h_2}{3G_2} = \frac{1.5}{3\times 892.7} = 56.0100\times 10^{-5}\ \text{mm}^3/\text{kg}$$

$$\kappa_0 = \frac{h_0}{G_0} = \frac{0.2}{1958.8} = 10.2103\times 10^{-5}\ \text{mm}^3/\text{kg}$$

$$\kappa = \kappa_0 + \kappa_1 + \kappa_2 = 86.0186\times 10^{-5}\ \text{mm}^3/\text{kg}$$

The parameter k of the interfacial shearing stress

$$k = \sqrt{\frac{\lambda}{\kappa}} = \sqrt{\frac{24.0912\times 10^{-5}}{86.0186\times 10^{-5}}} = 0.5292\ \text{mm}^{-1}$$

Hence, the parameter kl is $kl = 7.9382$. The maximum elastic interfacial shearing stress, if the bonding material behaves elastically under any level of loading, is

$$\tau_{\max}^{\infty} = k\frac{\Delta\alpha\Delta t}{\lambda} = 0.5292\frac{8.5\times 10^{-6}\times 130}{24.0912\times 10^{-5}} = 2.4273\ \text{kg/mm}^2$$

This result indicates that no inelastic strains could possibly occur if the yield stress of the solder material is equal or higher than the predicted elastic stress value. This value can be brought down by employing packages and PCBs with better thermal expansion match, by using thinner packages and PCBs, and by employing PCBs with lower Young's moduli. For the considered materials, the ratio $\frac{\tau_Y}{\tau_{\max}^{\infty}}$ is 0.6306, so that $\frac{l_*}{l} = 0.075$, that is, about 7.5% of the interface's length experiences inelastic deformations, and it is for these joints that the lifetime predictions should be conducted.

4.6.4 Numerical Example #2 (Ceramic Substrate)

The input data are the same as in example #1, but the PCB is substituted with a 1-mm thick ceramic substrate with effective Young's modulus 30612.2 kg/mm^2, Poisson's ratio 0.21, and shear modulus 12649.5 kg/mm^2. And here are the computed data.

Axial and interfacial compliances of the materials:

$$\lambda_2 = \frac{1-\nu_2}{E_2 h_2} = \frac{1-0.21}{30612.2\times 1.0} = 2.5807\times 10^{-5}\ \text{mm/kg};$$

$$\lambda = \lambda_1 + \lambda_2 = 6.5691\times 10^{-5}\ \text{mm/kg}$$

$$\kappa_2 = \frac{h_2}{3G_2} = \frac{1.0}{3\times 12649.5} = 2.6352\times 10^{-5}\ \text{mm}^3/\text{kg};$$

$$\kappa = \kappa_0 + \kappa_1 + \kappa_2 = 32.6438\times 10^{-5}\ \text{mm}^3/\text{kg}$$

The parameter k of the interfacial shearing stress: $k = \sqrt{\frac{\lambda}{\kappa}} = \sqrt{\frac{6.5691\times 10^{-5}}{32.6438\times 10^{-5}}} = 0.4486\ \text{mm}^{-1}$.

The product kl is $kl = 6.7289$, so that the assembly is long and stiff. Formula (4.16) yields

$$\tau_{max}^{\infty} = k\frac{\Delta\alpha\Delta t}{\lambda} = 0.4486\frac{1.6\times 10^{-6}\times 130}{6.5691\times 10^{-5}} = 1.4204\ \text{kg/mm}^2$$

Since the estimated yield stress in shear of the solder material is $\tau_Y = 1.5306\ \text{kg/mm}^2$, that is, higher than this value, no inelastic stresses are expected to occur.

Thus, the suggested model can be used to check if inelastic zones indeed exist in the design of interest, and, if they cannot be avoided, how large they are. Then one of the Coffin–Manson-types of equations aimed at the evaluation of the fatigue lifetime of the solder material could be applied. The numerical example carried out for a 30-mm long surface-mount package and a 200 μm thick lead-free solder indicated that, in the case of a high expansion PCB substrate, about 7.5% of the interface's size experienced inelastic strains, while no such strains could occur in the case of a low expansion ceramic substrate. The FEA computations confirmed the data obtained based on the developed analytical model.

4.7 PROBABILISTIC PALMGREN-MINER RULE FOR SOLDER MATERIALS EXPERIENCING ELASTIC DEFORMATIONS

4.7.1 Background/Incentive

It has been recently shown that there are effective ways not only to reduce the interfacial stresses in electronic packaging assemblies with solder joint arrays as the second level of interconnections, but to do that to an extent that inelastic strains in the peripheral joints, where the induced thermal stresses and strains are the highest, are avoided. While various and numerous modifications of the empirical Coffin–Manson relationship are used to predict the fatigue life of solder materials experiencing inelastic strains and operated in low-cycle fatigue conditions, the Palmgren-Miner rule of the linear accumulation of fatigue damages, although suggested many decades ago, is still viewed by many material scientists and reliability physicists as a suitable model that enables one to quantify the cumulative fatigue damage in metals experiencing elastic strains.

In this analysis, the Palmgren-Miner rule is extended for the case of random loading, and a simple formalism is suggested for the evaluation of the remaining useful lifetime (RUL) for a solder material subjected to random loading and experiencing elastic thermally induced shearing deformations. Special highly focused and highly cost-effective accelerated tests have to be conducted, of course, to establish the S-N curve for the given solder material. In future work, we intend to extend the suggested methodology to take into account various aspects of the physics-of-failure: the role of the growth kinetics of intermetalic compound (IMC) layers; the random number, size and orientation of grains in the joints; position of the joint with respect to the mid-cross section of the assembly (peripheral joints are more prone to elevated interfacial stresses); and assembly size. However, all this effort, important as it is, is beyond the scope of this analysis, which is aimed at the extension of the classical Palmgren-Miner rule for the case of random loading.

Solder joints employed in today's IC packaging engineering as second-level interconnections provide both electrical connection and mechanical support [12–14]. In the latter capacity, they are subjected to thermally induced stresses primarily caused by the (global) thermal expansion (contraction) mismatch of the dissimilar materials of the package and the substrate. The induced interfacial stresses concentrate at the assembly ends and could be quite high [15–17], thereby compromising the integrity of the peripheral joints. Therefore, there is an obvious incentive to minimize the induced stresses for the improved short- and long-term reliability of the package design.

It has been recently shown [18–26] that effective ways exist not only to reduce the interfacial stresses in packaging assemblies with solder joint interconnections, but to do that to an extent that inelastic strains in the peripheral joints, where the induced stresses and strains are the highest, are avoided [27–34]. This could be achieved, particularly, by using solder joints with elevated stand-offs and/or by employing inhomogeneous solder joint systems, in which the peripheral zones of the package are characterized by lower Young's moduli and/or lower soldering temperatures. If such an effort is successful, the high thermal stresses in the solder material, when the assembly is fabricated at the elevated temperature and is subsequently cooled down to a low (room or testing) temperature, and/or when the assembly experiences temperature cycling during reliability testing and actual operation, will be replaced by the temperature cycling within the elastic range.

It goes without saying that such a situation would be highly desirable and result in a considerably longer lifetime of the solder material. As is known (see, for example, [35–38]), the reliability of a material has to be checked and assured with respect to two major loading conditions: short-term high-level loading (ultimate strength) and long-term relatively-low-level, but repetitive, loading (fatigue strength). The ultimate strength (short-term reliability) is defined by the ability/capacity of the material to withstand a significant load applied just once. This strength is measured by the maximum ("ultimate") stress that a material is able to withstand in a typical loading condition (usually in tension or in shear) before breaking. Shear-off product development testing is a typical example of such a condition. As to the fatigue/endurance strength (long-term reliability) of an elastic material, it is defined by the highest stress that this material can withstand for the given number of cycles without breaking. Fatigue failure occurs because of the accumulation of micro-damages that result in the developing and propagation of fatigue cracks. In electronics reliability, the fatigue strength (lifetime, number-of-cycles-till-failure) is usually determined by temperature cycling.

The research literature in this field is enormous, especially when addressing situations, when the solder joints experience inelastic deformations. Since it has been shown that in many situations such deformations could be avoided by taking appropriate design measures, in the analysis that follows, a methodology for the evaluation of the remaining useful lifetime (RUL) for a solder material experiencing elastic thermally induced shearing deformations is addressed.

It is assumed that this could be done, in the first approximation and for the preliminary and tentative evaluations, on the basis of the simplest and well-known Palmgren-Miner rule of linear accumulation of fatigue damages [28, 29] and that this rule could be extended for the situation when thermal loading is applied in a random fashion, which is a typical situation in electronics and photonics systems operation. Certainly, special accelerated tests are required to establish the S-N curve for the given solder material in a situation when its yield stress is not exceeded. This should be done with consideration of the degradation effects of the solder material and its interfaces (see, for example, [41–48]).

4.7.2 Probabilistic Palmgren-Miner Rule

A typical stress versus number-of-cycles-to-failure (Wöhler) S-N curve/diagram in logarithmic coordinates is shown in Figure 4.7 for the case when a variable tensile stress is applied. When the applied stress is in shear, this diagram can be approximated by a power law (see Figure 4.8):

$$N = N_f \left(\frac{\tau_f}{\tau_a} \right)^m . \tag{4.55}$$

Here, N_f is the number of cycles corresponding to reaching the fatigue curve, τ_f is the level of the steady-state fatigue, τ_a is the amplitude of the variable shearing stress, and $m = \tan \alpha$ is the tangent of the angle that the limited-fatigue portion of the diagram forms with the vertical line that divides the limited fatigue and the steady-state fatigue regions.

In accordance with the Palmgren-Miner theory of the linear accumulation of fatigue damages, these damages are independent of the degree of the consumption of the fatigue lifetime at the given moment of time. The accumulated damages are also independent of the "prehistory" of loading, and therefore those that are caused by the current loading cycle can simply be added to previous damages. Such an assumption seems to be particularly justified in the case of random loading, when,

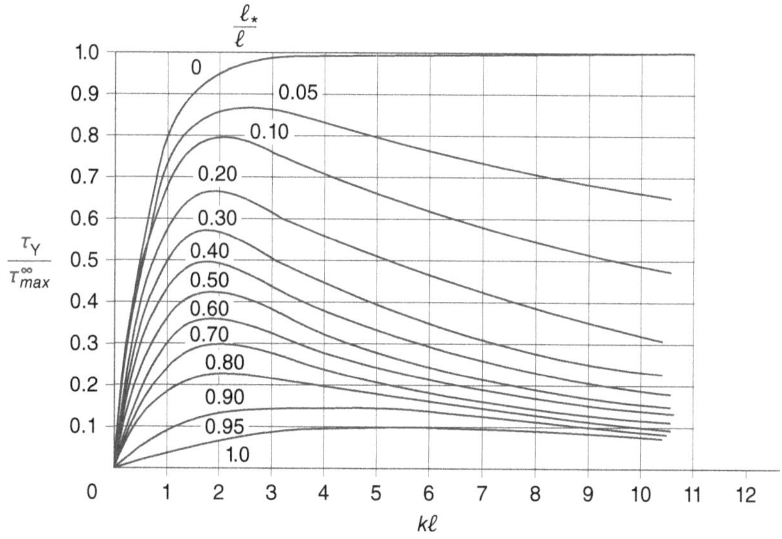

FIGURE 4.7 Yield-stress to maximum-elastic-stress ratios versus product of the parameter of the interfacial shearing stress and half-assembly length for different ratios of the length of the inelastic zone to half-assembly length.

because of the sequential action of cycles with high and low stresses, the material's weakenings caused by high-stress cycles interchange with its strengthenings caused by low-stress cycles, so that the effect of the "prehistory" of loading is being continuously smoothed down. The accumulated fatigue damage is assessed as

$$\xi = \sum_{i=1}^{k} \xi_i = \sum_{i=1}^{k} \frac{n_i}{N_i}. \tag{4.56}$$

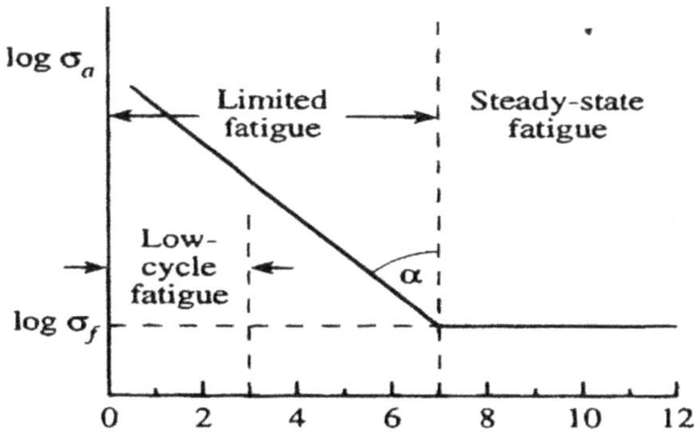

FIGURE 4.8 Stress versus number-of-cycles-to-failure (Wöhler) curve/diagram in logarithmic coordinates.

Here, $\xi_i = \frac{n_i}{N_i}$ is the damage from the ith level of loading, k is the total number of the loading level, n_i is the number of loading cycles of the ith level, and N_i is the number of loading cycles of the ith level leading to failure. Such a failure can be caused either by a single loading of the ith level, or by the entire spectrum of loadings of different levels. Assuming that the work W of the external loading leading to fatigue failure is the same in both cases, one has

$$W = \frac{N_i}{n_i} W_i = \sum_{i=1}^{k} W_i, \tag{4.57}$$

where W_i is the work of n_i cycles of a variable loading of the ith level. This relationship yields

$$W_i = \frac{n_i}{N_i} \sum_{i=1}^{k} W_i \tag{4.58}$$

By summing up the elementary works W_i in both parts of this equation, the following Palmgren-Miner formula for the linear accumulation of fatigue damages can be obtained:

$$\xi = \sum_{i=1}^{k} \frac{n_i}{N_i} = 1. \tag{4.59}$$

When the material experiences continuous random loading, this formula can be generalized as follows:

$$\int_0^{N_f} \frac{dN}{N} = 1. \tag{4.60}$$

Here, N is the number of cycles that corresponds to achieving the fatigue curve, and N_f is the number of cycles till fatigue failure.

4.7.3 Remaining Useful Life

The total number of cycles accumulated for the time $t = RUL$ can be naturally determined as $N_t = \frac{t}{t_e}$, where t_e is the effective period of random loading. Assuming that the loading cycles are distributed in a uniform fashion during the material's lifetime $t = RUL$ within the interval $0 \prec N \succ N_t$, the probability density function for the number N of cycles is

$$f_N(N) = \frac{1}{N_t} = \frac{t_e}{t}. \tag{4.61}$$

Using the condition

$$f_\tau(\tau_a)d\tau_a = f_N(N)dN \tag{4.62}$$

of the equality of the elementary probabilities for the random amplitudes τ_a of loading and the number N of cycles we have

$$dN = f_\tau(\tau_a)\frac{d\tau_a}{f_N(N)} = \frac{t}{t_e} f_\tau(\tau_a)d\tau_a. \tag{4.63}$$

In formulas (4.62) and (4.63), $f_\tau(\tau_a)$ is the probability density distribution function for the random amplitudes τ_a of loading. Introducing formulas (4.55) and (4.63) into condition (4.60), the following expression for the time-to-failure (RUL) can be obtained:

$$t = RUL = \beta t_e N_f, \tag{4.64}$$

where the factor

$$\beta = \frac{\tau_f^m}{\int_{\tau_f}^{\tau_u} \tau_a^m f_\tau(\tau_a)d\tau_a} = \frac{\tau_f^m}{\int_{\tau_f}^{\tau_Y} \tau_a^m f_\tau(\tau_a)d\tau_a} \tag{4.65}$$

considers the roles of the material properties (through the ultimate shearing stress τ_u) and the level of loading (through the level and distribution of random amplitudes of the shearing stress). The ultimate shearing stress τ_u could be accepted in this analysis, limited to the elastic deformations, equal to the yield stress τ_Y of the material.

Two important aspects of the fatigue limit τ_f should be pointed out.

The fatigue limit τ_f, even when it is low, as it is in the case in question, could still decrease, because of the material aging/degradation, with an increase in the number of cycles. In this connection, we would like to indicate that there is a way to separate the irreversible and unfavorable physics-of-failure related degradation process, which results in the increase in the failure rate with time and in a lower fatigue limit, from the also irreversible, but favorable, statistics-related process that results in the decreased failure rate with time [36, 37]. This aspect of the assessment of the fatigue lifetime is, however, beyond the scope of this analysis. Another aspect has to do with the fact that the fatigue limit τ_f depends on the loading spectrum: the larger the portion of this spectrum above the fatigue limit, the larger the rate of the decrease in this limit. In an approximate analysis, one could assume that the rate in the decrease in the fatigue limit depends on its initial value τ_f^0, the rate $\frac{n}{N}$ of the accumulation of fatigue damages and the total level N of the accumulated damages at the given moment of time. These considerations could be formalized in the following equation:

$$\frac{d\tau_f}{dn} = -\alpha \frac{\tau_f^0}{N}\left(1 - \frac{n}{N}\right)^{a-1}, \tag{4.66}$$

where τ_f is the fatigue limit after n loadings, τ_f^0 is the initial value of the fatigue limit, N is the number of cycles till fatigue failure, and α is the materials parameter. Equation (4.66) has the following solution:

$$\tau_f = \tau_f^0\left(1 - \frac{n}{N}\right)^{\alpha} \tag{4.67}$$

As evident from this solution, the fatigue limit τ_f depends, for the given material, characterized by the exponent α, on the accumulated fatigue damage $\frac{n}{N}$, and becomes zero at the moment of failure. The role of the variability of the fatigue limit is, however, beyond the scope of this analysis.

4.7.4 Rayleigh Law for the Random Amplitude of the Interfacial Shearing Stress

The long-term distributions of the stress amplitudes could be exponential, or distributed in accordance with the log-normal law, or with the Weibull law. Let us address, as a suitable example, the case when the random amplitude of the interfacial shearing stress is distributed in accordance with Rayleigh law

$$f(\tau_a) = \frac{\tau_a}{D_\tau}\exp\left[-\left(\frac{\tau_a^2}{2D_\tau}\right)\right] = \frac{\tau_a}{\tau_*^2}\exp\left[-\left(\frac{\tau_a^2}{2\tau_*^2}\right)\right]. \tag{4.68}$$

Here D_τ is the variance of the random amplitude τ_a and $\tau_* = \sqrt{D_\tau}$ is the most likely value (mode) of this amplitude. Introducing expression (4.68) into formula (11), we obtain

$$\beta = \frac{D_\tau \tau_f^m}{\int_{\tau_f}^{\tau_Y} \tau_a^{m+1} \exp\left(-\frac{\tau_a^2}{2D_\tau}\right) d\tau_a} \tag{4.69}$$

When the loading cycle is asymmetric, the random amplitudes τ_a should be multiplied by the factor k_* that can be determined from Goodman's law (see, for example, [49]) as

$$k_* = \left(1 - \frac{\tau_m}{\tau_Y}\right)^{-1} \tag{4.70}$$

Here,

$$\tau_m = \frac{\tau_{\max} + \tau_{\min}}{2} \tag{4.71}$$

is the mean value of the stress cycle. Then, formula (4.69) yields

$$\beta = \frac{x_0^m}{K_*^{m+2} \int_{x_0}^{x_1} x^{m+1} \exp\left(-\frac{x^2}{2}\right) dx} \tag{4.72}$$

where the following notations are used:

$$x = \frac{\tau_a}{\sqrt{D_\tau}}, \quad x_0 = \frac{\tau_f}{\sqrt{D_\tau}}, \quad x_1 = \frac{\tau_Y}{\sqrt{D_\tau}}. \tag{4.73}$$

The integral in formula (4.72) can be taken numerically, but could also be expressed through tabulated functions. Since the integrand in this integral does not make physical sense for the shearing stress amplitudes exceeding the yield stress, one could put the upper limit in this integral equal to infinity. Then,

$$\int_{x_0}^{x_1} x^{m+1} \exp\left(-\frac{x^2}{2}\right) dx = \int_{x_0}^{\infty} x^{m+1} \exp\left(-\frac{x^2}{2}\right) dx = 2^{m/2} \Gamma\left(\frac{m}{2}+1\right) P(x_0^2, m+2), \tag{4.74}$$

where

$$\Gamma(\alpha) = \int_0^{\infty} x^{\alpha-1} e^{-x} dx \tag{4.75}$$

is the gamma function and

$$P(\chi^2, n) = \frac{2^{1-n/2}}{\Gamma(n/2)} \int_{\chi}^{\infty} x^{n-1} e^{-x^2/2} dx \tag{4.76}$$

is the Pierson function. Both functions are tabulated (see, for example, [37]). The coefficient in front of the Pierson function in (4.74) can be computed as

$$2^{m/2} \Gamma\left(\frac{m}{2}+1\right) = \sqrt{\frac{\pi}{2}} m!! = \sqrt{\frac{\pi}{2}} (1x3x5x \ldots x(m-2)m) \tag{4.77}$$

for odd m numbers and as

$$2^{m/2} \Gamma\left(\frac{m}{2}+1\right) = 2^{m/2} \left(\frac{m}{2}\right)! \tag{4.78}$$

for even m numbers.

4.7.5 Numerical Example

Input data

Structural Element	Package (Comp. #1)	PCB (Comp. #2)	Solder (96.5%Ag3.5%Sn)
Effective Young's modulus, kg/mm^2	8775.5	2321.4	1939.0
Poisson's ratio	0.30	0.30	0.38
CTE, $1/°C$	6.5×10^{-6}	15.0×10^{-6}	x
Thickness, mm	2.0	1.5	0.2

Estimated yield stress of the solder material in shear $\tau_Y = 1.700$ kg/mm^2;
Soldering temperature 160°C; Lowest testing temperature –20°C; Highest testing temperature: +100°C;
Largest external thermal strain:
$\varepsilon_{max} = \Delta\alpha\Delta t_{max} = (15.0-6.5)10^{-6}\times180 = 153.0\times10^{-5}$;
Smallest external thermal strain:
$\varepsilon_{min} = \Delta\alpha\Delta t_{min} = (15.0-6.5)10^{-6}\times60 = 51.0\times10^{-5}$;
The fatigue limit is $\tau_f = 0.200$ kg/mm^2;
Most likely interfacial shearing stress in actual operation condition is $\tau_* = 1.00$ kg/mm^2;
Factor considering the slope of the limited fatigue region: $m = 10$;
Effective period of random loading: $t_e = 24$ hours;
Number of cycles to failure $N_f = 10^4$.

Calculated data

Axial compliances of the assembly components:

$$\lambda_1 = \frac{1-\nu_1}{E_1h_1} = \frac{1-0.3}{8775.5\times2.0} = 3.9884\times10^{-5} \text{ mm/kg};$$

$$\lambda_2 = \frac{1-\nu_2}{E_2h_2} = \frac{1-0.3}{2321.4\times1.5} = 20.1028\times10^{-5} \text{ mm/kg};$$

Flexural rigidities of the assembly components:

$$D_1 = \frac{E_1h_1^3}{12(1-\nu_1^2)} = \frac{8775.5\times8}{12\times0.91} = 6428.9 \text{ kg/mm};$$

$$D_2 = \frac{E_2h_2^3}{12(1-\nu_2^2)} = \frac{2321.4\times3.375}{12\times0.91} = 717.5 \text{ kg/mm};$$

Axial compliance of the assembly:

$$\lambda = \lambda_1 + \lambda_2 + \frac{h_1}{4D_1} = \frac{h_2}{4D_2} = (3.9884 + 20.1028 + 7.7774 + 52.2648)10^{-5}$$

$$= 84.1334 \times 1^{-5} \text{ mm/kg}$$

Shear moduli:

- of the package

$$G_1 = \frac{E_1}{2(1+\nu_1)} = \frac{8775.5}{2\times 1.3} = 3375.2 \text{ kg/mm}^2;$$

- of the PCB

$$G_2 = \frac{E_2}{2(1+\nu_2)} = \frac{2321.4}{2\times 1.3} = 892.8 \text{ kg/mm}^2;$$

- of the solder

$$G_0 = \frac{E_0}{2(1+\nu_0)} = \frac{1939.0}{2\times 1.38} = 702.5 \text{ kg/mm}^2;$$

Interfacial compliances:

- of the package

$$\kappa_1 = \frac{h_1}{3G_1} = \frac{2.0}{3\times 3375.2} = 19.752\times 10^{-5} \text{ mm}^3\text{/kg};$$

- of the PCB

$$\kappa_2 = \frac{h_2}{3G_2} = \frac{1.5}{3\times 892.8} = 56.004\times 10^{-5} \text{ mm}^3\text{/kg};$$

- of the solder

$$\kappa_0 = \frac{h_0}{G_0} = \frac{0.2}{702.5} = 28.469\times 10^{-5} \text{ mm}^3\text{/kg}$$

Total interfacial compliance of the assembly

$$\kappa = 19.752\times 10^{-5} + 56.004\times 10^{-5} + 28.469\times 10^{-5} = 104.230\times 10^{-5} \text{ mm}^3\text{/kg}$$

Parameter of the interfacial shearing stress

$$k = \sqrt{\frac{\lambda}{\kappa}} = \sqrt{\frac{84.1334 \times 10^{-5}}{104.230 \times 10^{-5}}} = 0.8984 \text{ mm}^{-1}$$

Highest interfacial shearing stress

$$\tau_{\max} = k\varepsilon_{\max} = 0.8984 \frac{153 \times 10^{-5}}{84.1334 \times 10^{-5}} = 1.6338 \text{ kg/mm}^2$$

Lowest interfacial shearing stress

$$\tau_{\min} = k\varepsilon_{\min} = 0.8984 \frac{51 \times 10^{-5}}{84.1334 \times 10^{-5}} = 0.5446 \text{ kg/mm}^2$$

Mean value of the stress cycle

$$\tau_m = \frac{\tau_{\max} + \tau_{\min}}{2} = \frac{1.6338 + 0,5446}{2} = 1.0892 \text{ kg/mm}^2$$

Factor accounting for the ratio of the fatigue limit for the asymmetric cycle to the fatigue limit for the symmetric cycle

$$k_* = \left(1 - \frac{\tau_m}{\tau_Y}\right)^{-1} = \left(1 - \frac{1.0892}{1.700}\right)^{-1} = 2.7832$$

The lower limit in the integral (4.74)

$$x_0 = \frac{\tau_f}{\tau_*} = \frac{0.2}{1.0} = 0.2$$

Pierson function [49]

$$P(x_0^2, m+2) = P(0.04, 12) = 4.90$$

Then, formula (4.72) yields

$$\beta = \frac{x_0^m}{K_*^{m+2} \int_{x_0}^{x_1} x^{m+1} \exp\left(-\frac{x^2}{2}\right) dx} = \frac{x_0^m}{K_*^{m+2} 2^{m/2} \left(\frac{m}{2}\right)!} \frac{0.2^{10}}{2.7832^{12} \times 2^5 \times 5!} = 2.4108$$

Formula (4.64) predicts that the time to the fatigue failure is

$$t = RUL = 2.4108 \times 24 \times 10^4 \text{ hours} = 5.7859 \times 10^5 \text{ hours} \approx 66.0 \text{ years}$$

We conclude that

- A simple and practically useful methodology is suggested for the evaluation of the RUL for a solder material experiencing elastic thermally induced shearing deformations.
- The classical Palmgren-Miner rule is extended for the case of random loading. Certainly, special accelerated tests are required to establish the S-N curve for the given solder material in a situation when its yield stress is not exceeded.
- In future work, we intend to extend the suggested methodology to take into account various aspects of the physics-of-failure: the role of the growth kinetics of IMC layers; the random number, size, and orientation of grains in the joints; position of the joint with respect to the mid-cross section of the assembly (peripheral joints are more prone to elevated interfacial stresses); and assembly size.

REFERENCES

1. H.D. Solomon, "Fatigue of 60/40 Solder," *IEEE CPMT Transactions*, vol. CHMT-9, No. 4, 1986.
2. H.S. Morgan, "Thermal Stresses in Layered Electrical Assemblies Bonded with Solder," *ASME Journal of Electronic Packaging*, vol. 113, No. 4, 1991.
3. T. Hatsuda, H. Doi, T. Hayasida, "Thermal Strains in Flip-Chip Joints of Die-Bonded Chip Packages," Procedures of the EPS Conference, San Diego, California, 1991.
4. R. Darveaux, K. Banerji, "Constitutive Relations for Tin-Based Solder Joints," 50th ECTC, 1992.
5. J. Lau (ed.), *Ball Grid Array Technology*, McGraw-Hill, 1995.
6. J.S. Hwang, *Modern Solder Technology for Competitive Electronics Manufacturing*, McGraw-Hill, 1996.
7. E. Suhir, "Interfacial Thermal Stresses in a Bi-Material Assembly with a Low-Yield-Stress Bonding Layer," *Journal of Applied Physics-D, Modeling and Simulation in Materials Science and Engineering*, vol. 14, 2006.
8. B.A. Boley, J.H. Weiner, "Theory of Thermal Stresses," *Dover Civil and Mechanical Engineering*, Nov. 2, 2011.
9. E. Suhir, L. Bechou, B. Levrier, D. Calvez, "Assessment of the Size of the Inelastic Zone in a BGA Assembly," 2013 IEEE Aerospace Conference, Big Sky, Montana, March 2013.
10. E. Suhir, S. Yi, R. Ghaffarian, "How Many Peripheral Solder Joints in a Surface Mounted Design Experience Inelastic Strains?" *Journal of Electronic Materials*, vol. 46, No. 3, 2017.
11. E. Suhir, R. Ghaffarian, "Electron Device Subjected to Temperature Cycling: Predicted Time-to Failure," *Journal of Electronic Materials*, vol. 48, No. 2, 2019.
12. J. Lau (ed.), *Solder Joint Reliability: Theory and Applications*, Van Nostrand Reinhold, 1990.
13. J.-P. Clech, F. M. Langerman, J. A. Augis, "Local CTE Mismatch in SM Leaded Packages: A Potential Reliability Concern," Procedures of the 40th Electronic Components and Technology Conference, Las Vegas, Nevada, May 1990.

14. W. Engelmaier, "Reliability for Surface Mount Solder Joints: Physics and Statistics of Failure," Procedures of Surface Mount International, vol. 1, San Jose, California, August 1992.
15. E. Suhir, "Stresses in Bi-Metal Thermostats," *ASME Journal of Applied Mechanics*, vol. 53, No. 3, Sept. 1986.
16. E. Suhir, "Interfacial Stresses in Bi-Metal Thermostats," *ASME Journal of Applied Mechanics*, vol. 56, No. 3, September 1989.
17. A.Y. Kuo, "Thermal Stress at the Edge of a Bi-Metallic Thermostat," *ASME Journal of Applied Mechanics*, vol. 57, 1990.
18. E. Suhir, "Predicted Thermal Mismatch Stresses in a Cylindrical Bi-Material Assembly Adhesively Bonded at the Ends," *ASME Journal of Applied Mechanics*, vol. 64, No. 1, 1997.
19. E. Suhir, "Adhesively Bonded Assemblies with Identical Nondeformable Adherends and Inhomogeneous Adhesive Layer: Predicted Thermal Stresses in the Adhesive," *Journal of Reinforced Plastics and Composites*, vol. 17, No. 14, 1998.
20. E. Suhir, "Thermal Stress in a Polymer Coated Optical Glass Fiber with a Low Modulus Coating at the Ends," *Journal of Material Research*, vol. 16, No. 10, 2001.
21. E. Suhir, "Thermal Stress in a Bi-Material Assembly Adhesively Bonded at the Ends," *Journal of Applied Physics*, vol. 89, No.1, 2001.
22. E. Suhir, "Thermal Stress in an Adhesively Bonded Joint with a Low Modulus Adhesive Layer at the Ends," *Journal of Applied Physics*, April 2003.
23. E. Suhir, "Interfacial Thermal Stresses in a Bi-Material Assembly with a Low-Yield-Stress Bonding Layer," *Modeling and Simulation in Materials Science and Engineering*, vol. 14, 2006.
24. E. Suhir, "On a Paradoxical Situation Related to Bonded Joints: Could Stiffer Mid-Portions of a Compliant Attachment Result in Lower Thermal Stress?" *Journal of Solid Mechanical and Materials Engineering (JSMME)*, vol. 3, No. 7, 2009.
25. E. Suhir, "Thermal Stress in a Bi-Material Assembly with a 'Piecewise-Continuous' Bonding Layer: Theorem of Three Axial Forces," *Journal of Physics D: Applied Physics*, vol. 42, 2009.
26. E. Suhir, A. Shakouri, "Assembly Bonded at the Ends: Could Thinner and Longer Legs Result in a Lower Thermal Stress in a Thermoelectric Module (TEM) Design?" *ASME Journal of Applied Mechanics*, vol. 79, No. 6, 2012.
27. E. Suhir, L. Bechou, B. Levrier, "Predicted Size of an Inelastic Zone in a Ball-Grid-Array Assembly," *ASME Journal of Applied Mechanics*, vol. 80, March 2013.
28. E. Suhir, "Predicted Stresses in a Ball-Grid-Array (BGA)/Column-Grid-Array (CGA) Assembly with a Low Modulus Solder at Its Ends," *Journal of Materials Science: Materials in Electronics*, vol. 26, No. 12, 2015.
29. E. Suhir, "Analysis of a Short Beam with Application to Solder Joints: Could Larger Stand-off Heights Relieve Stress?" *European Journal of Applied Physics (EPJAP)*, vol. 71, 2015.
30. E. Suhir, R. Ghaffarian, J. Nicolics, "Could Application of Column-Grid-Array Technology Result in Inelastic-Strain-Free State-of-Stress in Solder Material?" *Journal of Materials Science: Materials in Electronics*, vol. 26, No. 12, 2015.
31. E. Suhir, R. Ghaffarian, "Predicted Stresses in a Ball-Grid-Array (BGA)/Column-Grid-Array (CGA) Assembly with Epoxy Adhesive at Its Ends," *Journal of Materials Science: Materials in Electronics*, vol. 27, No. 5, 2016.
32. E. Suhir, "Bi-Material Assembly with a Low-Modulus-and/or-Low-Fabrication-Temperature Bonding Material at Its Ends: Optimized Stress Relief," *Journal of Materials Science: Materials in Electronics*, vol. 27, No. 5, 2016.

33. E. Suhir, "Expected Stress Relief in a Bi-Material Inhomogeneously Bonded Assembly with a Low-Modulus-and/or-Low-Fabrication-Temperature Bonding Material at the Ends," *Journal of Materials Science: Materials in Electronics*, vol. 27, No. 6, 2016.
34. E. Suhir, R. Ghaffarian, J. Nicolics, "Could Thermal Stresses in an Inhomogeneous BGA/CGA System be Predicted Using a Model for a Homogeneously Bonded Assembly?" *Journal of Materials Science: Materials in Electronics*, vol. 27, No. 1, 2016.
35. E. Suhir, "Avoiding Low-Cycle Fatigue in Solder Material Using Inhomogeneous Column-Grid-Array (CGA) Design," *ChipScale Reviews*, March–April 2016.
36. S. Timoshenko, D.H. Young, *Theory of Structures*, McGraw-Hill, 1945.
37. S. Timoshenko, *Strength of Materials*, Van-Nostrand, 1955.
38. E. Suhir, *Structural Analysis in Microelectronic and Fiber Optic Systems*, Van-Nostrand, 1991.
39. A. Palmgren, "Die Lebensdauer von Kuegellagern," *Zeitschrift des Vereins Deutscher Ingenieure*, vol. 68, 1924.
40. M.A. Miner, "Cumulative Damage in Fatigue," *ASME Journal of Applied Mechanics*, vol. 67, 1945.
41. K. Mishiro, "Issues Related to the Implementation of Pb-free Electronic Solders," *Microelectronics Reliability*, vol. 42, 2002.
42. D.R. Frear, "Emerging Reliability Challenges in Electronic Packaging," International Reliability Physics Symposium, Phoenix, Arizona, 2008.
43. G. Cuddalorepatta, A. Dasgupta, "Creep and Stress Relaxation of Hypo-Eutectic Sn3.0Ag0.5Cu Pb-free Alloy: Testing and Modeling," 2007 ASME International Mechanical Engineering Congress and Exposition, Seattle, Washington, 2007
44. M. Pecht, P. McCluskey, J. Evans, *Failures in Electronic Assemblies and Devices*, Springer-Verlag, London, 2001.
45. C. Andersson, D. Andersson, P. Tegehall, J. Liu, "Effect of Different Temperature Cycle Profiles on the Crack Propagation and Microstructural Evolution of Lead Free Solder Joints of Different Electronic Components," 5th International Conference on Thermal and Mechanical Simulation and Experiments in Micro-electronics and Micro-systems, Brussels, Belgium, 2004.
46. R.K. Ulrich, W.D. Brown, *Advanced Electronic Packaging*, Wiley-Interscience, 2006.
47. E. Suhir, "Statistics- and Reliability-Physics-Related Failure Processes," *Modern Physics Letters B (MPLB)*, Vol. 28, No. 13, 2014.
48. E. Suhir, "Degradation Related Failure Rate Determined from the Experimental Bathtub Curve," SAE Conference, Seattle, Washington, Sept. 2015.
49. E. Suhir, *Applied Probability for Engineers and Scientists*, McGraw-Hill, 1997.

5 Elevated Stand-Off Heights Can Relieve Thermal Stress in Solder Joints

"There are truths, which are like new lands: the best way to them becomes known only after trying many other ways."

—Denis Diderot, French philosopher, art critic, and writer

5.1 BACKGROUND/MOTIVATION

Various solder joint technologies, both their metallurgical and mechanical aspects, have been addressed for many years in various books (see, for example, [1–2]), with a natural emphasis on reliability issues (see, for example, [3–6]), as well as in numerous technical articles (see, for example, [7–27]). Ball-grid-array (BGA) (Figure 5.1) is a widely used IC packaging technology today [28–36]. It enables to permanently surface mount electronic components on a printed circuit board (PCB) with high mounting density (high pin count). In addition, the application of BGA technology leads to a short signal delay.

One important disadvantage of the BGA technology, however, is that the BGA solder balls are not mechanically compliant. They do not flex the way the longer leads of the previous generations of the second-level (package-to-PCB) interconnections were, and are therefore unable to effectively relieve stresses and strains in the joints. The reliability of BGA solder joint interconnections is a crucial bottleneck of the technology, especially if lead-free solder materials are considered. The maximum normal and shearing stresses and strains often lead the BGA solder joints to fracture. Several decades ago, in the general theory of thermal stresses in assemblies comprised of dissimilar materials (see, for example, [37–40]), it was shown that an increase in the interfacial compliance of the assembly could result in an appreciable relief in the interfacial stresses, even to an extent that the induced stresses and strains remain within the elastic range, that is, that no inelastic strains occur in the bonding solder. This consideration brought in column-grid array (CGA) technologies [41–48] (Figure 5.2) characterized by much larger stand-off heights of the joints.

It is noteworthy in this connection that, as has been shown earlier [8] in application to Bell Labs Si-on-Si flip-chip multi-chip packaging technology (see Chapter 3, Figure 3.5),

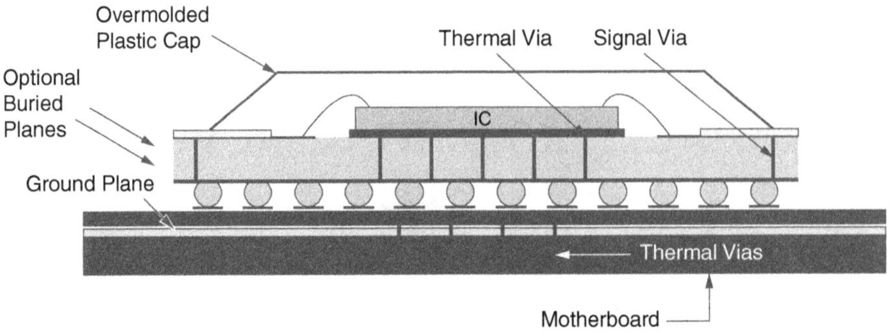

FIGURE 5.1 Ball grid array (BGA) technology. (From B. Guenin, "The Many Flavors of Ball Grid Array Packages," Electronics Cooling, Feb 1, 2002, http://www.electronics-cooling.com/2002/02/the-many-flavors-of-ball-grid-array-packages.)

solder joints configured as "pancakes" (i.e., those with large ratios of their diameter to the height) exhibit substantially higher stresses and strains than joints configured as "balls" (i.e., joints with lower aspect ratios). This observation is consistent with even more earlier findings [37–39], which indicated that by employing small-sized bonded assemblies with compliant interfaces and/or properly pre-engineered substrates with small "islands" at the interface with the vulnerable material [39] (Figure 5.3), one can considerably bring down the induced interfacial stresses. In these findings, however, the addressed assemblies and "islands" were characterized not by their height-to-width aspect ratios, but by the product kl of the parameter k of the interfacial shearing stress to half the" island" length l. As to the height-to-width aspect ratios of these assemblies/"islands," it was still expected that these ratios could be less than one; that is, that they were of a "pancake," rather than of a short cylinder, type. In addition, the aforementioned referenced papers dealt with "local" stresses, so that bending and "global" shear of the structural elements of interest did not take place and was

FIGURE 5.2 Column-grid array (CGA) technology. Left: General view. Right: Comparison with BGA geometry. (From https://www.topline.tv/CCGA.html.)

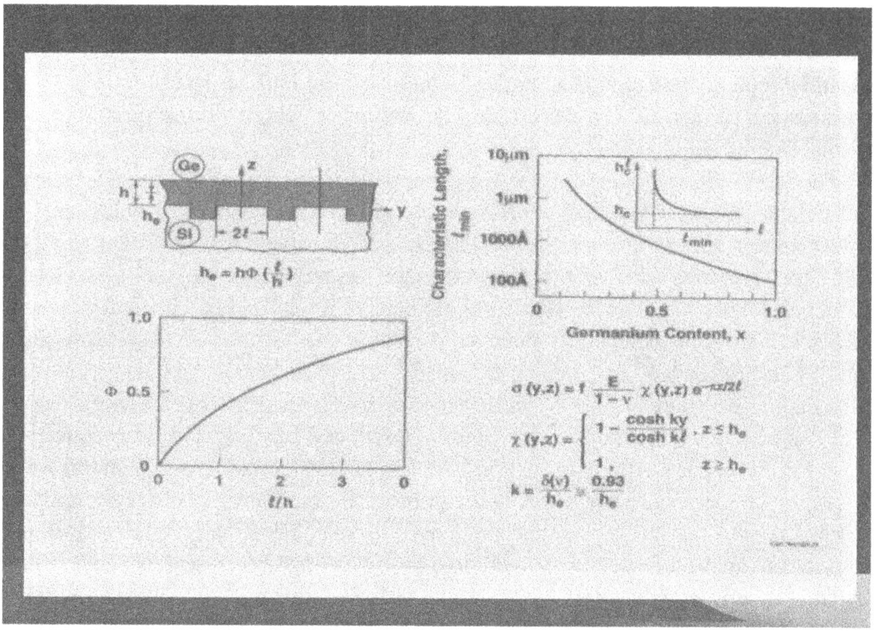

FIGURE 5.3 The suggested high-quality epitaxial growth of lattice mismatched materials. (Based on the findings of E. Suhir, "Interfacial Stresses in Bi-Metal Thermostats," ASME Journal of Applied Mechanics, vol. 56, No. 3, Sept. 1989.)

not addressed. In the analysis that follows, however, it is the effect of the "global" mismatch that results in bending of a beam-like structural element (solder joint) of interest that is addressed, and a situation when the ratio of the height to the width of such an element is typically higher than one, but still not high enough (lower than, say, 6), so that the effects of bending are small and could be neglected.

To demonstrate and to quantify the effectiveness of joints configured as short beams, an analytical (mathematical) stress model is developed. As to the offset, it is assumed to be always small compared to the beam's length (height) and even compared to its thickness. The analysis is limited to the elastic deformations. Unlike the classical Timoshenko beam theory [42], when the total deflections caused by the combined bending and shear deformations are sought for the given loading, we consider an inverse problem: the lateral force is sought for the given total displacement (end offset). In a short beam, because of its deformation in shear, this force is expected to be larger than in a long beam, since such a shear force has to overcome both the bending and the shear resistance of the beam to achieve the given end offset.

It is envisioned that short beams subjected to bending caused by the relative displacements of their ends could adequately mimic the state of stress in solder joint interconnections, including BGA and especially CGA systems, with large, compared to conventional joints, stand-off heights. When the package/PCB assembly (second level of interconnection) is subjected to the change in temperature, the thermal expansion (contraction) mismatch of the package and the PCB results in an easily predictable relative displacement (offset) of the ends of the solder joints. This offset can be determined from the known external thermal mismatch strain [evaluated as the product of the difference in the coefficients of thermal expansion (CTE) of the assembly components and the change in temperature] and the position of the joint with respect to the mid-cross section of the assembly, where the thermally induced displacements are always zero.

Therefore, there is an incentive for developing a stress model for a short beam with clamped and offset ends to adequately mimic the mechanical behavior of a solder joint in a thermally mismatched assembly. Note that a short cylinder clamped at its flat ends and subjected to outward radial loadings at these ends was viewed and considered as a suitable structural element to model the mechanical behavior of a solder joint in a thermally matched assembly (see Chapter 3). The subsequent analysis is limited to elastic deformations.

It should be pointed out that while the classical Timoshenko short-beam theory seeks the beam's deflections caused by the combined bending and shear deformations for the given loading, an inverse problem is considered here: the lateral force is sought for the given ends offset that can be easily predicted. In short beams, this force is larger than in long beams, since in order to achieve the given displacement (offset), the applied force has to overcome both bending and shear resistance of the beam. The normal and shearing stresses in such a beam could be viewed as suitable criteria of the beam's (joint's) material long-term reliability. We intend to show that these stresses can be brought down by employing beam-like joints; that is, joints with an increased stand-off height compared to conventional joints. It is imperative, of course, that if such joints are employed, there is still enough interfacial real estate, so that the BGA bonding strength is not compromised. On the other hand, owing to the lower stress level, reliability assurance might be much less of a challenge than in the case of conventional joint configuration. By employing beam-like solder joints resulting in lower stresses and strains, one could even try to avoid inelastic deformations of the joints altogether, thereby dramatically increasing their fatigue lifetime.

5.2 STRESSES IN A SHORT BEAM SUBJECTED TO BENDING CAUSED BY ITS END OFFSET

Consider a short beam of unit width, length (height), h, and thickness $2l$. The beam's ends are clamped and offset at the given distance Δ. The strain energy due to the beam's bending can be found as (see, for example, [40])

$$V_b = \frac{1}{2EI}\int_0^h M^2(z)dz, \tag{5.1}$$

Where EI is the beam's flexural rigidity, E is Young's modulus of the material, $I = \frac{2}{3}l^3$ is the moment of inertia of the beam's cross section (per unit width), l is half the beam's thickness, h is its length (height),

$$M(z) = M_0\left(1 - \frac{2z}{h}\right) \tag{5.2}$$

is the (linearly) distributed bending moment,

$$M_0 = \frac{1}{2} N_0 h \tag{5.3}$$

are the bending moments at the clamped ends, and N_0 is the lateral force (this force does not change along the beam). The origin of the coordinate z is at the beam's lower end.

The elastic curve $v(z)$ of the beam can be sought, in an approximate analysis, in the form of the method of initial parameters (see, for example, [40]):

$$v(z) = \frac{M_0 z^2}{2EI} - \frac{N_0 z^3}{6EI} = \frac{N_0 z^2}{6EI}\left(\frac{3}{2}h - z\right). \tag{5.4}$$

The condition $v(h) = \Delta$ yields:

$$M_0 = \frac{6\Delta EI}{h^2} = 4E\Delta\frac{l^3}{h^2}, \quad N_0 \frac{12\Delta EI}{h_3} = 8E\Delta\left(\frac{l}{h}\right)^3, \tag{5.5}$$

and the strain energy due to bending is

$$V_b = 4E\Delta^2\left(\frac{l}{h}\right)^3. \tag{5.6}$$

The strain energy due to shear (per unit volume and per unit beam's width) can be found as (see, for example, [40]):

$$V_s'' = \frac{3(1+v)}{2E}\tau_{zx}^2, \tag{5.7}$$

where v is Poisson's ratio of the material, and τ_{zx} is the shearing stress acting in the cross sections of the beam and associated with the distortion of the beam's form. Assuming that this stress is distributed over the beam's cross section in a parabolic fashion

$$\tau_{zx} = \tau_{\max}\left(1 - \frac{x^2}{l^2}\right), \tag{5.8}$$

we obtain the strain energy per unit beam length (height) by integrating this relationship over the beam's thickness:

$$V_s' = \frac{3(1+v)}{2E}\tau_{\max}^2 \int_{-l}^{l}\left(1 - \frac{x^2}{l^2}\right)^2 dx = \frac{8}{5}\frac{1+v}{E} l\tau_{\max}^2. \tag{5.9}$$

In formulas (5.8) and (5.9), τ_{max} is the maximum shearing stress at the origin ($x = 0$). From the obvious relationship

$$N_0 \int_{-l}^{l} \tau_{zx}(x)dx = \tau_{max} \int_{-l}^{l} \left(1 - \frac{x^2}{l^2}\right) dx = \frac{4}{3} l \tau_{max} \tag{5.10}$$

we have

$$\tau_{max} = \frac{3N_0}{4l} = 6E\Delta \frac{l^2}{h^3}, \tag{5.11}$$

and formula (5.9) yields

$$V_s' = \frac{288}{5}(1+\nu)E\frac{\Delta^2 l^5}{h^6}. \tag{5.12}$$

For the entire beam the strain energy (per unit beam width) due to shear is

$$V_s = \frac{288}{5}(1+\nu)E\Delta^2\left(\frac{l}{h}\right)^6. \tag{5.13}$$

Equating the total strain energy

$$V = V_b + V_s = 4E\Delta^2\left(\frac{l}{h}\right)^3\left[1 + \frac{72}{5}(1+\nu)\left(\frac{l}{h}\right)^3\right] \tag{5.14}$$

to the work $W = \frac{1}{2}N\Delta$ of the external lateral force N, we obtain the following formula for this force:

$$N = 8E\Delta\left(\frac{l}{h}\right)^3\left[1 + \frac{72}{5}(1+\nu)\left(\frac{l}{h}\right)^3\right]. \tag{5.15}$$

Comparing this formula with the second formula in (5.5), we conclude that the lateral force N in the presence of shear deformations is larger by the factor of

$$\chi = 1 + \frac{72}{5}(1+\nu)\left(\frac{l}{h}\right)^3 \tag{5.16}$$

than the force N_0 that does not consider these deformations. This is because the force N has to overcome not only bending, but also shear resistance of the beam in order to achieve the given offset of its ends. The factor (5.16) changes from 1.0, for very thin-and-tall beams characterized by the next-to-zero thickness-to-length (height) ratios, to 1.3, when this ratio is about 0.5 (for the Poisson's ratio of 0.33). From the relationships (5.11) and (5.15), we obtain the following expression for the maximum shearing stress:

$$\tau_{\max} = 6E\frac{\Delta}{l}\left(\frac{l}{h}\right)^3\left[1+\frac{72}{5}(1+v)\left(\frac{l}{h}\right)^3\right]. \tag{5.17}$$

This formula indicates particularly that the maximum shearing stress decreases rapidly with the decrease in the ratio of the beam's length (height) to its thickness. The normal stress

$$\sigma = \frac{3Nh}{4l} = \frac{h}{l}\tau_{\max} \tag{5.18}$$

in the beam, as long as the beam model is used, is always higher than the shearing stress. It is noteworthy that in an opposite extreme case of a joint with a very low length (height)-to-thickness ratio, the maximum shearing stress can be sought, using the method of interfacial compliance (see Chapter 2), as $\tau_{\max} = \frac{\Delta}{\kappa}$, where the interfacial compliance $\kappa = \frac{h}{G}$ in the denominator can be found as the ratio of the vertical dimension (height) of the joint h to the shear modulus G of the material. Thus, the maximum shearing stress is inversely proportional to the stand-off height of the joint in the case of a plate-like joint and becomes inversely proportional to the cube of the stand-off height in the case of a beam-like joint. The analysis of the shearing and peeling stresses in an entire BGA assembly is provided in the next section.

5.3 INTERFACIAL STRESSES IN ASSEMBLIES WITH SMALL STAND-OFF HEIGHTS

The analysis that follows is carried out under an assumption that the stresses in and the performance of a long enough BGA can be evaluated by replacing, when carrying out predictive stress modeling, an actual BGA with a continuous layer of solder of the same stand-off height as the BGA is. The interfacial longitudinal displacement can be sought, in an approximate analysis, using the interfacial compliance concept, in the form

$$\begin{aligned} u_1(x) &= -\alpha_1\Delta tx + \lambda_1\int_0^x N(\xi)d\xi - \kappa_1\tau(x) - \frac{h_1}{2}w_1'(x), \\ u_2(x) &= -\alpha_1\Delta tx + \lambda_2\int_0^x N(\xi)d\xi + \kappa_2\tau(x) - \frac{h_1}{2}w_2'(x). \end{aligned} \tag{5.19}$$

Here, α_1 and α_2 ($\alpha_1 \prec \alpha_2$) are the effective CTEs (contraction) of the package and the PCB materials; Δt is the change in temperature from the fabrication temperature to the low (room or testing) temperature; $\lambda_1 = \frac{1-\nu_1}{E_1 h_1}$ and $\lambda_2 = \frac{1-\nu_2}{E_2 h_2}$ are the axial compliances of the PCB and the package materials, respectively; E_1, ν_1, and E_2, h_2 are the elastic constants of the materials; h_1 and h_2 are the thicknesses of the PCB and the package;

$$N(x) = \int_{-L}^{x} \tau(\xi) d\xi \tag{5.20}$$

is the thermally induced force acting ion the cross sections of the package and the PCB; L is half assembly length; $\tau(x)$ is the so far unknown interfacial shearing stress; $\kappa_1 = \frac{h_1}{3G_1}$ and $\kappa_2 = \frac{h_2}{3G_2}$ are the effective longitudinal interfacial compliances of the PCB and the package; $G_1 = \frac{E_1}{2(1+\nu_1)}$ and $G_2 = \frac{E_2}{2(1+\nu_2)}$ are the effective shear moduli of the PCB and the package materials; and $w_1(x)$ and $w_2(x)$ are the deflection functions of the PCB and the package. The origin of the coordinate x is in the mid-cross section of the assembly. The condition of the displacement compatibility can be written as

$$u_1(x) - u_2(x) = \kappa_0 \tau(x), \tag{5.21}$$

where $\kappa_0 = \frac{h}{G}$ is the longitudinal interfacial compliance of the BGA (attachment). Substituting equations (5.19) into this condition, we obtain

$$\kappa\tau(x) - (\lambda_1 + \lambda_2) \int_0^x N(\xi) d\xi + \frac{h_1}{2} w_1'(x) + \frac{h_2}{2} w_2'(x) = \Delta\alpha\Delta t x, \tag{5.22}$$

where $\kappa = \kappa_0 + \kappa_1 + \kappa_2$ is the total longitudinal interfacial compliance of the assembly, and $\Delta\alpha = \alpha_1 - \alpha_2$ is the CTE difference between the PCB and the package.

Differentiating equation (5.22) with respect to the coordinate x we have

$$\kappa\tau'(x) - (\lambda_1 + \lambda_2) N(x) + \frac{h_1}{2} w_1'(x) + \frac{h_2}{2} w_2''(x) = \Delta\alpha\Delta t \tag{5.23}$$

Since no concentrated longitudinal external forces act at the assembly ends, the boundary condition $N(\pm L) = 0$ should be fulfilled, and since no concentrated bending moments act at the assembly ends, the conditions $w_1''(\pm L) = 0$ and $w_2''(\pm L) = 0$ for the curvatures should be fulfilled. Then, equation (5.23) results in the following condition for the shearing stress $\tau(x)$:

$$\tau(\pm L) = \pm \frac{\Delta\alpha\Delta t}{\kappa}. \tag{5.24}$$

Treating the package and its substrate as rectangular plates, we have the following equation of their bending (equilibrium):

$$D_1 w_1''(x) = -\frac{h_1}{2} N(x) - \int_{-L}^{x}\int_{-L}^{x} p(\xi) d\xi d\xi_1,$$

$$D_2 w_2''(x) = -\frac{h_2}{2} N(x) + \int_{-L}^{x}\int_{-L}^{x} p(\xi) d\xi d\xi_1. \quad (5.25)$$

Here, $D_1 = \frac{E_1 h_1^3}{12(1-\nu_1^2)}$ and $D_2 = \frac{E_2 h_2^3}{12(1-\nu_2^2)}$ are flexural rigidities of the PCB and the package, and $p(x)$ is the peeling stress. The package substrate (PCB) and the package curvatures are therefore

$$w_1''(x) = -\frac{h_1}{2D_1} N(x) - \frac{1}{D_1}\int_{-L}^{x}\int_{-L}^{x} p(\xi) d\xi d\xi_1,$$

$$w_2''(x) = -\frac{h_2}{2D_2} N(x) - \frac{1}{D_2}\int_{-L}^{x}\int_{-L}^{x} p(\xi) d\xi d\xi_1. \quad (5.26)$$

Note that since the force $N(x)$ and the moment $\int_{-L}^{x}\int_{-L}^{x} p(\xi) d\xi d\xi_1$ are zero at the assembly ends, the curvatures (5.26) are zero as well.

Introducing formulas (5.26) for the curvatures into equation (5.23), we obtain the following equation for the interfacial shearing stress function $\tau(x)$:

$$\kappa\tau'(x) - \lambda N(x) + \mu \int_{-L}^{x}\int_{-L}^{x} p(\xi) d\xi d\xi_1 = \Delta\alpha\Delta t. \quad (5.27)$$

Here,

$$\lambda = \lambda_1 + \lambda_2 + \frac{h_1^2}{4D_1} + \frac{h_2^2}{4D_2} \quad (5.28)$$

is the axial compliance of the assembly with consideration of its finite flexural rigidity and, hence, with consideration of bending, and

$$\mu = \frac{h_1}{2D_1} - \frac{h_2}{2D_2} \quad (5.29)$$

is the parameter that considers the difference in the flexural rigidities of the PCB and the package. Formula (5.28) indicates that finite flexural rigidities of the PCB and the package result in higher axial compliance of the assembly and, hence, in higher values of the parameter of the interfacial shearing stress. In an approximate analysis, aimed at the assessment of the role of the BGA compliance, the parameter μ can be put equal to zero, so that the interfacial shearing and peeling stresses are not coupled. Then, the shearing stress can be evaluated from the simplified equation

$$\kappa\tau'(x) - \lambda N(x) = \Delta\alpha\Delta t \tag{5.30}$$

Considering (5.20), this equation results in the following equation for the induced force $N(x)$:

$$N''(x) - k^2 N(x) = \frac{\Delta\alpha\Delta t}{\kappa}, \tag{5.31}$$

where $k = \sqrt{\frac{\lambda}{\kappa}}$ is the parameter of the interfacial shearing stress.

We seek the solution to equation (5.31) in the form:

$$N(x) = C_0 + C_2 \cosh kx. \tag{5.32}$$

Since the condition $N(\pm L) = 0$ has to be fulfilled, solution (5.32) yields

$$C_2 = -\frac{C_0}{\cosh kL}, \tag{5.33}$$

and can be written as

$$N(x) = C_0\left(1 - \frac{\cosh kx}{\cosh kL}\right). \tag{5.34}$$

Introducing this formula into equation (5.31), and carrying out the differentiation, we obtain

$$C_0 = -\frac{\Delta\alpha\Delta t}{\lambda}. \tag{5.35}$$

Thus, solution (5.34) results in the following expression for the lateral force:

$$N(x) = -\frac{\Delta\alpha\Delta t}{\lambda}\left(1 - \frac{\cosh kx}{\cosh kL}\right). \tag{5.36}$$

The interfacial shearing stress can be found, as follows from formula (5.20), by differentiation:

$$\tau(x) = N'(x) = k\frac{\Delta\alpha\Delta t}{\lambda}\frac{\sinh kx}{\cosh kL}. \tag{5.37}$$

This solution satisfies the boundary condition (5.24).

For a long assembly, when $kL \geq 2.5$, solution (5.37) can be written as

$$\tau(x) = k\frac{\Delta\alpha\Delta t}{\lambda}e^{-k(L-x)} = \tau_{max}e^{-k(L-x)}, \tag{5.38}$$

where

$$\tau_{max} = k\frac{\Delta\alpha\Delta t}{\lambda} = \frac{\Delta\alpha\Delta t}{\sqrt{\lambda\kappa}} \tag{5.39}$$

is the maximum value of the interfacial shearing stress. It takes place at the assembly ends and decreases exponentially with the increase in the distance of the given cross section from the assembly ends.

Assuming that the PCB and the package are very rigid compared to the BGA system, so that the interfacial compliance of this system is expressed as $\kappa \approx \kappa_0 = \frac{h}{G}$, we obtain formula (5.39) in the approximate form

$$\tau_{max} = \Delta\alpha\Delta t = \sqrt{\frac{G}{\lambda h}}. \tag{5.40}$$

Thus, the maximum interfacial shearing stress in a long enough assembly is assembly length independent and is inversely proportional to the square root of the joint's stand-off height. For a short assembly, when $kL \leq 0.25$, solution (5.37) yields

$$\tau(x) = \frac{\Delta\alpha\Delta t}{\kappa}x = \tau_{max}\frac{x}{L}, \tag{5.41}$$

where the maximum value

$$\tau_{max} = \frac{\Delta\alpha\Delta t}{\kappa}L \tag{5.42}$$

of the interfacial shearing stress takes place at the assembly ends. The interfacial shearing stress is linearly distributed along the assembly. Assuming that the PCB and the package are very rigid compared to the BGA system, so that the interfacial compliance of this system can be found as $\kappa \approx \kappa_0 = \frac{h}{G}$, we obtain formula (5.42) in the following approximate form:

$$\tau_{max} = G\frac{\Delta\alpha\Delta t}{h}L \tag{5.43}$$

Thus, the maximum interfacial shearing stress in a short enough assembly increases with an increase in the assembly length and is inversely proportional to the BGA stand-off height.

The interfacial peeling stress can be sought as

$$p(x) = K[w_1(x) - w_2(x)]. \tag{5.44}$$

Here, K is the spring constant of the BGA in the through-thickness direction. The relationship (5.44) reflects an obvious assumption that the deflections of the PCB and the package have to be different in the given cross section of the assembly to result in a nonzero peeling stress. By differentiation, we find

$$p''(x) = K[w_1''(x) - w_2''(x)], \quad p'''(x) = K[w_1'''(x) - w_2'''(x)]. \tag{5.45}$$

Since there are no concentrated bending moments, nor concentrated lateral forced at the assembly ends, the right parts of these equations should be zero at the ends, and the following boundary conditions should be fulfilled for the sought peeling stress:

$$p''(\pm L) = 0, \quad p'''(\pm L) = 0. \tag{5.46}$$

Introducing formulas (5.26) into the first formula in (5.45) and differentiating the obtain expression twice with respect to the coordinate x, we obtain the following equation of the fourth order for the peeling stress function $p(x)$:

$$p^{IV}(x) + 4\beta^4 p(x) = -\mu K \frac{\Delta\alpha\Delta t}{\kappa} \frac{\cosh kx}{\cosh kL}, \tag{5.47}$$

where

$$\beta = \sqrt[4]{K \frac{D_1 + D_2}{4D_1 D}} \tag{5.48}$$

is the parameter of the peeling stress. Considering (5.48) in the right part of equation (5.47), this equation can be written as

$$p^{IV}(x) + 4\beta^4 p(x) = -4\beta^4 p_0 \frac{\cosh kx}{\cosh kL}, \tag{5.49}$$

where the notation

$$p_0 = \mu \frac{\Delta\alpha\Delta t}{\kappa} \frac{D_1 D_2}{D_1 + D_2} \tag{5.50}$$

is used. It is noteworthy that equation (5.49) has the form of an equation of bending of beams supported by an elastic foundation (see, for example, [40]). In the theory of beams on elastic foundations, this equation is written, however, for the deflections, and not for the peeling stress. But it does not actually matter mathematically, and the solution to equation (5.49) can be sought in the form

$$p(x) = C_0 V_0(\beta x) + C_2 V_2(\beta x) - p_0 \frac{\eta^4}{1+\eta^4} \frac{\cosh kx}{\cosh kL}. \tag{5.51}$$

Here,

$$\eta = \frac{\beta\sqrt{2}}{k} \tag{5.52}$$

is the ratio of the parameters of the peeling and the shearing stresses. The first two terms in the right part of solution (5.51) represent the general solution to the homogeneous equation that corresponds to equation (5.49) and the last term is the particular solution to the inhomogeneous equation (5.49). The functions $V_i(\beta x), i = 0,1,2,3,$ are expressed as follows:

$$\begin{aligned} V_0(\beta x) &= \cosh\beta x \cos\beta x, \\ V_2(\beta x) &= \sinh\beta x \sin\beta x, \\ V_{1,3}(\beta x) &= \frac{1}{\sqrt{2}}(\cosh\beta x \sin\beta x \pm \sinh\beta x \cos\beta x), \end{aligned} \tag{5.53}$$

and obey the following simple and convenient rules of differentiation:

$$\begin{aligned} V_0'(\beta x) &= -\beta\sqrt{2V_3}(\beta x), \\ V_1'(\beta x) &= -\beta\sqrt{2V_0}(\beta x), \\ V_2'(\beta x) &= -\beta\sqrt{2V_1}(\beta x), \\ V_3'(\beta x) &= -\beta\sqrt{2V_2}(\beta x). \end{aligned} \tag{5.54}$$

Using the boundary conditions (5.46), we obtain the following equations for the constants C_0 and C_2 of integration:

$$\begin{aligned} V_2(u)C_0 - V_0(u)C_2 &= -\frac{\eta^2}{1+\eta^4} p_0, \\ V_1(u)C_0 + V_3(u)C_2 &= -\frac{\eta}{1+\eta^4} p_0 \tanh kL, \end{aligned} \tag{5.55}$$

where $u = \beta L$. Equations (5.55) have the following solutions:

$$\begin{aligned} C_0 &= -2\sqrt{2}\frac{\eta}{1+\eta^4} p_0 \frac{\eta V_3(u) + V_0(u)\tanh kL}{\sinh 2u + \sin 2u}, \\ C_2 &= -2\sqrt{2}\frac{\eta}{1+\eta^4} p_0 \frac{-\eta V_1(u) + V_2(u)\tanh kL}{\sinh 2u + \sin 2u}, \end{aligned} \tag{5.56}$$

and solution (5.51) results in the following expression for the peeling stress:

$$p(x) = -2\sqrt{2}\frac{\eta}{1+\eta^4}p_0$$

$$\left[\frac{\eta V_3(u) + V_0(u)\tanh kL}{\sinh 2u + \sin 2u}V_0(\beta x) + \frac{-\eta V_1(u) + V_2(u)\tanh kL}{\sinh 2u + \sin 2u}V_2(\beta x)\right]$$

$$-\frac{\eta^4}{1+\eta^4}p_0\frac{\cosh kx}{\cosh kL} \tag{5.57}$$

For a long and/or stiff assembly ($kL \geq 2.5$), this solution yields:

$$p(x) = -\frac{1}{4}\frac{\eta}{1+\eta^4}p_0 e^{-\beta(L-x)}$$

$$[\eta(\sin(\beta(L-x)) - \cos(\beta(L-x)) + \sqrt{2}\cos(\beta(L-x))]$$

$$-\frac{\eta^4}{1+\eta^4}p_0 e^{-k(L-x)} \tag{5.58}$$

At the assembly end $x = L$:

$$p(L) = -\frac{1}{4}p_0\frac{\eta}{1+\eta^4}(-\eta + \sqrt{2} + 4\eta^3). \tag{5.59}$$

For large $\eta = \frac{\beta\sqrt{2}}{k}$ ratios, this formula leads to the following simple result: $p(L) = -p_0$. This result explains the physical meaning of the p_0 value: it is the peeling stress at the end of a long and stiff assembly. Formula (5.50) indicates that this stress is inversely proportional to the interfacial compliance κ and, hence, to the stand-off height h of the attachment. Thus, for long and stiff assemblies, the peeling stress is even more sensitive to the increase in the BGA stand-off height than the interfacial shearing stress. The peeling stress is next to zero for short (kL) ≤ 0.25) assemblies.

Let, for example, the CTE of the package and the PCB are 12×10^{-6}1/°C and 18×10^{-6}1/°C, respectively, the change in temperature from the reflow soldering (fabrication) temperature to the room temperature is 275°C, and the distance from the package mid-cross section to the location of the given solder joint of the BGA system is 12.0 mm. The predicted thermally induced end offset of the solder joint is $\Delta = (18 \times 10^{-6} - 12 \times 10^{-6}) \times 275 \times 12 \approx 0.02$ mm. Let the elastic constants of the solder material be $E = 30$ GPa $= 3060$ kg/mm^2 and $\nu = 0.30$, the height of the solder joint be $h = 0.8$ mm, and half its thickness be $l = 0.2$ mm. Then, formulas (5.17) and (5.18) yield: $\tau_{max} = 37.1$ kg/mm^2 and $\sigma = 148.3$ kg/mm^2. If the stand-off height is, say, $h = 1.6$ mm, then the predicted stresses are $\tau_{max} = 3.7$ kg/mm^2 and $\sigma = 29.7$ kg/mm^2. The stress relief is indeed significant.

5.4 HEAD-IN-PILLOW PROBLEM

5.4.1 Motivation

The head-in-pillow (HnP) defects (Figures 5.4 and 5.5) in lead-free solder joint interconnections of IC packages with conventional (small) stand-off heights of the solder joints, and particularly in packages with fine pitches, are attributed by many electronic material scientists [51–57] to the three major causes: the manufacturing process, solder material properties, and design-related issues. The latter are thought to be caused primarily by elevated stresses in the solder material, as well as by the excessive warpage of the PCB-package assembly and particularly to the differences in the thermally induced curvatures of the PCB and the package. So, if it is the difference in the postfabrication deflections of the PCB-package assembly that is the root cause of the solder materials failures and particularly the HnP defects, could the replacement of the conventional BGA designs with designs with elevated stand-off heights of the solder joints, such as CGA, result in significant stress and warpage relief and, hopefully, in a lower propensity of the IC package to HnP defects as well? In the analysis that follows, the stress-and-warpage issue is addressed using analytical predictive stress model. It is assumed that it is the difference in the post-fabrication deflections of the PCB-package assembly that is the root cause of the solder materials failures and particularly, and perhaps, the HnP defects.

5.4.2 Background

The causes of the observed HnP soldering defects in BGA packages, and particularly those with lead-free solders [53–57], are attributed by many electronic materials

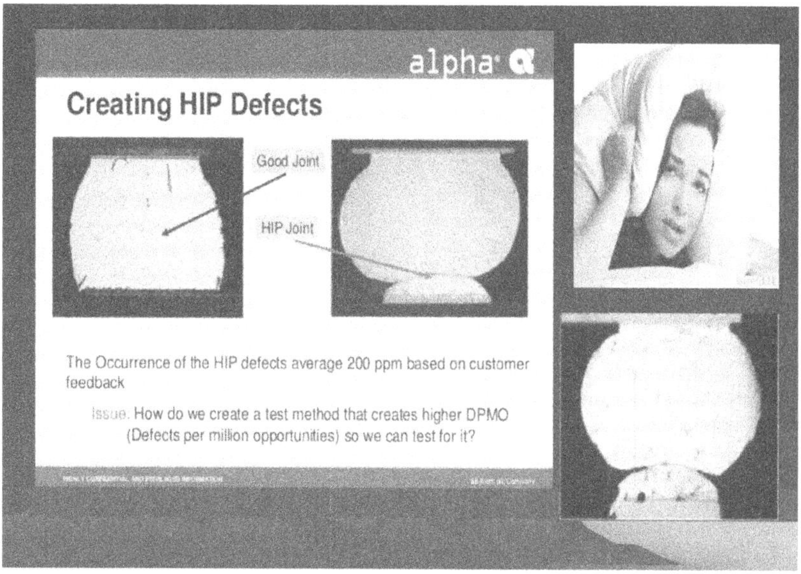

FIGURE 5.4 Head-in-pillow (HnP) defect: what is it?

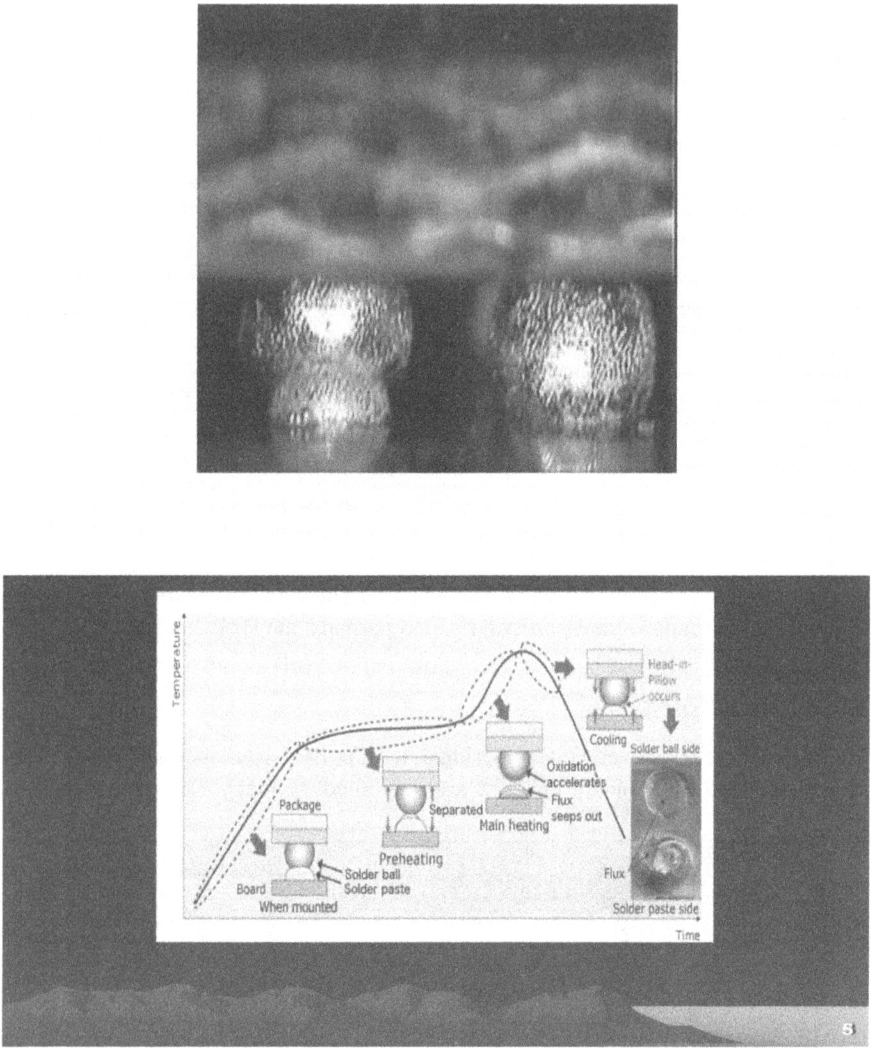

FIGURE 5.5 Head-in-pillow (HnP) defect mechanism.

specialists to process-related, solder material–related, and design-related issues. There is an indication that the design-related problems are caused by the elevated interfacial stresses and elevated warpage of the PCB-package assembly. A typical package structure is addressed in our analysis with an emphasis on the stresses and warpage issue. It is felt that it is the elevated interfacial thermal stresses and the difference in the postfabrication thermally induced deflections of the warped PCB and the warped package that might be the root cause of the possible HnP defects, as well as of an insufficient interfacial strength of the solder joint interconnections.

The advantages of the elevated stand-off heights of solder joint interconnections, as far as the thermal stress level is concerned, were first indicated in application to

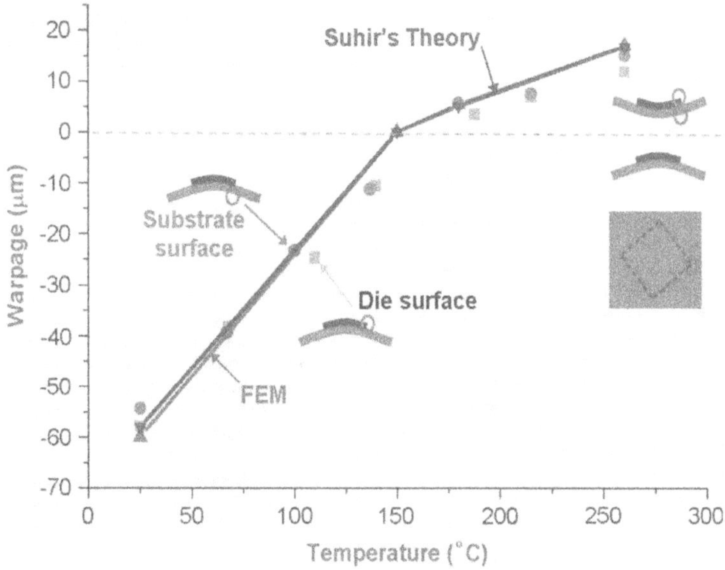

FIGURE 5.6 Warpages at the corner of the die and at the corresponding point on the substrate against the temperature for an FCBGA package under thermal cycling in the moiré experiment, FEA, and theory. (From M.-Y. Tsai, H.-Y. Chang, M. Pecht, "Warpage Analysis of Flip-Chip PBGA Packages Subject to Thermal Loading," IEEE Transactions on Device and Materials Reliability, vol. 9, No. 3, Sept. 2009.)

flip-chip solder joints about three decades ago (see Chapters 2 and 3) and have been recently addressed and modeled in application to the solder joints of the second level of interconnections [43], with an emphasis on the advantages of the CGA designs (Figure 5.2). The thermal stress model used in the analysis that follows is a modification and an extension of the previously developed constitutive thermal stress model [37, 38]. The model has been confirmed over the years by numerous FEA models, but recently it has also been confirmed by a thorough and comprehensive study carried out by Tsai, Chang, and Pecht [52] who employed not only finite element analysis (FEA), but also advanced moiré experimentation, and addressed both the induced stresses and the warpage (Figure 5.6).

5.4.3 Interfacial Stresses and Warpage

The distributed interfacial shearing and peeling stresses can be evaluated using formulas (5.41) and (5.58), respectively (see also Figures 5.7 and 5.8). The thermally induced bow (warpage) of the assembly as a whole can be determined from the approximate equation of equilibrium of the elastic moment (left part of this equation, containing flexural rigidities of the assembly components) and the thermally induced "external" bending moment at the right part of this equation:

$$(D_1 + D_2)w''(x) = \frac{h_1 + h_2}{2} T(x) = \frac{h_1 + h_2}{2} \frac{\Delta\alpha\Delta t}{\lambda} \left(1 - \frac{\cosh kx}{\cosh kl}\right) \tag{5.60}$$

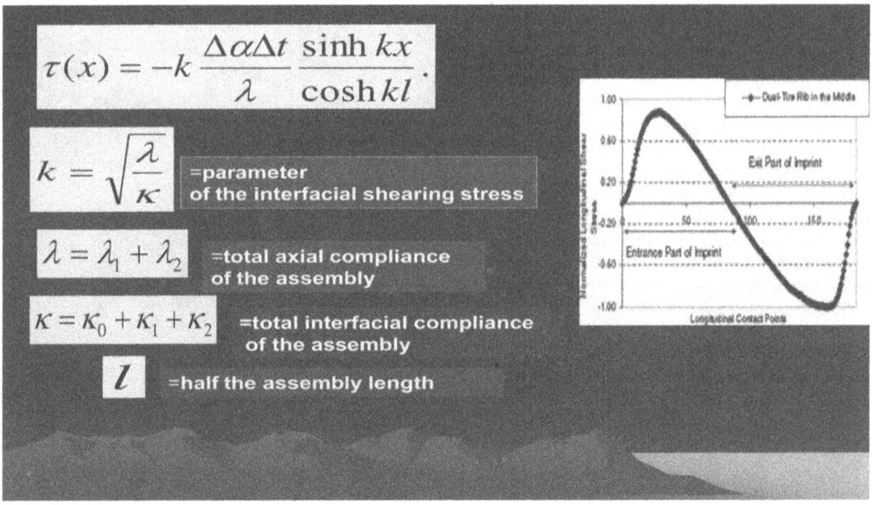

FIGURE 5.7 Interfacial shearing stress.

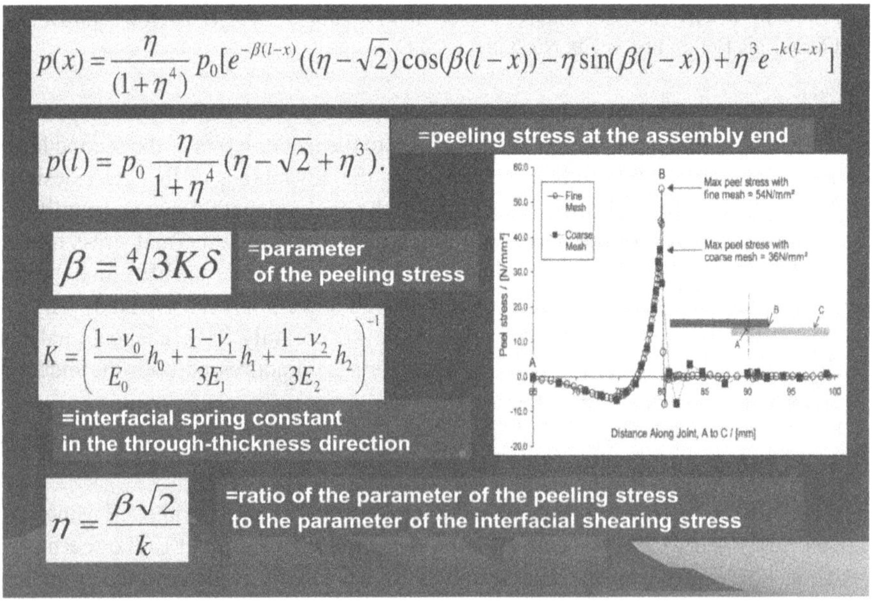

FIGURE 5.8 Peeling stress.

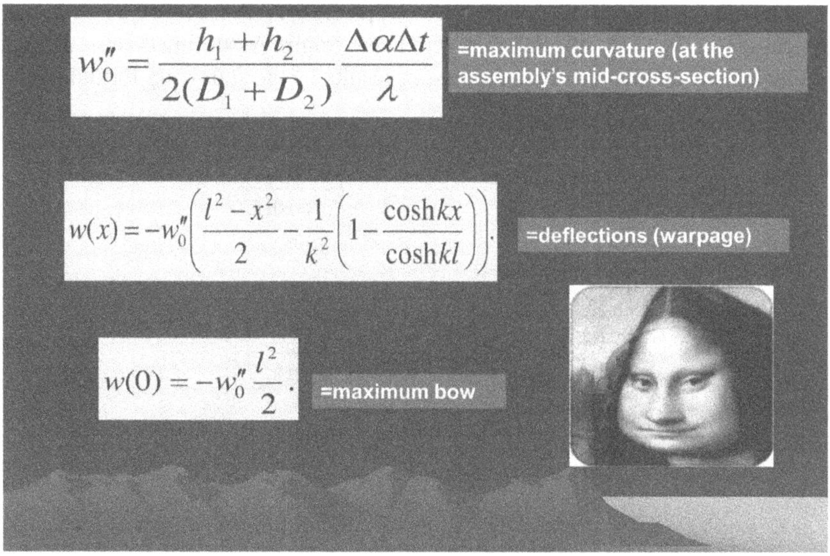
$w_0'' = \frac{h_1 + h_2}{2(D_1 + D_2)} \frac{\Delta\alpha\Delta t}{\lambda}$
=maximum curvature (at the assembly's mid-cross-section)
$w(x) = -w_0''\left(\frac{l^2 - x^2}{2} - \frac{1}{k^2}\left(1 - \frac{\cosh kx}{\cosh kl}\right)\right).$
=deflections (warpage)
$w(0) = -w_0''\frac{l^2}{2}.$
=maximum bow

FIGURE 5.9 Warpage.

Here, $D_1 = \frac{E_1 h_1^3}{12(1 - v_1^2)}$ and $D_2 = \frac{E_2 h_2^3}{12(1 - v_2^2)}$ are the flexural rigidities of the assembly components (PCB and package) treated here as thin elongated plates. From (5.60), the following formula for the assembly curvature $w''(x)$ can be obtained (see also Figure 5.9):

$$w''(x) = w_0''\left(1 - \frac{\cosh kx}{\cosh kl}\right), \tag{5.61}$$

where

$$w_0'' = \frac{h_1 + h_2}{2(D_1 + D_2)} \frac{\Delta\alpha\Delta t}{\lambda} \tag{5.62}$$

is the curvature in the middle of a long-and-stiff assembly. As evident from (5.61), the assembly curvature changes from its maximum value

$$w''_{\max} = w_0''\left(1 - \frac{1}{\cosh kl}\right) \tag{5.63}$$

in the middle of the assembly to zero at its ends. The angles of rotation of the assembly cross sections can be found from (5.63) as follows:

$$w'(x) = w_0''\left(x - \frac{1}{k}\frac{\sinh kx}{\cosh kl}\right). \tag{5.64}$$

The constant of integration is put to zero in this case, so that this expression makes physical sense: the angle of rotation is zero at the origin. Assuming that the assembly ends have zero deflections, this expression results in the following formula for the deflections:

$$w(x) = -w_0''\left[\frac{l^2 - x^2}{2} - \frac{1}{k^2}\left(1 - \frac{\cosh kx}{\cosh kl}\right)\right]. \tag{5.65}$$

In the middle of the assembly ($x = 0$)

$$w(0) = -w_0''\left[\frac{l^2}{2} - \frac{1}{k^2}\left(1 - \frac{1}{\cosh kl}\right)\right]. \tag{5.66}$$

In the case of a long and/or stiff assembly (large enough kl values),

$$w(0) = -w_0''\left(\frac{l^2}{2} - \frac{1}{k^2}\right). \tag{5.67}$$

For assemblies having very high k values of the parameter of the interfacial shearing stress (stiff PCB, stiff package, and stiff solder system) this formula yields

$$w(0) = -w_0''\frac{l^2}{2}. \tag{5.68}$$

5.4.4 Numerical Example

The numerical example is carried out for a hypothetical, but realistic, package/PCB soldered assembly. Two package designs are considered: BGA, characterized by small (conventional) stand-off heights of the solder joint interconnections, and CGA, characterized by elevated stand-off heights of the solder joints.

The calculated data suggest that the replacement of the conventional BGA designs with designs characterized by the elevated stand-off heights of the solder joints could result in significant stress and warpage relief and, supposedly, in lower propensity of the IC package to HnP defects as well. The computed data indicated that the effective stress in the solder material is relieved by about 40% and the difference between the maximum deflections of the PCB and the package is reduced by about 60% when the BGA design is replaced by the CGA system (Figure 5.10).

Thus, we conclude that although no convincing proof is provided that the use of solder joints with elevated stand-off heights will lessen the package propensity to the HnP defects, there is nevertheless a reason to believe that the application of solder joints with elevated stand-off heights could result in substantial improvement in the general IC package performance, including its propensity to HnP defects. Future work should include both FEA and experimental investigations. Particularly, it should be verified if indeed packages with elevated stand-off heights of the solder system are less prone to the observed HnP effect–related damages.

<u>*Coefficients of Thermal Expansion (CTE):*</u>
PCB $\alpha = 15x10^{-6} 1/^0 C$; **Package** $\alpha = 10x10^{-6} 1/^0 C$;
<u>*Young's moduli:*</u>
PCB $E = 17900 kg/mm^2$; **Package** $E = 10300 kg/mm^2$; **Solder** $E = 5510 kg/mm^2$;
<u>*Poisson's ratios:*</u>
PCB $\nu = 0.40$; **Package** $\nu = 0.35$; **Solder** $\nu = 0.35$;
<u>*Shear moduli:*</u>
PCB $G = 6393 kg/mm^2$; **Package** $G = 3815 kg/mm^2$; **Solder** $G = 2040.7 kg/mm^2$;
<u>*Thicknesses (stand-off heights):*</u>
PCB $h = 0.33mm$; **Package** $h = 0.33mm$;
<u>*Stand-off height:*</u>
BGA $h = 0.6mm$; **CGA** $h = 2.2mm$;
<u>*Assembly size (half-length)*</u>
$l = 20.0mm$;
<u>*Change in temperature*</u>
$\Delta t = 150^0 C$.

Solder System	BGA	CGA
Maxinum interfacial shearing stress $\tau_{max}, kg/mm^2$	**2.807**	**1.593**
Maximum peeling stress $p_{max}, kg/mm^2$	**0.1659**	**0.0212**
Effective interfacial stress $\sigma = \sqrt{p^2 + 3\tau^2}, kg/mm^2$	**4.8645**	**2.9324**
Difference in warpage between the package and the PCB $\Delta w, mm^2$	**14.8E-6**	**5.9E-6**

FIGURE 5.10 Calculated data.

REFERENCES

1. H.H. Manko, *Solders and Soldering: Materials, Design, Production and Analysis for Reliable Bonding*, McGraw-Hill, 1979.
2. J. Lau, *Ball Grid Array Technology*, McGraw-Hill, 1994.
3. J.H.L. Pang, *Solder Joint Reliability: Theory and Applications*, Van Nostrand Reinhold, 1991.
4. J. Lau, Y. Pao, *Solder Joint Reliability of BGA, CSP, Flip Chip, and Fine Pitch SMT Assemblies*, McGraw-Hill, 1997.
5. J. Lau, C. Wong, J. Prince, W. Nakayama, *Electronic Packaging: Design, Materials, Process, and Reliability*, McGraw-Hill, 1998.
6. E. Suhir, C.-P. Wong, Y.-C. Lee, (eds.), *Micro- and Opto-Electronic Materials and Structures: Physics, Mechanics, Design, Packaging, Reliability*, vol. 2, Springer, 2008.
7. R. Darveaux, K. Banerji, "Constitutive Relations for Tin-Based Solder Joints," *IEEE CHMT Transactions*, vol. 15, No. 6, 1992.
8. E. Suhir, "Mechanical Reliability of Flip-Chip Interconnections in Silicon-on-Silicon Multichip Modules," IEEE Conference, Santa Cruz, California, March 1993.
9. Y.H. Pao, S. Badgley, R. Govila, E. Jih, "Experimental and Modeling Study of Thermal Cyclic Behavior of Sn-Cu and Sn-Pb Solder Joints," MRS Symposium Proceeding, vol. 323, 1994.
10. R. Darveaux, K. Banerji, A. Mawer, G. Dody. *Reliability of Plastic Ball Grid Array Assemblies. Ball Grid Array Technology,* McGraw-Hill, 1995.
11. S. Wiese, et al., "Microstructural Dependence of Constitutive Properties of Eutectic SnAg and SnAgCu Solders," IEEE ECTC, 2003.
12. R. Schubert, E. Dudek, A. Auerswald, B. Gollhardt, B. Michel, H. Reichl. "Fatigue Life Models for SnAgCu and SnPb Solder Joints Evaluated by Experiments and Simulation," IEEE ECTC, 2003.

13. J. Lau, W. Dauksher, P. Vianco "Acceleration Models, Constitutive Equations and Reliability of Lead-free Solders and Joints," IEEE ECTC, 2003.
14. R. Syed, "Accumulated Creep Strain and Energy Density Based Thermal Fatigue Life Prediction Models for SnAgCu Solder Joints," IEEE ECTC, 2004.
15. Q. Xiao, L. Nguyen, W.D. Armstrong, "Aging and Creep Behavior of Sn3.9Ag0.6Cu Solder Alloy," IEEE ECTC, 2004.
16. Y. Kariya, W.J. Plumbridge, "Mechanical Properties of Sn3.0%Ag-0.5%Cu alloy," Proceedings of the 7th Symposium on Microjoining and Assembly Technology in Electronics, Yokohama, Japan, 2001.
17. J. Lau, Z. Mei, S. Pang, C. Amsden, J. Rayner, S. Pan, "Creep Analysis and Thermal-Fatigue Life Prediction of the Lead-Free Solder Sealing Ring of a Photonic Switch," *ASME Transactions, Journal of Electronic Packaging*, vol. 124, Dec. 2002.
18. H.J. Song, J.W. Morris, F. Hua, "The Creep Properties of Lead-Free Solder Joints," *Journal of Materials*, June 2002.
19. Q. Zhang, A. Dasgupta, P. Haswell. "Viscoplastic Constitutive Properties and Energy-Partitioning Model of Lead-Free Sn3.9Ag0.6Cu Solder Alloy," IEEE ECTC, 2003.
20. J. Lau, W. Dauksher, P. Vianco, "Acceleration Models, Constitutive Equations and Reliability of Lead-Free Solders and Joints," IEEE ECTC, 2003.
21. E. Suhir, "Axisymmetric Elastic Deformations of a Finite Circular Cylinder with Application to Low Temperature Strains and Stresses in Solder Joints," *ASME Journal of Applied Mechanics*, vol. 56, No. 2, 1989.
22. J. Lau, D. Shangguan, D. Lau, T. Kung, R. Lee, "Thermal-Fatigue Life Prediction Equation for Wafer-Level Chip Scale Package (WLCSP): LeadFree Solder Joints on Lead-Free Printed Circuit Board (PCB)," IEEE ECTC, 2004.
23. S. Ridout, C. Bailey, "Review of Methods to Predict Solder Joint Reliability Under Thermo-Mechanical Cycling," *Fatigue Engineering Material Structures*, vol. 30, 2006.
24. M. Osterman, A. Dasgupta, "Life Expectancies of Pb-Free SAC Solder in-Terconnects in Electronic Hardware," *Journal of Master Science*, vol. 18, 2007.
25. M. Vandevelde, M. Gonzalez, P. Limaye, P. Ratchev, E. Beyne, "Thermal Cycling Reliability of SnAgCu and SnPb Solder Joints: A Comparison for Several IC-Packages," *Microelectronics Reliability*, vol. 47, 2007.
26. M. Spraul, W. Nuchter, A. Moller, B. Wunderle, B. Michel, "Reliability of SnPb and Pb-free Flips Under Different Test Conditions," *Microelectronics Reliability*, vol. 47, 2007.
27. A.M. Lajimi, J. Cugnoni, J. Botsis, "Reliability Analysis of Lead-free Solders," Procedures of the World Congress on Engineering and Computer Science, WCECS 2008, San Francisco, California, 2008.
28. Q. Yu, M. Shiratori, Y. Oshima, "*Thermal Fatigue Reliability Assessment for Solder Joints of BGA Assembly*," The 11th Computational Mechanics Conference, JSME, No. 98–2, 1998.
29. Q. Yu, M. Shiratori, "Effects of BGA Solder Geometry on Fatigue Life and Reliability Assessment," *JIEP Journal*, vol. 1, No. 4, 1998.
30. C.B. Lee, S.B. Jung, Y.E. Shin, C.C. Shur, "Effect of Isothermal Ageing on Ball Shear Strength in BGA Joints with Sn-3.5Ag-0.75Cu Solder," *Materials Transactions*, vol. 43, No. 8, 2002.
31. M. Painaik, D.L. Santos, "*Effect of Flux Quantity on Sn-Pb and Pb-Free BGA Solder Shear Strength*," SEMI Technology Symposium: International Electronics Manufacturing Technology (IEMT) Symposium, San Jose, California, 2002.
32. M.-Y. Tsai, C.-H. Hsu, C.-T. Wang, "Investigation of Thermomechanical behaviors of Flip-Chip BGA Package During Manufacturing Process and Thermal Cycling," *IEEE CPMT Transactions*, vol. 27, No. 3, Sept. 2004.

33. B.T. Vaccaro, et al., "Plastic Ball Grid Array Package Warpage and Impact on Traditional MSL Classification for Pb-free Assembly," SMTAI, 2004.
34. K. Newman, "BGA Brittle Fracture–Alternative Solder Joint Integrity Test Methods," IEEE ECTC, 2005.
35. J.M. Koo, S.B. Jung, "Effect of Displacement Rate on Ball Shear Properties for Sn–37Pb and Sn–3.5Ag BGA Solder Joints During Isothermal Aging," *Microelectronics Reliability*, vol. 47, 2007.
36. B. Guenin, "The Many Flavors of Ball Grid Array Packages," Electronics Cooling, Feb 1, 2002, http://www.electronics-cooling.com/2002/02/the-many-flavors-of-ball-grid-array-packages.
37. E. Suhir, "Stresses in Bi-Metal Thermostats," *ASME Journal of Applied Mechanics*, vol. 53, No. 3, Sept. 1986.
38. E. Suhir, "Interfacial Stresses in Bi-Metal Thermostats," *ASME Journal of Applied Mechanics*, vol. 56, No. 3, Sept. 1989.
39. S. Luryi, E. Suhir, "A New Approach to the High-Quality Epitaxial Growth of Lattice-Mismatched Materials," *Applied Physics Letters*, vol. 49, No. 3, July 1986.
40. E. Suhir, *Structural Analysis in Microelectronic and Fiber Optic Systems, Basic Principles of Engineering Elasticity and Fundamentals of Structural Analysis*, van Nostrand Reinhold, 1991.
41. M.-Y. Tsai, C.-H. Ysu, C.-N. Han, "A Note on Suhir's Solution of Thermal Stresses for a Die-Substrate Assembly," *ASME Journal of Electronic Packaging*, vol. 12, No. 6, 2004.
42. S.P. Timoshenko, "On the Correction Factor for Shear of the Differential Equation for Transverse Vibrations of Bars of Uniform Cross-Section," Philosophical Magazine, 1921.
43. E. Suhir, "Predicted Stresses in a Ball-Grid-Array (BGA)/Column-Grid-Array (CGA) Assembly with a Low Modulus Solder at Its Ends," *Journal of Material Science: Materials in Electronics*, vol. 26, No. 12, 2015.
44. E. Suhir, R. Ghaffarian, J. Nicolics, "Could Application of Column-Grid-Array Technology Result in Inelastic-Strain-Free State-of-Stress in Solder Material?" *Journal of Material Science: Materials in Electronics*, vol. 26, No. 12, 2015.
45. E. Suhir, "Analysis of a Short Beam with Application to Solder Joints: Could Larger Stand-off Heights Relieve Stress?" *European Journal of Applied Physics (EPJAP)*, vol. 71, 2015.
46. E. Suhir, R. Ghaffarian, J. Nicolics, "Could Thermal Stresses in an Inhomogeneous BGA/CGA System be Predicted Using a Model for a Homogeneously Bonded Assembly?" *Journal of Material Science: Materials in Electronics*, vol. 27, No. 1, 2016.
47. E. Suhir, R. Ghaffarian, J. Nicolics, "Predicted Stresses in Ball-Grid-Array (BGA) and Column-Grid-Array (CGA) Interconnections in a Mirror-like Package Design," *Journal of Material Science: Materials in Electronics*, vol. 27, No. 3, 2016.
48. E. Suhir, R. Ghaffarian, "Predicted Stresses in a Ball-Grid-Array (BGA)/Column-Grid-Array (CGA) Assembly with Epoxy Adhesive at Its Ends," *Journal of Material Science: Materials in Electronics*, vol. 27, No. 5, 2016.
49. E. Suhir, R. Ghaffarian, "Column-Grid-Array (CGA) vs. Ball-Grid-Array (BGA): Board-Level Drop Test and the Expected Dynamic Stress in the Solder Material," *Journal of Material Science: Materials in Electronics*, vol. 27, No. 11, 2016.
50. E. Suhir, "Avoiding Low-Cycle Fatigue in Solder Material Using Inhomogeneous Column-Grid-Array (CGA) Design," ChipScale Reviews, March–April 2016.
51. A. Dudi, R. Aspaniar, S. Buttars, W.-W. Chin, P. Gill, "Head-on-Pillow SMT Failure Modes," SMTAI Conf., 2009.
52. M.-Y. Tsai, H.-Y. Chang, M. Pecht, "Warpage Analysis of Flip-Chip PBGA Packages Subject to Thermal Loading," *IEEE Transactions on Device and Materials Reliability*, vol. 9, No. 3, Sept. 2009.

53. Y. Liu, P. Fiacco, N. Lee, "Testing and Prevention of Head-in-Pillow," ECTC, 2010.
54. J. Savic, W. Xie, N. Islam, P.-G. Oh, R. Pendse, K.-O. Kim, "Warpage Mitigation Processes in the Assembly Large Body Size Mixed Pitch BGA Coreless Packages for Use in High Speed Network Applications," SMTAI Conference, 2013.
55. "IPC-9641: High Temperature Printed Board Flatness Guideline," IPC, June 2013, https://www.ipc.org/TOC/IPC-9641.pdf.
56. H. Rekers, P. Eng, "Case Study–Head in Pillow Defect on a Plastic BGA," SMTAI Conference on Soldering and Reliability, 2015.
57. E. Suhir, S. Yi, J.S. Hwang, R. Ghaffarian, "Elevated Stand-Off Heights of Solder Joint Interconnections Can Result in Appreciable Stress and Warpage Relief," *IMAPS Journal of Microelectronics and Electronic Packaging*, vol. 16, No. 1, Jan. 2019 (best paper of session award).

6 Stress Relief in Soldered Assemblies by Using Inhomogeneous Bonds

"The only real voyage of discovery consists not in seeing new landscapes, but in having new eyes."

—Marcel Proust, French author and critic

6.1 BACKGROUND/INCENTIVE

Solder materials, providing mechanical support to surface-mounted devices and subjected to elevated thermally induced interfacial shearing stresses, are prone to inelastic deformations. This considerably shortens their fatigue lifetime. Therefore, there exists a crucial need for stress reduction in solder joint interconnections. One effective way for doing that is the use of inhomogeneous bonding systems. Various assemblies with inhomogeneous bonding layers were addressed in application to assemblies bonded at the ends [1–10] to predict the size of an inelastic zone in ball-grid array (BGA) assemblies [11, 12], to explain a paradoxical situation when stiffer midportions of compliant bonds could result in an appreciably lower stresses at the assembly ends [13], to assess the possible stress relief when solder joints with elevated stand-off heights are employed [14–17], to evaluate the peculiarities associated with the use of identical adherends [18–20], the possibility of using the model developed for a homogeneously bonded assembly (of the type suggested in [21]) for assemblies of the BGA or column-grid array (CGA) type, in which the solder supports are "spaced" at certain distances from each other [22], and in a number of other electronic and photonic reliability–related problems (see, for example, [23, 24]).

Identical bonded components and "piecewise-continuous" bonding layers were considered in application to a new generation of low-cost memory storages, in which the bonding layer played the role of the memory storing medium [18–20]. The emphasis was on the evaluation of the conditions that could lead to plane boundaries between the "pieces" of the bonding layer, rather than to low level stresses. It has been recently shown [14–17] that significant stress relief can be obtained owing to the application of BGAs with the increased stand-off of the solder joint interconnections, not to mention the use of CGAs.

The objective of this chapter is to show that significant thermal stress relief can be achieved in an optimized design of an inhomogeneously bonded bimaterial assembly, if its bonding system is designed and fabricated in such a way that the interfacial shearing stress at the ends of the high-modulus and high-bonding temperature midportion of the assembly at its boundary with the low-modulus and low-bonding temperature peripheral portion is made equal to the stress at the assembly ends. In such an assembly, the interfacial shearing stress increases first from zero at the assembly mid-cross section to its maximum value at the end of the high-modulus and/or high-fabrication temperature midportion at its boundary with the low-modulus and/or low-fabrication temperature peripheral portion(s) of the assembly, drops to a low value at this boundary and then increases again to the same maximum value at the assembly end(s). Each of these maximum values is well below the maximum interfacial stress in a regular, homogeneously bonded, assembly.

6.2 ASSEMBLY'S MIDPORTION

Consider first the midportion of an in-homogeneously bonded assembly (having different bonding materials at the midportion and the peripheral portions of the assembly), experiencing the given change Δt_m in temperature and subjected, because of its mismatch with the peripheral portions, to external forces $\hat{T}$ applied to the midportion in a symmetric fashion (Figure 6.1). The interfacial longitudinal displacements of the assembly components can be sought, in an approximate analysis, using the concept of the interfacial compliances (see Chapter 2), as follows:

$$\left.\begin{aligned} u_1(x) &= -\alpha_1 \Delta t_m x + \lambda_1 \int_0^x T(\xi)d\xi - \kappa_1 \tau(x) - \frac{h_1}{2} w_1'(x) \\ u_2(x) &= -\alpha_2 \Delta t_m x - \lambda_2 \int_0^x T(\xi)d\xi + \kappa_2 \tau(x) + \frac{h_2}{2} w_2'(x) \end{aligned}\right\}. \tag{6.1}$$

Here, α_1 and α_2 are the coefficients of thermal expansion (CTE) of the component materials, Δt_m is the change in temperature (say, between the bonding temperature and the room or testing temperature),

$$\lambda_1 = \frac{1-\nu_1}{E_1 h_1}, \quad \lambda_2 = \frac{1-\nu_2}{E_2 h_2} \tag{6.2}$$

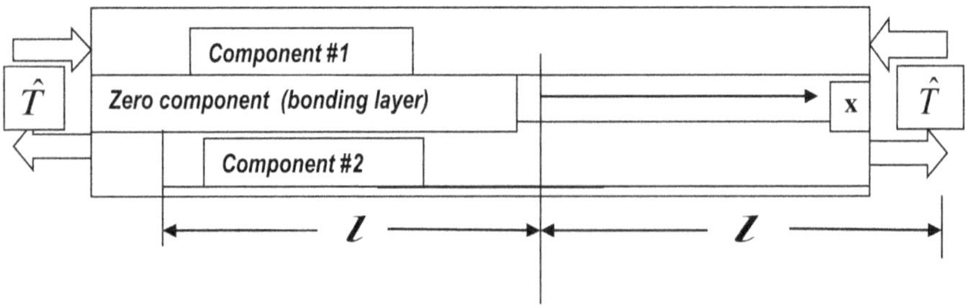

FIGURE 6.1 The midportion of an inhomogeneously bonded bimaterial assembly.

are the axial compliances of the assembly components, h_1 and h_2 are the component thicknesses, E_1 and E_2 are Young's moduli of the materials, v_1 and v_2 are Poisson's ratios of materials,

$$T(x) = \int_{-l}^{x} \tau(\xi)d\xi + \hat{T} \tag{6.3}$$

are the distributed forces acting in the x cross section of the assembly, $\tau(x)$ is the interfacial shearing stress, $\hat{T}$ is the thus far unknown force applied to the midportion from the peripheral portions of the assembly, l is half the assembly length, κ_1 and κ_2 are the interfacial compliances of the assembly components, and $w_1(x)$ and $w_2(x)$ are the components' deflections, so that their derivatives are angles of rotation. The origin of the coordinate x is in the mid-cross section of the assembly.

The first terms in (6.1) are stress-free thermal contractions. The second terms determine the displacements caused by the induced thermal forces. These displacements are evaluated in accordance with Hooke's law and do not consider that the interfacial displacements are somewhat larger than the displacements of the inner points of the cross section. The third terms account for that. They are, in effect, corrections to the displacements evaluated in accordance with the second terms. It is assumed that these corrections can be evaluated as the product of the interfacial compliances [15]

$$\kappa_1 = \frac{h_1}{3G_1}, \quad \kappa_2 = \frac{h_2}{3G_2} \tag{6.4}$$

of the bonded component of interest and the interfacial shearing stress acting in this cross section.

Here,

$$G_1 = \frac{E_1}{2(1+v_1)}, \quad G_2 = \frac{E_2}{2(1+v_2)}, \tag{6.5}$$

are the shear moduli of the bonded component materials. The fourth terms in (6.1) are due to bending. Since the case of cooling is considered here, and the CTE of component #1 is assumed to be lower than that of component #2, the interfacial surface of component #1 is configured in the concave fashion and the surface of component #2 is configured in the convex fashion. This circumstance is reflected by the signs in front of the corresponding displacement-related terms.

The condition of the compatibility of the interfacial displacements (6.1) can be written, in the presence of a compliant bond, if any, as

$$u_1(x) = u_2(x) - \kappa_0\tau(x), \tag{6.6}$$

where

$$\kappa_0 = \frac{h_0}{G_0} \tag{6.7}$$

is the longitudinal interfacial compliance of the bonding layer, and

$$G_0 = \frac{E_0}{2(1+v_0)} \tag{6.8}$$

is the shear modulus of the bonding material. The second term in the right part of condition (6.6) is due to the interfacial compliance of the bond.

Introducing formulas (6.1) into condition (6.6), the following equation for the interfacial shearing stress function $\tau(x)$ can be obtained:

$$\kappa\tau(x) - (\lambda_1 + \lambda_2)\int_0^x T(\xi)d\xi - \frac{h_1}{2}w_1'(x) - \frac{h_2}{2}w_2'(x) = \Delta\alpha\Delta t_m x. \tag{6.9}$$

Here, $\kappa = \kappa_0 + \kappa_1 + \kappa_2$ is the total interfacial compliance of the assembly, and $\Delta\alpha = \alpha_2 - \alpha_1$ is the difference in the CTEs of the component materials. By differentiating equation (6.9) with respect to the coordinate x we have

$$\kappa\tau'(x) - (\lambda_1 + \lambda_2)T(x) - \frac{h_1}{2}w_1''(x) - \frac{h_2}{2}w_2''(x) = \Delta\alpha\Delta t_m. \tag{6.10}$$

The curvatures $w_1''(x)$ and $w_2''(x)$ of the assembly components can be determined from the equilibrium equations

$$D_1 w_1''(x) = -\frac{h_1}{2}T(x) - \int_{-l}^x\int_{-l}^x p(\xi)d\xi d\xi, \quad D_2 w_2''(x) = -\frac{h_2}{2}T(x) + \int_{-l}^x\int_{-l}^x p(\xi)d\xi d\xi, \tag{6.11}$$

where

$$D_1 = \frac{E_1 h_1^3}{12(1-v_1^2)}, \quad D_2 = \frac{E_2 h_2^3}{12(1-v_2^2)} \tag{6.12}$$

are the flexural rigidities of the assembly components treated here as elongated rectangular plates, and $p(x)$ is the peeling stress. The left parts of equations (6.11) are elastic bending moments, the first terms in the right parts are the bending moments caused by the induced thermal forces $T(x)$, and the second terms are the bending moments due to the induced peeling stress $p(x)$.

Solving equations (11) for the component curvatures, we have

$$w_1''(x) = -\frac{h_1}{2D_1}T(x) - \frac{1}{D_1}\int_{-l}^x\int_{-l}^x p(\xi)d\xi d\xi, \quad w_2''(x) = -\frac{h_2}{2D_2}T(x) + \frac{1}{D_2}\int_{-l}^x\int_{-l}^x p(\xi)d\xi d\xi. \tag{6.13}$$

Introducing these expressions into equation (6.10), we obtain

$$\kappa\tau'(x) - \lambda_* T(x) + \mu\int_{-l}^x\int_{-l}^x p(\xi)d\xi d\xi = \Delta\alpha\Delta t_m. \tag{6.14}$$

Here,

$$\lambda_* = \lambda_1 + \lambda_2 + \frac{h_1^2}{4D_1} + \frac{h_2^2}{4D_2} \tag{6.15}$$

is the axial compliance of the assembly with consideration of the finite flexural rigidities of the assembly components, and

$$\mu = \frac{h_1}{2D_1} - \frac{h_2}{2D_2} \tag{6.16}$$

is the factor that considers the role of the components' geometrical dissimilarity. This role can be neglected in an approximate analysis, so that equation (6.14) can be replaced with the following simplified equation:

$$\kappa\tau'(x) - \lambda_* T(x) = \Delta\alpha\Delta t_m, \tag{6.17}$$

in which the shearing stress only is considered.

The force $T(x)$ should be symmetric with respect to the mid-cross section of the assembly and could be sought as

$$T(x) = C_0 + C_2 \cosh kx. \tag{6.18}$$

Introducing this solution into equation (17) and considering that, in accordance with formula (6.3),

$$\tau'(x) = T''(x) = k^2 C_2 \cosh kx, \tag{6.19}$$

we conclude that equation (6.17) is fulfilled, if the relationships

$$k = \sqrt{\frac{\lambda_*}{\kappa}}, \quad C_0 = -\frac{\Delta\alpha\Delta t_m}{\lambda_*} \tag{6.20}$$

take place. The constant C_2 can be found from the boundary condition

$$T(\pm l) = \hat{T} \tag{6.21}$$

as

$$C_2 = \left(\frac{\Delta\alpha\Delta t_m}{\lambda_*} + \hat{T} \right) \frac{1}{\cosh kl}. \tag{6.22}$$

Then, solution (6.18) results in the following expression for the induced force $T(x)$:

$$T(x) = -\frac{\Delta\alpha\Delta t_m}{\lambda_*} \left(1 - \frac{\cosh kx}{\cosh kl} \right) + \hat{T} \frac{\cosh kx}{\cosh kl}. \tag{6.23}$$

The first term in this expression is due to the local thermal mismatch of the assembly components, and the second term is caused by the external force $\hat{T}$ applied from the peripheral portions of the assembly. The interfacial shearing stress can be determined from (6.23) by differentiation

$$\tau(x) = T'(x) = k\left(\frac{\Delta\alpha\Delta t_m}{\lambda_*} + \hat{T}\right)\frac{\sinh kx}{\cosh kl} \tag{6.24}$$

The shearing stress is zero at the mid-cross section and increases to

$$\tau(x) = k\left(\frac{\Delta\alpha\Delta t_m}{\lambda_*} + \hat{T}\right)\tanh kl \tag{6.25}$$

at the end of the midportion.

6.3 ASSEMBLY'S PERIPHERAL PORTION(S) AND FORCES AT THE BOUNDARIES

Consider now the peripheral portion of the assembly (Figure 6.2). Unlike the midportion, the peripheral portion is subjected to an external loading applied to only one side of the assembly, but, even more importantly, its modulus and the application (soldering, curing) temperature could be much different from those in the midportion. The boundary conditions for the induced forces are also different:

$$T(-l) = \hat{T}, \quad T(l) = 0. \tag{6.26}$$

To satisfy these two conditions, the force $T(x)$ should be sought in the form

$$T(x) = C_0 + C_1 \sinh kx + C_2 \cosh kx. \tag{6.27}$$

This solution contains two constants of integration and is not symmetric with respect to its mid-cross section. Introducing the sought solution (6.27) into equation (6.17), we conclude that formulas (20) are still valid, provided, of course, that the appropriate

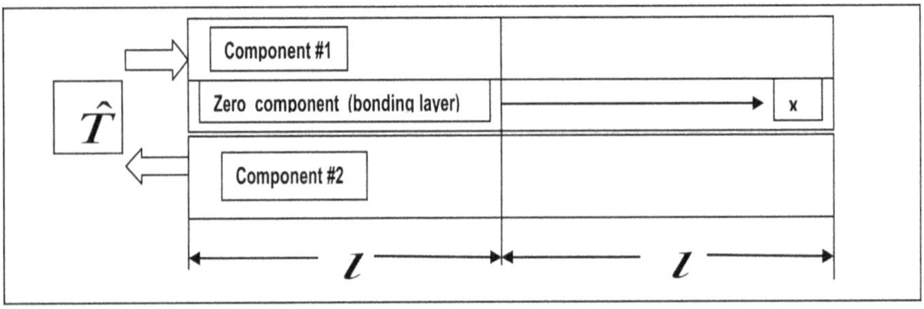

FIGURE 6.2 The peripheral portion of an inhomogeneously bonded bimaterial assembly subjected to thermal loading.

change in temperature is considered (Δt_p instead of Δt_m). The constants of integration obtained from the two conditions in (6.26) are

$$C_1 = -\frac{\hat{T}}{2\sinh kl}, \quad C_2 = \left(\frac{\Delta\alpha\Delta t_p}{\lambda} + \frac{\hat{T}}{2}\right)\frac{1}{\cosh kl}. \tag{6.28}$$

Introducing the second formula in (6.20) and formulas (6.28) into sought solution (6.27), we obtain the following formula for the induced force:

$$T(x) = -\frac{\Delta\alpha\Delta t_p}{\lambda_*}\left(1 - \frac{\cosh kx}{\cosh kl}\right) + \hat{T}\frac{\sinh[k(l-x)]}{\sinh 2kl}. \tag{6.29}$$

The first ("thermal") term in this expression is not different of the first term in (6.23), but the second term is quite different, because of the differently applied thermomechanical loading.

The interfacial shearing stress can be found from (6.29) by differentiation

$$\tau(x) = T'(x) = k\left[\frac{\Delta\alpha\Delta t_p}{\lambda_*}\frac{\sinh kx}{\cosh kl} - \hat{T}\frac{\cosh[k(l-x)]}{\sinh 2kl}\right]. \tag{6.30}$$

It changes from

$$\tau(-l) = -k\left(\frac{\Delta\alpha\Delta t_p}{\lambda_*}\tanh kl + \hat{T}\coth 2kl\right) \tag{6.31}$$

at the left end of the assembly to

$$\tau(l) = k\left(\frac{\Delta\alpha\Delta t_p}{\lambda_*}\tanh kl - \hat{T}\frac{1}{\sinh 2kl}\right) \tag{6.32}$$

at its right end. Formula (6.32) particularly indicates that if the peripheral portion is characterized by a large enough product of the parameter of the interfacial shearing stress and half the assembly length, the force at the inner end of the assembly where the external force is applied will not affect the shearing stress at the outer (free) end of the assembly, and this stress can be found as

$$\tau(l) = k\frac{\Delta\alpha\Delta t_p}{\lambda_*} \tag{6.33}$$

Formulas (6.25) and (6.31) yield

$$\tau(L-2l) = K\left(\frac{\Delta\alpha\Delta t_m}{\lambda_*} + \hat{T}\right)\tanh[K(L-2l)],$$

$$\tau(-l) = -k\left(\frac{\Delta\alpha\Delta t_p}{\lambda_*}\tanh kl + \hat{T}\coth 2kl\right), \tag{6.34}$$

where the notation in the first formula has been changed to account, because for the possibly different parameter of the interfacial shearing stress (K instead of k) and to indicate that half the length $L-2l$ of the midportion of the assembly can be found as the difference between half the assembly length L and the length $2l$ of one of its peripheral portions. The force $\hat{T}$ at the boundary between the midportion and the peripheral portions of the assembly with an inhomogeneous bonding layer can be determined from the condition of the compatibility of the longitudinal interfacial displacements of the two portions of the assembly at their boundary. Since the interfacial displacements can be found as products of the interfacial compliances and the interfacial shearing stresses, the condition of interest can be written for a long enough midportion (for such a midportion, the hyperbolic tangent in the first formula in (6.34) can be put equal to one), as

$$K\kappa_m\left(\frac{\Delta\alpha\Delta t_m}{\lambda_*}+\hat{T}\right)=-k\kappa_p\left(\frac{\Delta\alpha\Delta t_p}{\lambda_*}\tanh kl+\hat{T}\coth 2kl\right). \tag{6.35}$$

This condition yields

$$\hat{T}=-\frac{\Delta\alpha\Delta t_m}{\lambda_*}\frac{\eta+\delta\tanh kl}{\eta+\coth 2kl}=-2\frac{\Delta\alpha\Delta t_m}{\lambda_*}t\frac{\eta+\delta t}{1+t^2+2t\eta}, \tag{6.36}$$

where the notations

$$t=\tanh kl,\quad \eta=\frac{k}{K}=\sqrt{\frac{\kappa_m}{\kappa_p}},\quad \delta=\frac{\Delta t_p}{\Delta t_m} \tag{6.37}$$

and the formulas

$$\sinh 2kl=\frac{2t}{1-t^2};\quad \cosh 2kl=\frac{1+t^2}{1-t^2};\quad \coth 2kl=\frac{1+t^2}{2t}; \tag{6.38}$$

are used.

6.4 INTERFACIAL STRESSES

Introducing formula (6.36) for the induced force at the boundary between the midportion and the peripheral portions of the assembly into formulas (6.32) and (6.34) and assuming a long enough midportion, we obtain the following expressions for the interfacial shearing stresses:

$$\tau_m(L-2l)=K\frac{\Delta\alpha\Delta t_m}{\lambda_*}\frac{\coth 2kl+\delta\tanh kl}{\eta+\coth 2kl}=K\frac{\Delta\alpha\Delta t_m}{\lambda_*}\frac{1+(1+2\delta)t^2}{1+t^2+2t\eta}, \tag{6.39}$$

at the assembly's midportion at its boundary with the peripheral portion,

$$\tau_p(-l)=-k\frac{\Delta\alpha\Delta t_m}{\lambda_*}\frac{\delta\tanh kl(\eta+\coth 2kl-1)-\eta}{\eta+\coth 2kl}=-k\frac{\Delta\alpha\Delta t_m}{\lambda_*}t\frac{\delta[(1-t)^2+2t\eta]-2}{1+t^2+2t\eta}, \tag{6.40}$$

at the peripheral portion at its boundary with the midportion, and

$$\tau_p(l) = k\frac{\Delta\alpha\Delta t_m}{\lambda_*}\frac{\delta\tanh kl(\eta\sinh 2kl + \cosh 2kl + 1) + \eta}{\eta\sinh 2kl + \cosh 2kl}$$

$$= k\frac{\Delta\alpha\Delta t_m}{\lambda_*}\frac{2\delta t(1 + t\eta) + (1 - t^2)\eta}{1 + t^2 + 2t\eta}, \quad (6.41)$$

at the assembly end (free end of the peripheral portion). The shearing stress increases from zero at the mid-cross section of the assembly to the magnitude determined by formula (6.39) at the end of the midportion (at its boundary with the peripheral portion), then drops to the magnitude defined by formula (6.40) at the same cross section, then increases again, and reaches the magnitude (6.41) at the assembly ends.

6.5 NUMERICAL EXAMPLE

The numerical example is carried out for a typical package (component #1)/PCB (component #2) assembly with either BGA or CGA high-modulus solder joint interconnection at the midportion of the assembly, and with an epoxy adhesive at its peripheral portion(s), so that both the fabrication/curing temperature and Young's modulus of the peripheral material are low compared to the material in the midportion of the assembly. Our objective is to assess the level of the induced interfacial shearing stresses.

Input data:

Structural Element	Package	PCB	3-4%Ag0.5-1%Cu Solder at the Midportion	Epoxy Adhesive at the Peripheral Portions
Young's modulus, E, kg/mm^2	8775.5	2321.4	5510.0	415.3
Poisson's ratio, ν	0.25	0.40	0.35	0.35
CTE α, 1/°C	6.5×10^{-6}	15.0×10^{-6}	Not important	Not important
Thickness, h, mm	2.0	1.5	0.60 for BGA 2.20 for CGA	0.60 in the case of BGA 2.20 in the case of CGA
Shear modulus, G, kg/mm^2	3367.3	892.7	2040.7	153.8
Axial compliance, λ, mm/kg	3.9884×10^{-5}	20.1028×10^{-5}	Not important	Not important
Interfacial compliance, κ, mm^3/kg	19.7982×10^{-5}	56.0099×10^{-5}	29.4017×10^{-5} for BGA 107.806×10^{-5} for CGA	390.1170×10^{-5} in the case of BGA 1430.4291×10^{-5} in the case of CGA

Estimated yield stress of the solder material in shear is: $\tau_Y = 1.85$ kg/mm^2 for the solder; temperature change $\Delta t_m = 200$°C for solder, and $\Delta t_p = 100$°C for the epoxy, so that their ratio is $\frac{\Delta t_p}{\Delta t_m} = 0.5$. Half assembly length is $L = 15$ mm. Lengths of the peripheral portions are $2l = 2.0$ mm. Thermally induced force in a long midportion of the assembly is $\frac{\Delta\alpha\Delta t_m}{\lambda_*} = \frac{0.0017}{72.3701\times10^{-5}} = 2.3490$ kg/mm.

Calculated data

Axial compliances of the assembly components:

$$\lambda_1 = \frac{1-\nu_1}{E_1 h_1} = \frac{1-0.3}{8775.5 \times 2.0} = 3.9884 \times 10^{-5}\ \text{mm/kg};$$

$$\lambda_2 = \frac{1-\nu_2}{E_2 h_2} = \frac{1-0.3}{2321.4 \times 1.5} = 20.1028 \times 10^{-5}\ \text{mm/kg};$$

$$\lambda = \lambda_1 + \lambda_2 = 24.0912 \times 10^{-5}\ \text{mm/kg};$$

Interfacial compliances of the assembly components:

$$\kappa_1 = \frac{h_1}{3G_1} = \frac{2.0}{3 \times 3367.3} = 19.7983 \times 10^{-5}\ \text{mm}^3/\text{kg},$$

$$\kappa_2 = \frac{h_2}{3G_2} = \frac{1.5}{3 \times 892.7} = 56.100 \times 10^{-5}\ \text{mm}^3/\text{kg}.$$

Interfacial compliances of the solder:

$$\kappa_0 = \frac{h_0}{G_0} = \frac{0.6}{2040.7} = 29.4017 \times 10^{-5}\ \text{mm}^3/\text{kg}$$

in the case of BGA, and

$$\kappa_0 = \frac{h_0}{G_0} = \frac{2.2}{2040.7} = 107.8061 \times 10^{-5}\ \text{mm}^3/\text{kg}$$

in the case of for CGA. Interfacial compliances of the epoxy:

$$\kappa_0 = \frac{h_0}{G_0} = \frac{0.6}{153.8} = 390.1170 \times 10^{-5}\ \text{mm}^3/\text{kg}$$

in the case of BGA in its midportion, and

$$\kappa_0 = \frac{h_0}{G_0} = \frac{2.2}{153.8} = 1430.4291 \times 10^{-5}\ \text{mm}^3/\text{kg}$$

in the case of for CGA in the midportion. Total interfacial compliance at the midportion of the assembly:

$$\kappa_m = \kappa_0 + \kappa_1 + \kappa_2 = 105.2100 \times 10^{-5}\ \text{mm}^3/\text{kg}$$

in the case of BGA, and

$$\kappa_m = \kappa_0 + \kappa_1 + \kappa_2 = 183.6144 \times 10^{-5}\ \text{mm}^3/\text{kg}$$

in the case of CGA. Total interfacial compliance at the peripheral portions of the assembly:

$$\kappa_p = \kappa_0 + \kappa_1 + \kappa_2 = 465.9253 \times 10^{-5}\ \text{mm}^3/\text{kg}$$

in the case of BGA, and

$$\kappa_p = \kappa_0 + \kappa_1 + \kappa_2 = 1506.2374 \times 10^{-5}\ \text{mm}^3/\text{kg}$$

in the case of CGA. Flexural rigidities of the assembly components:

$$D_1 = \frac{E_1 h_1^3}{12(1-\nu_1^2)} = \frac{8775.5 \times 2.0^3}{12(1-0.25^2)} = 6240.3556\ \text{kgmm}$$

$$D_2 = \frac{E_2 h_2^3}{12(1-\nu_2^2)} = \frac{2321.4 \times 1.5^3}{12(1-0.40^2)} = 777.2545\ \text{kgmm}$$

Total axial compliance of the assembly with consideration of the finite flexural rigidities of its components:

$$\lambda_* = \lambda + \frac{h_1^2}{4D_1} + \frac{h_2^2}{4D_2} = 24.0912 \times 10^{-5} + \frac{2.0^2}{4 \times 6240.3556} + \frac{1.5^2}{4 \times 777.2545}$$
$$= 72.3701 \times 10^{-5}\ \text{mm/kg}$$

The parameter of the interfacial shearing stress at the midportion of the assembly:

$$K = \sqrt{\frac{\lambda_*}{\kappa_m}} = \sqrt{\frac{72.3701 \times 10^{-5}}{105.2100 \times 10^{-5}}} = 0.8294\ \text{mm}^{-1}$$

in the case of BGA, and

$$K = \sqrt{\frac{\lambda_*}{\kappa_m}} = \sqrt{\frac{72.3701 \times 10^{-5}}{183.6144 \times 10^{-5}}} = 0.6278\ \text{mm}^{-1}$$

in the case of CGA. The parameter of the interfacial shearing stress at the peripheral portions of the assembly:

$$k = \sqrt{\frac{\lambda_*}{\kappa_p}} = \sqrt{\frac{72.3701 \times 10^{-5}}{465.9253 \times 10^{-5}}} = 0.3941\ \text{mm}^{-1}$$

in the case of BGA, and

$$k = \sqrt{\frac{\lambda_*}{\kappa_p}} = \sqrt{\frac{72.3701 \times 10^{-5}}{1506.2374 \times 10^{-5}}} = 0.2192\ \text{mm}^{-1}$$

in the case of CGA. The ratio of these parameters of the interfacial shearing stress:

$$\eta = \frac{k}{K} = \frac{0.3941}{0.8294} = 0.4752$$

in the case of BGA, and

$$\eta = \frac{k}{K} = \frac{0.2192}{0.6278} = 0.3492$$

in the case of CGA. The force at the boundary is

$$\hat{T} = -2\frac{\Delta\alpha\Delta t_m}{\lambda_*}\frac{t(\eta+\delta t)}{1+t^2+2t\eta} = -2\times 2.3490\frac{0.3749(0.4752+0.5\times 0.3749)}{1.1405+0.3563} = -0.7797\,\text{kg/mm}$$

in the case of BGA, and

$$\hat{T} = -2\frac{\Delta\alpha\Delta t_m}{\lambda_*}\frac{t(\eta+\delta t)}{1+t^2+2t\eta} = -2\times 2.3490\frac{0.2158(0.3492+0.5\times 0.2158)}{1.0465+0.1507} = -0.3871\,\text{kg/mm}$$

in the case of CGA. The interfacial shearing stress in the midportion of the assembly at its boundary with the peripheral portion is

$$\tau_m(L-2l) = K\frac{\Delta\alpha\Delta t_m}{\lambda_*}\frac{\coth 2kl+\delta\tanh kl}{\eta+\coth 2kl} = K\frac{\Delta\alpha\Delta t_m}{\lambda_*}\frac{1+(1+2\delta)t^2}{1+t^2+2t\eta} =$$
$$= 0.8294\times 2.3490\frac{1+(1+2\times 0.5)\times 0.3749^2}{1+0.3749^2+2\times 0.3749\times 0.4752} = 1.6674\,\text{kg/mm}^2$$

in the case if BGA and

$$\tau_m(L-2l) = K\frac{\Delta\alpha\Delta t_m}{\lambda_*}\frac{\coth 2kl+\delta\tanh kl}{\eta+\coth 2kl} = K\frac{\Delta\alpha\Delta t_m}{\lambda_*}\frac{1+(1+2\delta)t^2}{1+t^2+2t\eta} =$$
$$= 0.6278\times 2.3490\frac{1+(1+2\times 0.5)\times 0.2158^2}{1+0.2158^2+2\times 0.2158\times 0.3492} = 1.3464\,\text{kg/mm}^2$$

in the case of CGA. The interfacial shearing stress in the peripheral portion of the assembly at its boundary with the midportion is

$$\tau_p(-l) = -k\frac{\Delta\alpha\Delta t_m}{\lambda_*}\frac{\delta\tanh kl(\eta+\coth 2kl-1)-\eta}{\eta+\coth 2kl}$$
$$= -k\frac{\Delta\alpha\Delta t_m}{\lambda_*}t\frac{\delta[1+t^2-2(1-\eta)t]-2\eta}{1+t^2+2t\eta} =$$
$$= -0.3941\times 2.3490\times 0.3749\times\frac{0.5[1+0.3749^2-2(1-0.4752)\times 0.3749]-2\times 0.4752}{1+0.3749^2+2\times 0.3749\times 0.4752}$$
$$= 0.1338\,\text{kg/mm}^2,$$

in the case if a BGA and

$$\tau_p(-l) = -k\frac{\Delta\alpha\Delta t_m}{\lambda_*}\frac{\delta\tanh kl(\eta + \coth 2kl - 1) - \eta}{\eta + \coth 2kl}$$

$$= -k\frac{\Delta\alpha\Delta t_m}{\lambda_*}t\frac{\delta[1 + t^2 - 2(1-\eta)t] - 2\eta}{1 + t^2 + 2t\eta} =$$

$$= -0.2192 \times 2.3490 \times 0.2158$$

$$\times\frac{0.5[1 + 0.2158^2 - 2(1 - 0.3492) \times 0.2158] - 2 \times 0.3492}{1.0466 + 2 \times 0.2158 \times 0.3492}$$

$$= 0.0293\,\text{kg/mm}^2$$

in the case of a CGA. The stress at the assembly end is

$$\tau_p(l) = k\frac{\Delta\alpha\Delta t_m}{\lambda_*}\frac{\delta\tanh kl(\eta\sinh 2kl + \cosh 2kl + 1) + \eta}{\eta\sinh 2kl + \cosh 2kl}$$

$$= k\frac{\Delta\alpha\Delta t_m}{\lambda_*}\frac{2\delta t(1 + t\eta) + (1 - t^2)\eta}{1 + t^2 + 2t\eta} =$$

$$= 0.3941 \times 2.3490\frac{2 \times 0.5 \times 0.3749 \times (1 + 0.3749 \times 0.4752) + (1 - 0.3749^2) \times 0.4752}{1 + 0.3749^2 + 2 \times 0.3749 \times 0.4752}$$

$$= 0.5258\,\text{kg/mm}^2$$

in the case of BGA and

$$\tau_p(l) = k\frac{\Delta\alpha\Delta t_m}{\lambda_*}\frac{\delta\tanh kl(\eta\sinh 2kl + \cosh 2kl + 1) + \eta}{\eta\sinh 2kl + \cosh 2kl}$$

$$= k\frac{\Delta\alpha\Delta t_m}{\lambda_*}\frac{2\delta t(1 + t\eta) + (1 - t^2)\eta}{1 + t^2 + 2t\eta} =$$

$$= 0.2192 \times 2.3490\frac{2 \times 0.5 \times 0.2158(1 + 0.2158 \times 0.3492) + (1 - 0.2158^2) \times 0.3492}{1 + 0.2158^2 + 2 \times 0.2158 \times 0.3492}$$

$$= 0.2430\,\text{kg/mm}^2$$

in the case of CGA.

The calculated data are summarized in the following table:

Solder System	BGA	CGA
Stress in the midportion at its boundary with the peripheral portion, kg/mm^2	1.6674	1.3464
Stress in the peripheral portion at its boundary with the midportion, kg/mm^2	0.1338	0.0293
Stress at the assembly end, kg/mm^2	0.5258	0.2430

When a homogeneous bonding layer with the characteristics of the midportion in the carried out example were applied, the maximum interfacial shearing stress would be

$$\tau_{\max} = k_* \frac{\Delta\alpha\Delta t}{\lambda_*} = 0.8294 \times 2.3490 = 1.9483\,\text{kg/mm}^2$$

in the case of BGA, and

$$\tau_{\max} = k_* \frac{\Delta\alpha\Delta t}{\lambda_*} = 0.6278 \times 2.3490 = 1.4747\,\text{kg/mm}^2$$

in the case of CGA. Thus, the application of the epoxy adhesive at the assembly ends resulted in this example in 14.42% stress reduction in the case of BGA and in 8.70% reduction in the case of CGA.

6.6 OPTIMIZED DESIGN

6.6.1 Optimization Condition

Let us define an optimized design as the one in which the interfacial shearing stress at the ends of the midportion at its boundaries with the peripheral portions is equal to the stress at the assembly ends. It is anticipated that by doing that, a significant stress relief could be obtained. By equating expressions (6.39) and (6.41), we obtain the following condition of optimization that could be written in the form of a quadratic equation for the t value:

$$[1+\eta^2+2\delta(1-\eta^2)]t^2 - 2\delta\eta t + 1 - \eta^2 = 0. \tag{6.42}$$

This equation has the following solution:

$$t = \frac{\delta\eta}{1+\eta^2+2\delta(1-\eta^2)}\left[1+\sqrt{1-(1-\eta^2)\frac{1+\eta^2+2\delta(1-\eta^2)}{\delta^2\eta^2}}\right] \tag{6.43}$$

After the t value is determined, the lengths of the peripheral zones can be evaluated as

$$2l = -\frac{1}{k}\ln\frac{1-t}{1+t} \tag{6.44}$$

6.6.2 Peripheral Material with a Low Fabrication Temperature

For an assembly in which the parameter of the interfacial stress is the same for the midportion and the peripheral portions ($\eta = 1$), the condition (6.42) yields ($\delta = t$). This results in the following interfacial shearing stresses:

$$\tau_m(L-2l) = \tau_p(l) = k_* \frac{\Delta\alpha\Delta t_m}{\lambda_*}\frac{1-\delta+2\delta^2}{1+\delta},$$
$$\tau_p(-l) = -k_* \frac{\Delta\alpha\Delta t_m}{\lambda_*}\delta\frac{1-\delta+2\delta^2}{1+\delta}. \tag{6.45}$$

The function $f(\delta) = \frac{1-\delta+2\delta^2}{1+\delta}$ has its minimum $f_{\min} = \frac{8}{\sqrt{2}} - 5 = 0.6569$ at $\delta = \sqrt{2} - 1 = 0.4142$. Then, as follows from formulas (6.45), the shearing stresses at the ends of the midportion and the peripheral portions are

$$\tau_m(L-2l) = \tau_p(l) = 0.6569K\frac{\Delta\alpha\Delta t_m}{\lambda_*}. \tag{6.46}$$

Hence, stress relief as significant as 34.3% can be obtained in this case.

6.6.3 Peripheral Material with a Low Parameter of the Interfacial Shearing Stress

For an assembly in which only the difference in the parameters of the interfacial shearing stresses is considered ($\delta = t$), the condition (6.43) yields

$$t = \frac{\eta}{3-\eta^2}\left[1+\sqrt{1-(1-\eta^2)\frac{3-\eta^2}{\eta^2}}\right] \tag{6.47}$$

This equation and, hence, requirement (6.42) cannot be fulfilled for η values below $\eta = 0.8350$. The relationship (6.48) is tabulated for this and higher η values in Table 6.1:

TABLE 6.1

η	0.8350	0.8500	0.9000	0.9500	1.0000
$t = \tanh kl$	0.3626	0.5053	0.6975	0.8512	1.0000
$\bar{\tau}_m(L-2l) = \eta\bar{\tau}_p(l)$	0.8028	0.8352	0.8970	0.9496	1.0000
$\bar{\tau}_p(l)$	0.9614	0.9826	0.9967	0.9997	1.0000

The calculated values

$$\bar{\tau}_m(L-2l) = \frac{\tau_m(L-2l)}{K\dfrac{\Delta\alpha\Delta t}{\lambda_*}} = \frac{1+3t^2}{1+t^2+2t\eta},$$
$$\bar{\tau}_p(l) = \frac{\tau_p(l)}{k\dfrac{\Delta\alpha\Delta t}{\lambda_*}} = \frac{2t+\eta(1+t^2)}{1+t^2+2t\eta}, \tag{6.48}$$

of the dimensionless stresses at the end of the midportion and at the end of the assembly are also shown in this table. Clearly, $\bar{\tau}_m(L-2l) = \eta\bar{\tau}_p(l)$. The table data show particularly that the length of the peripheral area should be small enough so that the induced stresses become sufficiently low. The data for the $\bar{\tau}_m(L-2l)$ stresses indicate that depending on the η ratio, the stress relief could be as high as 19.7%. Note that this relief, although significant, is lower than in the case of a peripheral material with a low fabrication temperature.

6.6.4 Peripheral Material with a Low Parameter of the Interfacial Shearing Stress and Low Fabrication Temperature

From (6.42), we have

$$2\delta = \frac{1-\eta^2+(1+\eta^2)t^2}{t[\eta-(1-\eta^2)t]}. \tag{6.49}$$

Then, formula (6.39) yields

$$\bar{\tau}_m = \frac{\tau_m(L-2l)}{K\dfrac{\Delta\alpha\Delta t_m}{\lambda_*}} = \frac{\coth 2kl+\delta\tanh kl}{\eta+\coth 2kl} = \frac{1+(1+2\delta)t^2}{1+t^2+2t\eta} = \eta\frac{1+t^2+2\eta t^3}{(1+t^2+2t\eta)[\eta-(1-\eta^2)t]}. \tag{6.50}$$

For assemblies, in which the peripheral portions are characterized with the same parameter of the interfacial stress as the assembly midportion ($\eta = 1$), formula (6.50) yields

$$\bar{\tau}_m = 1-2t\frac{1-t^2}{(1+t)^2} \tag{6.51}$$

For an assembly with very short peripheral areas ($t = 0$) or long enough peripheral areas ($kl \geq 2.5$), when $t = \tanh kl$ can be put equal to one, formula (6.51) leads to the same result: $\bar{\tau}_m = 1$, which means that no relief in the interfacial stress could be expected. But what about the intermediate lengths?

Equation (6.50) and the condition $\frac{d\bar{\tau}_m}{dt} = 0$ lead to the following equation for the sought t value that minimizes the stress (6.50):

$$(1-3\eta^2+4\eta^4)t^4 - 4\eta(1-3\eta^2)t^3 + 2(1+3\eta^2)t^2 + 4\eta(1-\eta^2)t + 1 - 3\eta^2 = 0. \quad (6.52)$$

In the extreme case $\eta = 1$, this condition yields

$$t^4 + 4t^3 + 4t^2 - 1 = 0. \quad (6.53)$$

Its solution is $t = \sqrt{2} - 1 = 0.4142$. Then, formula (51) yields $\bar{\tau}_m = 0.6569$, so stress relief as high as 34.3% can be expected in this case.

The general minimization condition (6.52) could and should be used to determine the optimum t value for the given (accepted) η ratio. Then, the appropriate δ ratio could be determined on the basis of formula (6.49). It should be pointed out that one cannot simplify formula (6.50) first and then use the minimization condition $\frac{d\bar{\tau}_m}{dt} = 0$ to determine the corresponding t value. Using as an example the simplified relationship (6.51), let us show that this approach is erroneous. Applying the condition $\frac{d\bar{\tau}_m}{dt} = 0$ to formula (6.51), we obtain the following equation for the t value: $t^2 - \frac{2}{5}t - \frac{1}{5} = 0$. Than we have: $t = \frac{\sqrt{6}+1}{5} = 0.6899$. This result is quite different of the result $t = 0.4142$ obtained using the general stress minimization condition (6.52), and so is the stress relief of $\bar{\tau}_m = 0.8525$ predicted by formula (6.51) for the $t = 0.6899$ value. This relief is only 14.7%, while the actual stress relief predicted based on the general stress minimization condition (6.52) is as high as 34.3%.

As another suitable example, let us consider a hypothetical case of $\eta = 0.8350$. Then, the condition (6.52) yields

$$f(t) = t^4 + 4.2756t^3 + 7.2507t^2 + 1.1859t - 1.2801 = 0.$$

This equation has the following solution: $t = 0.3210$, and formulas (6.49) and (6.50) yield $\delta \approx 1$ and $\bar{\tau}_m = 0.7987$. Hence, stress relief of about 20.1% is expected in this case.

6.7 CONCLUSIONS

The following conclusions can be drawn from the carried out analysis:

- Simple and physically meaningful predictive analytical models are developed for the assessment of thermal stresses in a BGA or in a CGA with a low-modulus solder material at the peripheral portions of the assembly.

- It is shown that significant thermal stress relief can be achieved if the bonding system is designed in such a way that the interfacial shearing stress at the ends of the high-modulus and high-bonding temperature midportion of the assembly at its boundary with the low-modulus and low-bonding temperature peripheral portion is made equal to the stress at the assembly ends. It is shown that stress relief as high as 34.3% can be obtained in the optimized design.
- If such a design is employed, there is a possibility that no inelastic stresses and strains in the solder joints will occur, and the fatigue life of the vulnerable solder material will be dramatically increased.
- Further work will include systematic computations to evaluate the role of the material and geometric characteristics of the design on the level of the induced stresses; FEAs and experimental investigations.

REFERENCES

1. E. Suhir, "'Global' and 'Local' Thermal Mismatch Stresses in an Elongated Bi-material Assembly Bonded at the Ends," in E. Suhir, (ed.), *Structural Analysis in Microelectronic and Fiber-Optic Systems*, symposium proceedings, ASME Press, 1995.
2. E. Suhir, "Predicted Thermal Mismatch Stresses in a Cylindrical Bi-Material Assembly Adhesively Bonded at the Ends," *ASME Journal of Applied Mechanics*, vol. 64, No. 1, 1997.
3. E. Suhir, "Thermal Stress in a Polymer Coated Optical Glass Fiber with a Low Modulus Coating at the Ends," *Journal of Material Research*, vol. 16, No. 10, 2001.
4. E. Suhir, "Thermal Stress in a Bi-Material Assembly Adhesively Bonded at the Ends," *Journal of Applied Physics*, vol. 89, No. 1, 2001.
5. E. Suhir, "Thermal Stress in an Adhesively Bonded Joint with a Low Modulus Adhesive Layer at the Ends," *Journal of Applied Physics*, vol. 55, 2003.
6. E. Suhir, A. Shakouri, "Assembly Bonded at the Ends: Could Thinner and Longer Legs Result in a Lower Thermal Stress in a Thermoelectric Module (TEM) Design?" *ASME Journal of Applied Mechanics*, vol. 79, No. 6, 2012.
7. E. Suhir, "Predicted Stresses in a Ball-Grid-Array (BGA)/Column-Grid-Array (CGA) Assembly with a Low Modulus Solder at Its Ends," *Journal of Materials Science: Materials in Electronics*, vol. 26, No. 12, 2015.
8. E. Suhir, "Expected Stress Relief in a Bi-Material Inhomogeneously Bonded Assembly with a Low-Modulus-and/or-Low-Fabrication-Temperature Bonding Material at the Ends," *Journal of Materials Science: Materials in Electronics*, vol. 27, No. 6, 2016.
9. E. Suhir, R. Ghaffarian, "Predicted Stresses in a Ball-Grid-Array (BGA)/Column-Grid-Array (CGA) Assembly with an Epoxy Adhesive at Its Ends," *Journal of Materials Science: Materials in Electronics*, vol. 27, No. 5, 2016.
10. E. Suhir, "Bi-Material Assembly with a Low-Modulus-and/or-Low-Fabrication-Temperature Bonding Material at Its Ends: Optimized Stress Relief," *Journal of Materials Science: Materials in Electronics*, vol. 27, No. 5, 2016.
11. E. Suhir, "Interfacial Thermal Stresses in a Bi-Material Assembly with a Low-Yield-Stress Bonding Layer," *Modeling and Simulation in Materials Science and Engineering*, vol. 14, 2006.
12. E. Suhir, L. Bechou, B. Levrier, "Predicted Size of an Inelastic Zone in a Ball-Grid-Array Assembly," *ASME Journal of Applied Mechanics*, vol. 80, 2013.

13. E. Suhir, "On a Paradoxical Situation Related to Bonded Joints: Could Stiffer Mid-Portions of a Compliant Attachment Result in Lower Thermal Stresses?" *Journal of Solid Mechanics and Materials Engineering (JSMME)*, vol. 3, No. 7, 2009.
14. E. Suhir, "Analysis of a Short Beam with Application to Solder Joints: Could Larger Stand-off Heights Relieve Stress?" *European Physical Journal of Applied Physics (EPJAP)*, vol. 71, 2015.
15. E. Suhir, R. Ghaffarian, J. Nicolics, "Could Application of Column-Grid-Array Technology Result in Inelastic-Strain-Free State-of-Stress in Solder Material?" *Journal of Materials Science: Materials in Electronics*, Sept. 2015.
16. E. Suhir, "Avoiding Low-Cycle Fatigue in Solder Material Using Inhomogeneous Column-Grid-Array (CGA) Design," ChipScale Reviews, March–April 2016.
17. E. Suhir, R. Ghaffarian, J. Nicolics, "Predicted Stresses in Ball-Grid-Array (BGA) and Column-Grid-Array (CGA) Interconnections in a Mirror-like Package Design," *Journal of Materials Science: Materials in Electronics*, vol. 27, No. 3, 2016.
18. E. Suhir, "Adhesively Bonded Assemblies with Identical Nondeformable Adherends and Inhomogeneous Adhesive Layer: Predicted Thermal Stresses in the Adhesive," *Journal of Reinforced Plastic Composites*, vol. 17, No. 14, 1998.
19. E. Suhir, "Adhesively Bonded Assemblies with Identical Nondeformable Adherends and 'Piecewise Continuous' Adhesive Layer: Predicted Thermal Stresses and Displacements in the Adhesive," *International Journal of Solids and Structures*, vol. 37, 2000.
20. E. Suhir, "Thermal Stress in a Bi-Material Assembly with a ''Piecewise-Continuous'' Bonding Layer: Theorem of Three Axial Forces," *Journal of Applied Physics D*, vol. 42, 2009.
21. E. Suhir, "Stresses in Bi-Metal Thermostats," *ASME Journal of Applied Mechanics*, vol. 53, No. 3, 1986.
22. E. Suhir, R. Ghaffarian, J. Nicolics, "Could Thermal Stresses in an Inhomogeneous BGA/CGA System be Predicted Using a Model for a Homogeneously Bonded Assembly?" *Journal of Materials Science: Materials in Electronics*, vol. 27, No. 1, 2016.
23. E. Suhir, R. Ghaffarian, "Reliability Physics Behind the QFN State of Stress," *Journal of Materials Science: Materials in Electronics*, vol. 28, No. 2, 2017.
24. E. Suhir, "Solder Joint Interconnections in Automotive Electronics: Design-for-Reliability and Accelerated Testing," SIITME Proceedings, Jesse, Romania, 2018.

7 Thermal Stresses in a Flip-Chip Design

"The truth is rarely pure and never simple."

—Oscar Wilde, Irish dramatist and novelist

7.1 BACKGROUND/INCENTIVE

Flip-chip (FC) technology (Figure 7.1), suggested about 50 years ago by the General Electric Light Military Electronics Dept. in Utica, New York, rapidly became widespread in the electronic industry owing to its ability to minimize the real estate occupied by an IC package. FC solder joints provide, in addition to electrical connection, mechanical support for the chips. This makes FC designs prone to physical failures. Various measures, including "dummy" solder joints (i.e., joints not needed to provide electrical connections) were suggested over the years to improve the situation. Particularly, a variety of encapsulation ("underfill") technologies were investigated and applied. It has been established that FC and fine-pitch ball-grid array (FPBGA) package designs are an attractive way to pursue as far as small-size and low-cost of IC packages are concerned. There are, however, situations when the acceptable reliability level in FC and FPBGA technologies cannot be achieved without bringing in surrogate materials, s.c. underfills, between the chip and its substrate to strengthen the solder joint interconnections. Underfill technology was suggested, among other possible encapsulation technologies, about three decades ago [1], but has become the technology of choice in today's IC packaging. Solder material fatigue failures caused by the inelastic thermal strains in the peripheral portions of the solder joint interconnections and delaminations (typically at the chip–solder interface) were observed and established as the main failure modes in FC and FPBGA structures with underfills. It is nonetheless expected that the appropriately chosen underfill material and the adequate application technology will improve the situation. Of course, one should determine, first of all, if it would be possible to get away with an unencapsulated design ("to underfill or not to underfill: that's the question?") and still meet the reliability requirements. If, however, an underfill decision is made, one should select the adequate underfill material and technology, with consideration of numerous factors associated with the application of this technology: encapsulation process, time-to-market (completion), cost, manufacturability, testability, reparability, and so on. The published experimental and modeling work on the mechanical performance and reliability of FC designs is enormous (see, for example, [1–31]). As to the modeling work, it was conducted by mostly using finite element analysis (FEA). It is, however,

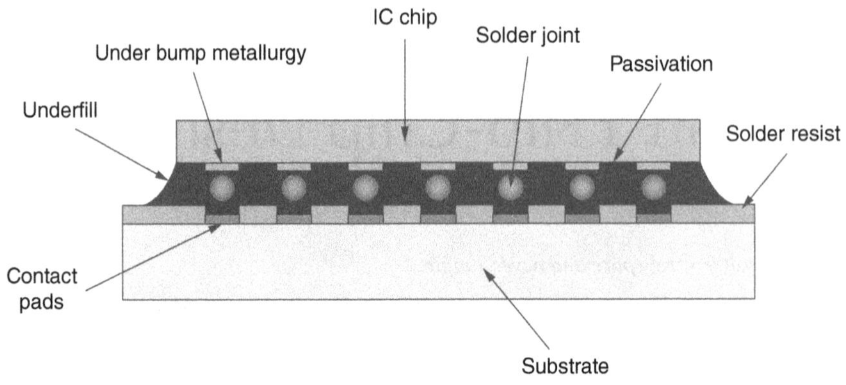

FIGURE 7.1 Flip-chip (FC) assembly.

the analytical technique (see Chapter 1) based on the concept of the interfacial compliance (see Chapter 2) that is employed in this chapter analyses.

It has recently been shown [32–33] that a real opportunity exists for bringing down and, in many cases, even avoiding inelastic strains in the second-level (package-to-printed circuit board) solder joint interconnections in IC packages. This could be done by employing low expansion (such as, say, ceramic) substrate, instead of a printed circuit board (PCB), and/or by increasing the compliance of the solder joint system by employing joints with elevated stand-offs (such as column-grid arrays). Appreciable stress relief could also be achieved, in addition to other measures, by implementing inhomogeneous solder joint systems, in which the solder material at the peripheral portions of the assembly has a lower modulus and/or a higher yield stress (see Chapter 6), and/or a lower melting temperature, than the solder material in its midportion, or by using an epoxy adhesive at the assembly corners. If such efforts turn out to be successful to an extent that the inelastic strains and, hence, the low cycle fatigue conditions in solder material are avoided, its fatigue lifetime will improve dramatically: elastic strains are much less damaging than inelastic strains. If the material performs within the elastic range, the fatigue lifetime could be assessed assuming linear accumulation of fatigue damages. This could be done by using the well-known Palmgren-Miner rule (see, for example, [34, 35]), rather than one of the numerous empirical Coffin–Manson models. But what about the first-level (chip-to-substrate) interconnections and particularly FC designs including encapsulated ones [1–31]? What could possibly be done to relieve the thermal stresses in FC solder joints? Could this be achieved to an extent that inelastic strains in solder material are avoided, or, if this is impossible, could at least the sizes of the inelastic peripheral zones of the soldered assemblies be predicted and minimized? This will certainly increase the lifetime of the interconnection.

Some results obtained in connection with these questions are set forth in this chapter. The emphasis is on the quantitative assessment of the possible stress relief. Particularly, we intend to find out to what extent a bimaterial predictive model could be employed in the analysis of FC designs [36–39]. We start with an analytical thermal stress model for a typical FC package design.

7.2 THERMAL STRESS MODEL FOR A TYPICAL FLIP-CHIP PACKAGE DESIGN

7.2.1 Approach

The addressed design consists of a silicon FC bonded to an organic substrate and covered by a lid (Figure 7.2). Organic or copper lids are considered. The lid is configured in such a way that its midportion is bonded to the back side of the chip using a thermal interface material (a heat sink is intended to be subsequently mounted on the outer surface of the lid) and its peripheral portions are adhesively bonded to the same substrate using compliant attachments (such as solder) around the lid's perimeter. The in-plane compliances of all the attachments, including the effective compliance of the encapsulated solder joint interconnections, are taken into account. The effective mechanical properties [coefficient of thermal expansion (CTE) and elastic constants] of the encapsulated FC bumps are taken into account by averaging these properties for this composite material.

The following approach (sequence of actions) is employed in the analysis. First, the midportion (Figure 7.3) of the design shown in Figure 7.2 is considered. This portion is a trimaterial assembly. It consists of the FC (zero component), the midportion of the substrate (component #1) located below the chip, and the midportion of the lid (component #2) located above the chip. The assembly is subjected to the combined action of the internal thermally induced forces caused by the dissimilar materials of the assembly (the chip, the substrate and the lid), as well as by the "external" forces

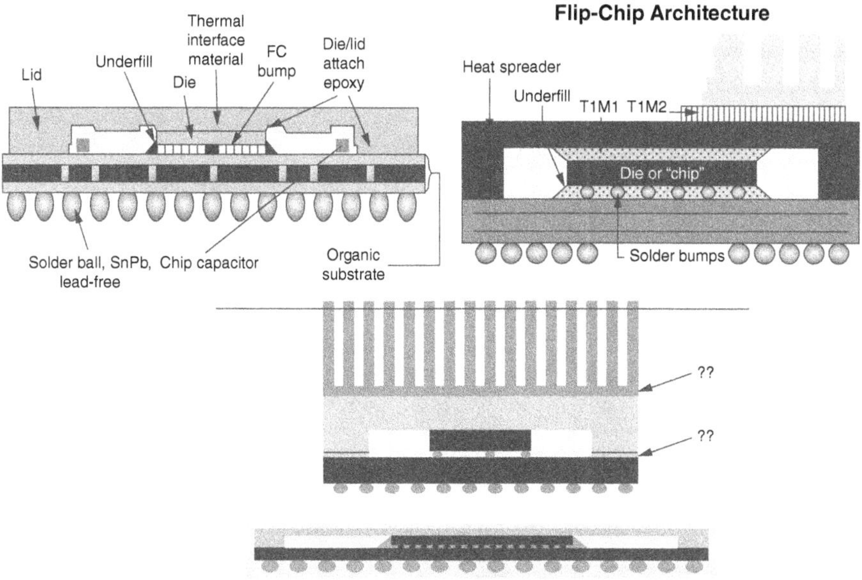

FIGURE 7.2 Flip-chip design with a heat-sink mounted on the outer surface of the lid. (Top right from https://alsicthermalmanagement.blogspot.com/2010/05/alsic-for-microprocessors.html.)

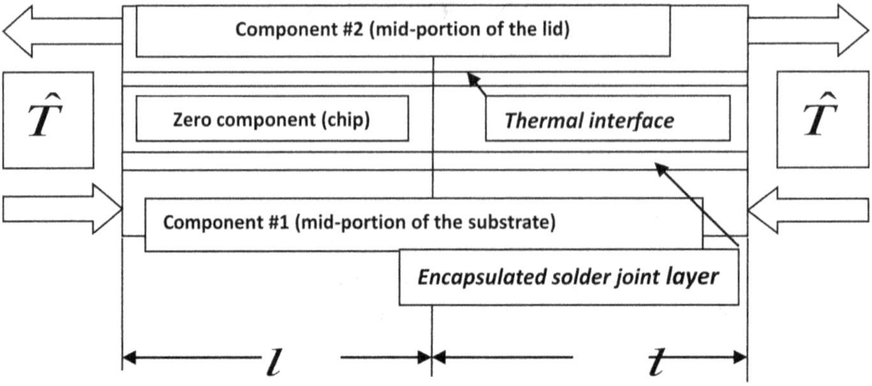

FIGURE 7.3 Midportion of the design.

$\hat{T}$ applied to this assembly from the peripheral portions of the FC design. These peripheral portions (Figure 7.4) consist of the thick peripheral portions of the lid and the peripheral portions of the substrate. The forces $\hat{T}$, "external" with respect to the assembly under consideration, but internal for the package as a whole, should be, of course, the same for the substrate and the lid components of the trimaterial assembly located in the midportion of the design, and for the substrate and the lid components of the bimaterial assemblies located at the design's peripheral portions. The magnitude of these forces could be determined based on the condition of the compatibility of the displacements at the extreme cross sections of the design's midportion and its peripheral portions.

It should be pointed out that the assembly in the midportion of the design is identified here as a trimaterial assembly, although it is comprised, in effect, of five materials: in addition to the three major components, it contains also materials of two relatively thin layers: encapsulated (by using underfill material) solder joint layer and thermal interface layer. These two layers are assumed, however, to be thin and/or low modulus, so that they do not generate thermally induced forces acting in their cross sections. These layers experience thermally induced interfacial stresses only, and affect the magnitude and the distribution of these stresses, but not the next-to-zero normal stresses acting in the bonding layer cross sections. This is also true as far as the assemblies at the peripheral portions of the FC design are concerned: the adhesive bonds in them do not experience tension or compression, but are subjected to

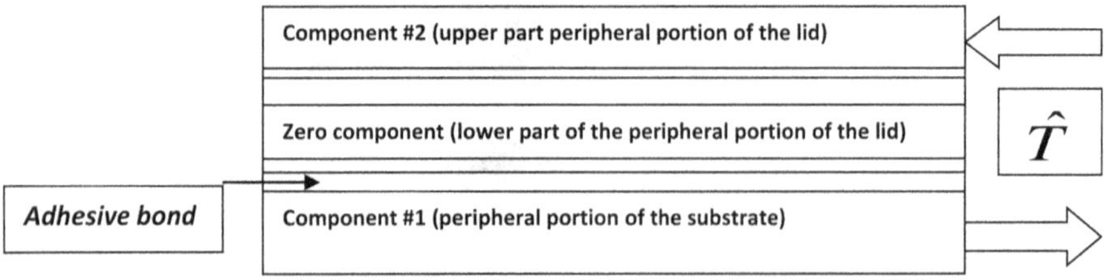

FIGURE 7.4 Peripheral portion of the design.

shearing deformations only, and do not have to be considered when the equilibrium conditions for the axial thermally induced forces are addressed.

7.2.2 Forces Acting in the Midportion of the Assembly Located at the Design's Midportion

The thermally induced forces, $T_i^*, i = 0,1,2$, acting in the midportion of the assembly in Figure 7.3, can be determined from the strain compatibility conditions

$$-\alpha_0 \Delta t + \lambda_0 T_0^* = -\alpha_1 \Delta t + \lambda_1 T_1^* = -\alpha_2 \Delta t + \lambda_2 T_2^* \tag{7.1}$$

and the equilibrium condition

$$T_0^* + T_1^* + T_2^* = 0 \tag{7.2}$$

for the induced forces as follows:

$$\begin{aligned} T_0^* &= [(\alpha_0 - \alpha_1)\lambda_2 + (\alpha_0 - \alpha_2)\lambda_1]\frac{\Delta t}{D}; \\ T_1^* &= [(\alpha_1 - \alpha_0)\lambda_2 + (\alpha_1 - \alpha_2)\lambda_0]\frac{\Delta t}{D}; \\ T_2^* &= [(\alpha_2 - \alpha_1)\lambda_0 + (\alpha_2 - \alpha_0)\lambda_1]\frac{\Delta t}{D}; \end{aligned} \tag{7.3}$$

Here,

$$D = \lambda_0\lambda_1 + \lambda_1\lambda_2 + \lambda_2\lambda_0, \tag{7.4}$$

is the determinant of the system of equations (7.3) for the sought forces; Δt is the change in temperature from the manufacturing (high) temperature to the operation (low) temperature; $\alpha_i, i = 0,1,2$, are the effective (temperature independent) CTEs of the component materials;

$$\lambda_i = \frac{1 - v_1}{E_i h_i}, i = 0,1,2 \tag{7.5}$$

are the axial compliances of the components; E_i and v_i are effective elastic constants of the materials; and h_i are their thicknesses.

The thermal interfacial stresses are zero in the regions where the forces acting in the assembly component cross sections are constant. These forces start to change at the end portions of the assembly, and the interfacial shearing stresses at these portions depend not only on the variable forces acting in the cross sections of the assembly's peripheral portions, but also on the parameter of the interfacial stresses. To determine this parameter for the midportion of the assembly, let us consider the longitudinal interfacial displacements of the assembly components in Figure 7.3.

These displacements can be evaluated, using the concept of the interfacial compliance, by the approximate formulas:

$$u_{10}(x) = -\alpha_1 \Delta t x + \lambda_1 \int_0^x T_1(\xi)d\xi - \kappa_1\tau_1(x), \quad u_{01}(x) = -\alpha_0 \Delta t x + \lambda_0 \int_0^x T_0(\xi)d\xi + \kappa_0\tau_1(x),$$

$$u_{20}(x) = -\alpha_2 \Delta t x + \lambda_2 \int_0^x T_2(\xi)d\xi - \kappa_2\tau_2(x), \quad u_{02}(x) = -\alpha_0 \Delta t x + \lambda_0 \int_0^x T_0(\xi)d\xi + \kappa_0\tau_2(x). \tag{7.6}$$

In these formulas, $u_{10}(x)$ are the longitudinal interfacial displacements of component #1 (substrate) at its interface with the zero component (chip), $u_{01}(x)$ are the longitudinal interfacial displacements of the zero component (chip) at its interface with component #1 (substrate), $u_{20}(x)$ are the longitudinal interfacial displacements of component #2 (lid) at its interface with the zero component (chip), $u_{02}(x)$ are the longitudinal interfacial displacements of the zero component (chip) at its interface with component #2 (lid),

$$\kappa_0 = \frac{h_0}{G_0} + \frac{h_{01}}{G_{01}} + \frac{h_{02}}{G_{02}}, \quad \kappa_1 = \frac{h_1}{3G_1}, \quad \kappa_2 = \frac{h_2}{3G_2}, \tag{7.7}$$

are the longitudinal interfacial compliances of the assembly components, $\frac{h_{01}}{G_{01}}$ is the longitudinal interfacial compliance of the thermal interface layer, $\frac{h_{02}}{G_{02}}$ is the effective longitudinal interfacial compliance of the encapsulated solder joint layer,

$$G_i = \frac{E_i}{2(1+\nu_i)}, \; i = 0,1,2, \tag{7.8}$$

are the shear moduli of the component materials, $\tau_1(x)$ and $\tau_2(x)$ are the thus far unknown shearing stresses at the interfaces of the zero component (chip) with components #1 (substrate) and #2 (lid), respectively,

$$T_1(x) = \int_{-l}^{x} \tau_1(\xi)d\xi - \hat{T}, \tag{7.9}$$

is the force acting in the cross section of component #1 (substrate), and

$$T_2(x) = \int_{-l}^{x} \tau_2(\xi)d\xi + \hat{T}, \tag{7.10}$$

is the force acting in the cross section x of component #2 (substrate). Here, l is half of the assembly length. The origin of the coordinate x is in the mid-cross section of the assembly. The first terms in equations (7.6) are unrestricted (stress-free) thermal displacements of the components of the assembly fabricated at an elevated temperature and subsequently cooled down to a low (room, testing, or operation) temperature. The second terms are the displacements caused by the forces due to both thermal and mechanical ("external") loading. These terms were obtained using Hooke's law under an assumption that all the longitudinal displacements are the same for the given cross section. The last terms are corrections to this assumption. They consider that the interfacial displacements are somewhat larger than the displacements of the inner points of the given cross section. In accordance with the concept of the interfacial compliance, it is assumed that such deviations from planarity could be evaluated as the product of the location independent interfacial compliance of the given component, evaluated by formulas (7.7), and the thus far unknown distributed interfacial shearing stress acting in this cross section. It is also assumed that the states of stress in the adjacent cross sections do not affect the state of stress in the given cross section.

Formulas (7.7) were obtained using the theory-of-elasticity solution (Ribière solution) for a long and narrow strip [36–39]. The first formula in (7.7) was obtained for a strip loaded in an antisymmetric fashion along both longitudinal edges, while the second and the third formulas were obtained for a strip loaded just on one long edge. This explains the difference in the denominators in these formulas: while the zero component of the assembly experiences loading along both its longitudinal edges, components #1 and #2 are loaded only along their inner strips and experience, for the same shear loading, lower longitudinal displacements and, because of that, are less flexible.

Clearly,

$$T_1(l) = \int_{-l}^{l} \tau_1(\xi)d\xi = -\hat{T}, \quad T_2(l) = \int_{-l}^{l} \tau_1(\xi)d\xi = 0. \tag{7.11}$$

The force $T_0(x)$ acting in the cross sections of the zero component can be found from the condition of equilibrium

$$T_0(x) = -\hat{T} - T_1(x) - T_2(x). \tag{7.12}$$

The conditions

$$u_{10}(x) = u_{01}(x), \quad u_{20}(x) = u_{02}(x). \tag{7.13}$$

of the displacements compatibility and equation (7.12) lead to the following equations for the interfacial shearing stresses $\tau_1(x)$ and $\tau_2(x)$:

$$\begin{aligned}
&(\kappa_0 + \kappa_1)\tau_1(x) - (\lambda_0 + \lambda_1)\int_0^x T_1(\xi)d\xi - \lambda_0 \int_0^x T_2(\xi)d\xi = (\alpha_0 - \alpha_1)\Delta tx + \lambda_0 \hat{T}x \\
&(\kappa_0 + \kappa_2)\tau_2(x) - (\lambda_0 + \lambda_2)\int_0^x T_2(\xi)d\xi - \lambda_0 \int_0^x T_1(\xi)d\xi = (\alpha_0 - \alpha_2)\Delta tx + \lambda_0 \hat{T}x
\end{aligned} \tag{7.14}$$

By differentiating, we find:

$$(\kappa_0+\kappa_1)\tau_1'(x)-(\lambda_0+\lambda_1)T_1(x)-\lambda_0 T_2(x)=(\alpha_0-\alpha_1)\Delta t+\lambda_0\hat{T}$$
$$(\kappa_0+\kappa_2)\tau_2'(x)-(\lambda_0+\lambda_2)T_2(x)-\lambda_0 T_1(x)=(\alpha_0-\alpha_2)\Delta t+\lambda_0\hat{T} \tag{7.15}$$

The next differentiation yields:

$$(\kappa_0+\kappa_1)\tau_1''(x)-(\lambda_0+\lambda_1)\tau_1(x)-\lambda_0\tau_2(x)=0,$$
$$(\kappa_0+\kappa_2)\tau_2''(x)-(\lambda_0+\lambda_2)\tau_2(x)-\lambda_0\tau_1(x)=0 \tag{7.16}$$

Because of the symmetry of the structure with respect to its mid-cross section, the interfacial thermal stresses are antisymmetric with respect to this cross section and can be sought in the form

$$\tau_1(x)=C_1\sinh kx,\quad \tau_2(x)=C_2\sinh kx, \tag{7.17}$$

where k is the thus far unknown parameter of the interfacial shearing stress. Introducing solutions (7.17) into equations (7.16) and requiring that the determinant of the obtained algebraic equations

$$\begin{aligned}&[(\kappa_0+\kappa_1)k^2-(\lambda_0+\lambda_1)]C_1-\lambda_0 C_2=0,\\ &-\lambda_0 C_1+[(\kappa_0+\kappa_2)k^2-(\lambda_0+\lambda_2)C_2=0\end{aligned} \tag{7.18}$$

is zero (otherwise the constants C_1 and C_2 will also be zero), the following biquadratic equation for the parameter k can be obtained:

$$k^4-(k_1^2+k_2^2)k^2-\delta k_1^2k_2^2=0, \tag{7.19}$$

where the notations

$$k_{1,2}=\sqrt{\frac{\lambda_0+\lambda_{1,2}}{\kappa_0+\kappa_{1,2}}},\quad \delta=\frac{\lambda_0^2}{(\lambda_0+\lambda_1)(\lambda_0+\lambda_2)} \tag{7.20}$$

are used. The δ value characterizes the deviation of the parameter of the interfacial shearing stress in a trimaterial assembly from the magnitude of this parameter in a bimaterial assembly. This value changes from zero to one, when the axial compliance λ_0 of the inner (zero) component changes from zero to infinity, or, practically, to a large enough number exceeding the axial compliances of the two other components.

Biquadratic equation (7.19) results in the following formula for the parameter of the interfacial shearing stress:

$$k=\sqrt{\frac{k_1^2+k_2^2}{2}\left[1+\sqrt{1+\delta\left(\frac{2k_1k_2}{k_1^2+k_2^2}\right)^2}\right]} \tag{7.21}$$

When $\delta = 0$, then $k = \sqrt{k_1^2 + k_2^2}$. This result indicates that when the axial compliance of the inner component (such as a silicon chip) is considerably smaller than the axial compliance of the two other components (the substrate and the lid), the parameter k of the interfacial shearing stresses in the trimaterial assembly in question can be evaluated as the square root of the parameters squared of the interfacial shearing stresses of the upper and the lower bimaterial assemblies (the zero component is included in such a case in both the upper and the lower assemblies). If, in addition, the two outer components of the assembly are identical and characterized by the same interfacial stress parameter k_*, then the interfacial stress parameter is $k = k_*\sqrt{2} = 1.4142k_*$. This result indicates that the parameter of the interfacial stress in a trimaterial assembly is always higher than in a bimaterial assembly. If all the three components in the trimaterial assembly are identical, then, with $k = k_1 = k_2 = k_*$ and $\delta = \frac{1}{4}$, solution (7.20) yields

$$k = k_*\sqrt{1 + \frac{\sqrt{5}}{2}} = 1.4553k_*.$$

Thus, the interfacial stress parameter is always higher for a trimaterial assembly, than for a bimaterial one.

The induced force $T_1(x)$ acting in component #1 (substrate) can be sought in the form

$$T_1(x) = C_0 + C_1 \sinh kx + C_2 \cosh kx. \tag{7.22}$$

The boundary conditions $T_1(\pm l) = \mp\hat{T}$ yield

$$C_1 = -\frac{\hat{T}}{\sinh kl}, \quad C_0 + C_2 \cosh kl = 0. \tag{7.23}$$

In the mid-cross section of component #1 ($x = 0$), the sought solution (7.22) yields

$$C_0 + C_2 = T_1(0). \tag{7.24}$$

The second equation in (7.23) and equation (7.24) yield

$$C_0 = T_1(0)\frac{\cosh kl}{\cosh kl - 1}, \quad C_2 = -T_1(0)\frac{1}{\cosh kl - 1}. \tag{7.25}$$

For large enough kl values (long and stiff assemblies), these formulas can be simplified as

$$C_0 = T_1(0) = T_1^*, \quad C_2 = -T_1^*\frac{1}{\cosh kl}. \tag{7.26}$$

Introducing these formulas and the first formula in (7.23) into the sought solution (7.22), we obtain

$$T_1(x) = T_1^*\left(1 - \frac{\cosh kx}{\cosh kl}\right) - \hat{T}\frac{\sinh kx}{\sinh kl}. \tag{7.27}$$

Similarly, we find

$$T_2(x) = T_2^*\left(1 - \frac{\cosh kx}{\cosh kl}\right) + \hat{T}\frac{\sinh kx}{\sinh kl}. \tag{7.28}$$

These two solutions could be written as

$$T_{1,2}(x) = T_{1,2}^*\left(1 - \frac{\cosh kx}{\cosh kl}\right) \mp \hat{T}\frac{\sinh kx}{\sinh kl}. \tag{7.29}$$

These expressions satisfy the zero boundary conditions at the assembly ends and the conditions $T_{1,2}(0) = T_{1,2}^*$ in the mid-cross section ($x = 0$) of sufficiently long assemblies ($kl \to \infty$). The interfacial shearing stresses can be determined by differentiation

$$\tau_{1,2}(x) = T'_{1,2}(x) = -kT_{1,2}^*\frac{\sinh kx}{\cosh kl} \mp k\hat{T}\frac{\cosh kx}{\sinh kl}. \tag{7.30}$$

At the assembly ends

$$\tau_{1,2}(\pm l) = \mp k(T_{1,2}^* \tanh kl + \hat{T}\coth kl). \tag{7.31}$$

For long assemblies with stiff interfaces (large kl values),

$$\tau_{1,2}(\pm l) = \mp k(T_{1,2}^* + \hat{T}). \tag{7.32}$$

7.2.3 Peripheral Portions of the Design

The thermally induced forces acting in the cross sections of the assembly components of the peripheral portions of the design (Figure 7.4) can be evaluated using the same approach as the one used for the its midportion (Figure 7.3). The only difference is that in the peripheral portions of the design, the zero component and component #2 (lid) are made of the same material, and that there is just a single bonding material: the adhesive layer between component #1 (substrate) and the zero component, which is just part of the lid. The total force $T_1(x)$ acting in component #1 (substrate) can be sought in the form (7.22), and the following three equations can be obtain for the constants C_0, C_1, and C_2:

$$C_0 + C_1 \sinh kl + C_2 \cosh kl = -\hat{T};$$

$$C_0 - C_1 \sinh kl + C_2 \cosh kl = 0;$$

$$C_0 + C_1 = T_2(0) = T_1^*; \tag{7.33}$$

The first equation determines the boundary condition for component #1 at its right end. The second equation is the boundary condition for this component at its left end. The third equation determines the force acting in the mid-cross section of component #1, assuming that the assembly is sufficiently long. Then, for large enough kl values, we find

$$C_0 = T_1^* + \frac{\hat{T}}{2\cosh kl} \approx T_1^*; \quad C_1 = -\frac{\hat{T}}{2\sinh kl}; \quad C_2 = -\frac{2T_1^* + \hat{T}}{2\cosh kl}; \tag{7.34}$$

and therefore

$$T_1(x) = T_1^*\left(1 - \frac{\cosh kx}{\cosh kl}\right) - \hat{T}\frac{\sinh[k(l+x)]}{\sinh 2kl}. \tag{7.35}$$

Similarly, we find

$$T_2(x) = T_2^*\left(1 - \frac{\cosh kx}{\cosh kl}\right) + \hat{T}\frac{\sinh[k(l+x)]}{\sinh 2kl}. \tag{7.36}$$

The two solutions can be combined as follows:

$$T_{1,2}(x) = T_{1,2}^*\left(1 - \frac{\cosh kx}{\cosh kl}\right) \mp \hat{T}\frac{\sinh[k(l+x)]}{\sinh 2kl}. \tag{7.37}$$

The interfacial shearing stresses can be determined by differentiation

$$\tau_{1,2}(x) = T'_{1,2}(x) = -kT_{1,2}^*\frac{\sinh kx}{\cosh kl} \mp k\hat{T}\frac{\cosh[k(l+x)]}{\sinh 2kl}. \tag{7.38}$$

At the assembly ends

$$\begin{aligned} \tau_{1,2}(l) &= -k(T_{1,2}^* \tanh kl + \hat{T}\coth 2kl), \\ \tau_{1,2}(-l) &= k\left(T_{1,2}^* \tanh kl + \frac{\hat{T}}{\sinh 2kl}\right). \end{aligned} \tag{7.39}$$

For long assemblies with stiff interfaces, these formulas yield

$$\tau_{1,2}(l) = \mp k(T_{1,2}^* + \hat{T}), \quad \tau_{1,2}(-l) = kT_{1,2}^*. \tag{7.40}$$

Clearly, when the force $\hat{T}$ is applied at the right end of a sufficiently long assembly, only this end will experience elevated stress caused by this force, while the remote end will not be affected at all.

The force $\hat{T}$ is "external" with respect to the midportion and the peripheral portions of the design, but is, as has been indicated, "internal," as far as the design as a whole is concerned. This force can be found from the following condition of the compatibility of the displacements at the left boundary of the midportion of the design at its contact with the left peripheral portion:

$$(\kappa_{0p}+\kappa_{1p})\tau_{1p}(l_p)+(\kappa_{0m}+\kappa_{1m})\tau_{1m}(-l_m)+\lambda_{1m}\hat{T}\Delta=0 \tag{7.41}$$

The first term is the interfacial displacement at the right end of the peripheral portion of the design for component #1 (substrate), the second term is the interfacial displacement at the left end of the midportion of the design for the same component, the third term, evaluated using Hooke's law, is the axial displacement of the portion of component #1 that is located between the design's peripheral and the midportions; κ_{0p} is the interfacial compliance of the zero component in the peripheral portion of the design (actually, of the lower part of the lid, whose thickness is the same as the thickness of the chip and the interfacial compliance of the bonding layer should be considered and included into this compliance); κ_{1p} is the interfacial compliance of component #1 (substrate) at the peripheral portion of the design; κ_{0m} is the interfacial compliance of the chip (the interfacial compliances of the encapsulated solder joint layer and the thermal interface layer should be included into this compliance); $\kappa_{1m}=\kappa_{1p}$ is the interfacial compliance of component #1 (substrate) at the midportion of the design;

$$\tau_{1p}(l_p)=-k_p(T_{1p}^*+\hat{T}) \tag{7.42}$$

is the interfacial shearing stress at the right end of the peripheral portion of the design; l_p is half-the-length of the peripheral portion of the design; k_p is the parameter of the interfacial shearing stresses at this portion; T_{1p}^* is the thermally induced force acting in component #1 (substrate) in the peripheral portion of the design;

$$\tau_{1m}(-l_m)=k_m(T_{1m}^*+\hat{T}) \tag{7.43}$$

is the interfacial shearing stress at the left end of the midportion of the design; l_m is half-the-length of the midportion of the design; k_m is the parameter of the interfacial shearing stresses at this portion; T_{1m}^* is the thermally induced force acting in component #1 (substrate) in the midportion of the design;

$$\lambda_{1m}=\lambda_{1p}=\frac{1-\nu_1}{E_1h_1} \tag{7.44}$$

is the axial compliance of component #1 (substrate); E_1 is the effective Young's modulus of the substrate material; ν_1 is its Poisson's ratio; h_1 is the substrate's thickness; and Δ is the length of the substrate's portion between the chip and the lid. Equation (7.41) is the condition of the compatibility of the three displacements: the two interfacial displacements of the peripheral and the midportions of the design and the axial displacement of the gap portion of the substrate. The interfacial displacements are determined in accordance with the concept of the interfacial compliance, and the third one, in accordance with Hooke's law.

Introducing expressions (7.42) and (7.43) into condition (7.41) and solving the obtained equation for the force $\hat{T}$, the following formula for this force can be obtained:

$$\hat{T} = \frac{(\kappa_{0m} + \kappa_{1m})k_m T_{1m}^* - (\kappa_{0p} + \kappa_{1p})k_p T_{1p}^*}{(\kappa_{0p} + \kappa_{1p})k_p - (\kappa_{0m} + \kappa_{1m})k_m + \lambda_1 \Delta} \tag{7.45}$$

One could certainly introduce this formula into equations (7.32) and (7.40) and obtain the analytical expressions that would consider the role of this force in a hidden fashion. We will, however, compute this force as part of the subsequent numerical examples. This will enable comparing the level of this force in comparison with the thermally induced forces acting in the cross sections of the assembly components.

7.3 NUMERICAL EXAMPLES

7.3.1 Design with an Organic Lid: Midportion of the Design

Input data

Structural Element	#1 (Substrate)	#2 (Lid)	0 (Chip)	#01 (TIL)	#02 (ESJL)
Thickness, mm	1.0	1.0	0.5	0.05	0.1
Young's modulus, kg/mm^2	2300	800	13000	200	10000
Poisson's ratio	0.28	0.35	0.24	0.4	0.38
Shear modulus, kg/mm^2	900	300	5200	70	3600
CTExE6, 1/°C	15.0	18.0	2.4	–	–

Half assembly length, $l_m = 2.5$ mm; Change in temperature: $\Delta t = 180$°C;
TIL = Thermal interface layer; ESJL = Encapsulated solder joint layer

Calculated data

Axial compliances:

- Substrate:

$$\lambda_1 = \frac{1-\nu_1}{E_1 h_1} = \frac{1-0.28}{2300 \times 1.0} = 31.3043 \times 10^{-5} \text{ mm/kg},$$

- Lid:

$$\lambda_2 = \frac{1-\nu_2}{E_2 h_2} = \frac{1-0.35}{800 \times 1.0} = 81.2500 \times 10^{-5} \text{ mm/kg}$$

- Chip:

$$\lambda_0 = \frac{1-\nu_0}{E_0 h_0} = \frac{1-0.24}{13000 \times 0.5} = 11.6923 \times 10^{-5} \text{ mm/kg}$$

- Determinant:

$$D = \lambda_0\lambda_1 + \lambda_1\lambda_2 + \lambda_2\lambda_0 = (366.0195 + 2543.4744 + 949.9994)\times 10^{-10}$$
$$= 3859.4933\times 10^{-10}\ \text{mm}^2/\text{kg}^2$$

- Ratio:

$$\frac{\Delta t}{D} = \frac{180}{3859.4933\times 10^{-10}} = 0.046638\times 10^{10}\,\frac{^\circ\text{C}\times\text{kg}^2}{\text{mm}^2}$$

Forces acting in mid-cross section of a long assembly:

- In the chip:

$$T_0^* = [(\alpha_0 - \alpha_1)\lambda_2 + (\alpha_0 - \alpha_2)\lambda_1]\frac{\Delta t}{D} = (-12.6\times 81.2500 - 15.6\times 31.3043)$$
$$\times 0.0046638 = -7.0521\ \text{kg/mm}$$

- In the substrate:

$$T_2^* = [(\alpha_2 - \alpha_1)\lambda_0 + (\alpha_2 - \alpha_0)\lambda_1]\frac{\Delta t}{D} = (3.0\times 11.6923 + 15.6\times 31.3043)$$
$$\times 0.0046638 = 2.4411\ \text{kg/mm}$$

- In the lid:

$$T_1^* = [(\alpha_1 - \alpha_0)\lambda_2 + (\alpha_1 - \alpha_2)\lambda_0]\frac{\Delta t}{D} = (-3.0\times 11.6923 + 12.6\times 81.2500)$$
$$\times 0.0046638 = 4.6110\ \text{kg/mm}$$

Thus, the low expansion chip is in compression, while the substrate and the lid are in tension. The structure is in equilibrium so that the total force acting on it is zero.

Interfacial compliances:

- The chip, with the thermal interface layer (TIL) and the encapsulated solder joint layer (ESJL):

$$\kappa_0 = \frac{h_0}{G_0} + \frac{h_{01}}{G_{01}} + \frac{h_{02}}{G_{02}} = \frac{0.5}{5200} + \frac{0.05}{70} + \frac{0.1}{3600} = (9.6154 + 71.4286 + 2.7778)\times 10^{-5}$$
$$= 83.8218\times 10^{-5}\ \text{mm}^3/\text{kg}$$

- The substrate:

$$\kappa_1 = \frac{h_1}{3G_1} = \frac{1.0}{3\times 900} = 37.0370\times 10^{-5}\ \text{mm}^3/\text{kg},$$

- The lid:

$$\kappa_2 = \frac{h_2}{3G_2} = \frac{1.0}{3 \times 300} = 111.1111 \times 10^{-5} \text{ mm}^3/\text{kg},$$

Partial parameters of the interfacial shearing stress (for bimaterial assemblies):

$$k_1 = \sqrt{\frac{\lambda_0 + \lambda_1}{\kappa_0 + \kappa_1}} = \sqrt{\frac{11.6923 \times 10^{-5} + 31.3043 \times 10^{-5}}{83.8218 \times 10^{-5} + 37.0370 \times 10^{-5}}} = \sqrt{\frac{42.9966}{120.8588}} = 0.5965 \text{ mm}^{-1}$$

$$k_2 = \sqrt{\frac{\lambda_0 + \lambda_2}{\kappa_0 + \kappa_2}} = \sqrt{\frac{11.6923 \times 10^{-5} + 81.2500 \times 10^{-5}}{83.8218 \times 10^{-5} + 111.1111 \times 10^{-5}}} = \sqrt{\frac{92.9423}{194.9329}} = 0.6905 \text{ mm}^{-1}$$

Parameter of the trimaterial assembly:

$$\delta = \frac{\lambda_0^2}{(\lambda_0 + \lambda_1)(\lambda_0 + \lambda_2)} = \frac{(11.6923 \times 10^{-5})^2}{(42.9966 \times 10^{-5})(92.9423 \times 10^{-5})} = 0.034210.$$

This parameter is small, because the zero component (high-modulus chip) is characterized by a very low axial compliance.

Parameter of the interfacial shearing stress:

$$k = \sqrt{\frac{k_1^2 + k_2^2}{2}\left[1 + \sqrt{1 + \delta\left(\frac{2k_1k_2}{k_1^2 + k_2^2}\right)^2}\right]} =$$

$$= \sqrt{\frac{0.3558 + 0.4768}{2}\left[1 + \sqrt{1 + 0.03421\left(\frac{2 \times 0.5965 \times 0.6905}{0.3558 + 0.4768}\right)^2}\right]} = 0.9162 \text{mm}^{-1}$$

7.3.2 Design with an Organic Lid: Peripheral Portion of the Design

Input data

Structural Element	#1 (Substrate)	#2 (Lid's Upper Part)	0 (Lid's Lower Part)	#01 (Adhesive Layer)
Thickness, mm	1.0	1.0	0.5	0.05
Young's modulus, kg/mm^2	2300	800	800	200
Poisson's ratio	0.28	0.35	0.35	0.40
Shear modulus, kg/mm^2	900	300	300	70
CTExE6, 1/deg.C	15.0	18.0	18.0	–

Half assembly length

$$l_m = 2.5 \text{ mm};$$

Change in temperature:

$$\Delta t = 180°\text{C}.$$

Calculated data

Axial compliances:

- Substrate:

$$\lambda_1 = \frac{1-v_1}{E_1 h_1} = \frac{1-0.28}{2300 \times 1.0} = 31.3043 \times 10^{-5} \text{ mm/kg},$$

- Lid's upper part:

$$\lambda_2 = \frac{1-v_2}{E_2 h_2} = \frac{1-0.35}{800 \times 1.0} = 81.2500 \times 10^{-5} \text{ mm/kg},$$

- Lid's lower part:

$$\lambda_0 = \frac{1-v_0}{E_0 h_0} = \frac{1-0.35}{800 \times 1.0} = 162.5000 \times 10^{-5} \text{ mm/kg},$$

- Determinant:

$$D = \lambda_0\lambda_1 + \lambda_1\lambda_2 + \lambda_2\lambda_0 = (5086.9488 + 2543.4744 + 13203.1250) \times 10^{-10} =$$
$$= 20833.5482 \times 10^{-10} \text{ mm}^2/\text{kg}^2$$

- Ratio:

$$\frac{\Delta t}{D} = \frac{180}{20833.5482 \times 10^{-10}} = 0.008640 \times 10^{-10} \frac{°\text{C} \times \text{kg}^2}{\text{mm}^2}$$

Forces acting in mid-cross section of a long assembly:

- In the chip:

$$T_0^* = [(\alpha_0 - \alpha_1)\lambda_2 + (\alpha_0 - \alpha_2)\lambda_1]\frac{\Delta t}{D} = 3.00 \times 81.2500 \times 0.0008640 = 0.2106 \text{ kg/mm}$$

- In the substrate:

$$T_1^* = [(\alpha_1 - \alpha_2)\lambda_0 + (\alpha_1 - \alpha_0)\lambda_2]\frac{\Delta t}{D} = (-3.0 \times 162.5000 - 3.0 \times 81.2500) \times 0.008640 =$$
$$= -0.6318 \text{ kg/mm}$$

- In the lid:

$$T_2^* = [(\alpha_2 - \alpha_1)\lambda_0 + (\alpha_2 - \alpha_0)\lambda_1]\frac{\Delta t}{D} = 3.0 \times 162.5000 \times 0.0008640 = 0.4212\,\text{kg/mm}$$

Thus, the substrate is in compression, and the lid is in tension.
Interfacial compliances:

- The lid's lower part (with the bonding layer):

$$\kappa_0 + \kappa_{01} = \frac{h_0}{G_0} + \frac{h_{01}}{G_{01}} = \frac{0.5}{300} + \frac{0.05}{70} = (166.6667 + 71.4286) \times 10^{-5}$$

$$= 238.0953 \times 10^{-5}\,\text{mm}^3/\text{kg}$$

- The substrate:

$$\kappa_1 = \frac{h_1}{3G_1} = \frac{1.0}{3 \times 900} = 37.0370 \times 10^{-5}\,\text{mm}^3/\text{kg},$$

- The lid's upper part:

$$\kappa_2 = \frac{h_2}{3G_2} = \frac{1.0}{3 \times 300} = 111.1111 \times 10^{-5}\,\text{mm}^3/\text{kg},$$

- Partial parameters of the interfacial shearing stress:

$$k_1 = \sqrt{\frac{\lambda_0 + \lambda_1}{\kappa_0 + \kappa_1}} = \sqrt{\frac{162.5000 \times 10^{-5} + 31.3043 \times 10^{-5}}{238.0953 \times 10^{-5} + 37.0370 \times 10^{-5}}} = \sqrt{\frac{193.8043}{275.1323}} = 0.8393\,\text{mm}^{-1}$$

$$k_2 = \sqrt{\frac{\lambda_0 + \lambda_2}{\kappa_0 + \kappa_2}} = \sqrt{\frac{162.5000 \times 10^{-5} + 81.2500 \times 10^{-5}}{238.0953 \times 10^{-5} + 111.1111 \times 10^{-5}}} = \sqrt{\frac{243.7500}{349.2064}} = 0.8355\,\text{mm}^{-1}$$

- Parameter of the trimaterial assembly:

$$\delta = \frac{\lambda_0^2}{(\lambda_0 + \lambda_1)(\lambda_0 + \lambda_2)} = \frac{(162.5000 \times 10^{-5})^2}{(193.8043 \times 10^{-5})(243.7500 \times 10^{-5})} = 0.5590.$$

This parameter is not small because all the assembly components are characterized by relatively high axial compliances.

- Parameter of the interfacial shearing stress:

$$k = \sqrt{\frac{k_1^2 + k_2^2}{2}\left[1 + \sqrt{1 + \delta\left(\frac{2k_1k_2}{k_1^2 + k_2^2}\right)^2}\right]} =$$

$$= \sqrt{\frac{0.7044 + 0.6981}{2}\left[1 + \sqrt{1 + 0.5560\left(\frac{2 \times 0.8393 \times 0.8355}{0.7044 + 0.6981}\right)^2}\right]} = 1.2558\,\text{mm}^{-1}$$

- "External" force

The length of the substrate's portion between the chip and the lid is $\Delta = 7.5$ mm Then, with

$$\kappa_{0m} = 83.8218 \times 10^{-5}\ \text{mm}^3/\text{kg};\ \kappa_{1m} = 37.0370 \times 10^{-5}\ \text{mm}^3/\text{kg};$$
$$k_m = 0.9162\,\text{mm}^{-1};\ T_{1m}^* = 4.6110\ \text{kg/mm};$$
$$\kappa_{0p} = 238.0953 \times 10^{-5}\ \text{mm}^3/\text{kg};\ \kappa_{1p} = 37.0370 \times 10^{-5}\ \text{mm}^3/\text{kg};$$
$$k_p = 1.2558\,\text{mm}^{-1};\ T_{1p}^* = -0.6318\ \text{kg/mm};\ \lambda_1 = 31.3043 \times 10^{-5}\ \text{mm/kg},$$

we obtain:

$$\hat{T} = \frac{(\kappa_{0m} + \kappa_{1m})k_m T_{1m}^* - (\kappa_{0p} + \kappa_{1p})k_p T_{1p}^*}{(\kappa_{0p} + \kappa_{1p})k_p - (\kappa_{0m} + \kappa_{1m})k_m + \lambda_1\Delta} =$$

$$= \frac{510.5799 \times 10^{-5} + 218.2939 \times 10^{-5}}{345.5111 \times 10^{-5} - 110.7308 \times 10^{-5} + 234.7823 \times 10^{-5}} = 1.5522\ \text{kg/mm}$$

This force is considerably larger than the thermally induced forces in the assembly components.

Stresses:

Normal stresses in the assembly components:

- Maximum normal (compressive) stress in the chip

$$\sigma_0 = \frac{T_0^*}{h_0} = \frac{-7.0521}{0.5} = -14.1042\ \text{kg/mm}^2$$

- Maximum normal (tensile) stress in the substrate in its midportion

$$\sigma_1 = \frac{T_0^* - \hat{T}}{h_1} = \frac{4.6110 - 1.5522}{1.0} = 3.0588\ \text{kg/mm}^2$$

- Maximum normal (tensile) stress in the lid in its midportion

$$\sigma_2 = \frac{T_2^* + \hat{T}}{h_1} = \frac{2.4411 + 1.5522}{1.0} = 3.9933\ \text{kg/mm}^2$$

Interfacial shearing stresses:

- Maximum shearing stresses in the design's midportion at the chip–substrate interface

$$\tau_1(\pm l_m) = \mp k(T_1^* + \hat{T}) = \mp 0.9162(4.6110 + 1.5522) = \mp 5.6468 \text{ kg/mm}^2$$

- Maximum shearing stresses in the design midportion at the chip–lid interface

$$\tau_1(\pm l_m) = \mp k(T_2^* + \hat{T}) = \mp 0.9162(2.4411 + 1.5522) = \mp 3.9933 \text{ kg/mm}^2$$

- Maximum shearing stresses in the design's peripheral portion at the chip–substrate interface at its boundary with the midportion

$$\tau_1(l_m) = -k(T_1^* + \hat{T}) = -1.2558(-0.6318 + 1.5522) = -1.1558 \text{ kg/mm}^2$$

- Maximum shearing stresses in the design's peripheral portion at the chip–substrate interface and at its free boundary

$$\tau_1(-l) = kT_1^* = 1.2558 \times (-0.6318) = -0.7934 \text{ kg/mm}^2.$$

- Maximum shearing stresses in the design peripheral portion at the chip–lid interface at its boundary with the midportion

$$\tau_2(l) = -k(T_2^* + \hat{T}) = -1.2558(0.4212 + 1.5522) = -2.4782 \text{ kg/mm}^2,$$

- Maximum shearing stresses in the design's peripheral portion at the chip–substrate interface at its free boundary

$$\tau_1(-l) = kT_2^* = 1.2558 \times 0.4212 = 0.5289 \text{ kg/mm}^2.$$

7.3.3 Design with a Copper Lid: Midportion of the Design

Input data

Structural Element	#1 (Substrate)	#2 (Lid)	0 (Chip)	#01 (TIL)	#02 (ESJL)
Thickness, mm	1.0	1.0	0.5	0.05	0.1
Young's modulus, kg/mm^2	2300	12200	13000	200	10000
Poisson's ratio	0.28	0.33	0.24	0.4	0.38
Shear modulus, kg/mm^2	900	4586	5200	70	3600
CTExE6, 1/°C	15.0	16.5	2.4	–	–

Half assembly length, $l_m = 5.0$ mm Change in temperature: $\Delta t = 180$°C
TIL = Thermal interface layer ESJL = Encapsulated solder joint layer

Calculated data

Axial compliances:

- Substrate:

$$\lambda_1 = \frac{1-\nu_1}{E_1 h_1} = \frac{1-0.28}{2300 \times 1.0} = 31.3043 \times 10^{-5}\ \text{mm/kg},$$

- Lid:

$$\lambda_2 = \frac{1-\nu_2}{E_2 h_2} = \frac{1-0.33}{12200 \times 1.0} = 5.4918 \times 10^{-5}\ \text{mm/kg}$$

- Chip:

$$\lambda_0 = \frac{1-\nu_0}{E_0 h_0} = \frac{1-0.24}{13000 \times 0.5} = 11.6923 \times 10^{-5}\ \text{mm/kg}$$

- Determinant:

$$D = \lambda_0\lambda_1 + \lambda_1\lambda_2 + \lambda_2\lambda_0 = (366.0193 + 171.9170 + 64.2118) \times 10^{-10}$$
$$= 602.1481 \times 10^{-10}\ \text{mm}^2/\text{kg}^2$$

- Ratio:

$$\frac{\Delta t}{D} = \frac{180}{602.1481 \times 10^{-10}} = 0.29893 \times 10^{-10}\ \frac{°\text{C} \times \text{kg}^2}{\text{mm}^2}$$

Forces acting in mid-cross section of a long assembly:

- In the chip:

$$T_0^* = [(\alpha_0 - \alpha_1)\lambda_2 + (\alpha_0 - \alpha_2)\lambda_1]\frac{\Delta t}{D} = (-12.6 \times 5.4918 - 14.1 \times 31.3043) \times 0.029893$$
$$= -15.2630\ \text{kg/mm}$$

- In the substrate:

$$T_1^* = [(\alpha_1 - \alpha_0)\lambda_2 + (\alpha_1 - \alpha_2)\lambda_0]\frac{\Delta t}{D} = (12.6 \times 5.4918 - 1.5 \times 11.6923) \times 0.029893$$
$$= 1.5442\ \text{kg/mm}$$

- In the lid:

$$T_2^* = [(\alpha_2 - \alpha_1)\lambda_0 + (\alpha_2 - \alpha_0)\lambda_1]\frac{\Delta t}{D} = (1.5 \times 11.6923 + 14.1 \times 31.3043) \times 0.029893$$
$$= 13.7188\ \text{kg/mm}$$

Thus, at low temperature condition, the low expansion chip is in compression, while the high expansion substrate and the lid are in tension. The structure is in equilibrium, so that the total force acting on it is zero. The thermally induced forces in the chip and in the lid of the design with a high-modulus lid are considerably higher than in a design with a low modulus lid: by the factor of about 2.2 for the chip and of about 5.6 for the lid. The induced force in the substrate went down considerably: there is only about 33% of the thermal force in it in the case of a low modulus lid. In effect, the forces in the chip and in the lid approximately equilibrate each other, while the substrate is understressed.

Interfacial compliances:

- The chip, with the TIL and the ESJL

$$\kappa_0 = \frac{h_0}{G_0} + \frac{h_{01}}{G_{01}} + \frac{h_{02}}{G_{02}} = \frac{0.5}{5200} + \frac{0.05}{70} + \frac{0.1}{3600} = (9.6154 + 71.4286 + 2.7778) \times 10^{-5} =$$
$$= 83.8218 \times 10^{-5}\ \text{mm}^3/\text{kg}$$

- The substrate:

$$\kappa_1 = \frac{h_1}{3G_1} = \frac{1.0}{3 \times 900} = 37.0370 \times 10^{-5}\ \text{mm}^3/\text{kg},$$

- The lid:

$$\kappa_2 = \frac{h_2}{3G_2} = \frac{1.0}{3 \times 4586} = 7.2685 \times 10^{-5}\ \text{mm}^3/\text{kg},$$

- Partial parameters of the interfacial shearing stress:

$$k_1 = \sqrt{\frac{\lambda_0 + \lambda_1}{\kappa_0 + \kappa_1}} = \sqrt{\frac{11.6923 \times 10^{-5} + 31.3043 \times 10^{-5}}{83.8218 \times 10^{-5} + 37.0370 \times 10^{-5}}} = \sqrt{\frac{42.9966}{120.8588}} = 0.5965\ \text{mm}^{-1}$$

$$k_2 = \sqrt{\frac{\lambda_0 + \lambda_2}{\kappa_0 + \kappa_2}} = \sqrt{\frac{11.6923 \times 10^{-5} + 5.4918 \times 10^{-5}}{83.8218 \times 10^{-5} + 7.2685 \times 10^{-5}}} = \sqrt{\frac{17.1841}{91.0903}} = 0.4343\ \text{mm}^{-1}$$

- Parameter of the trimaterial assembly:

$$\delta = \frac{\lambda_0^2}{(\lambda_0 + \lambda_1)(\lambda_0 + \lambda_2)} = \frac{(11.6923 \times 10^{-5})^2}{(42.9966 \times 10^{-5})(17.1841 \times 10^{-5})} = 0.1850$$

- Parameter of the interfacial shearing stress:

$$k = \sqrt{\frac{k_1^2 + k_2^2}{2}\left[1 + \sqrt{1 + \delta\left(\frac{2k_1k_2}{k_1^2 + k_2^2}\right)^2}\,\right]} =$$

$$= \sqrt{\frac{0.3558 + 0.1886}{2}\left[1 + \sqrt{1 + 0.1850\left(\frac{2 \times 0.5965 \times 0.4343}{0.3558 + 0.4768}\right)^2}\,\right]} = 0.7525\ \text{mm}^{-1}$$

7.3.4 Design with a Copper Lid: Peripheral Portions of the Design

Input data

Structural Element	#1 (Substrate)	#2 (Lid's Upper Part)	0 (Lid's Lower Part)	#01 (Adhesive Layer)
Thickness, mm	1.0	1.0	0.5	0.05
Young's modulus, kg/mm^2	2300	12200	12200	200
Poisson's ratio	0.28	0.33	0.33	0.40
Shear modulus, kg/mm^2	900	4586	4586	70
CTExE6, 1/deg.C	15.0	16.5	16.5	–

Half assembly length

$$l_m = 2.5 \text{ mm};$$

Change in temperature:

$$\Delta t = 180°\text{C}.$$

Calculated data

Axial compliances:

- Substrate:

$$\lambda_1 = \frac{1-\nu_1}{E_1 h_1} = \frac{1-0.28}{2300 \times 1.0} = 31.3043 \times 10^{-5} \text{ mm/kg},$$

- Lid's upper part:

$$\lambda_2 = \frac{1-\nu_2}{E_2 h_2} = \frac{1-0.33}{12200 \times 1.0} = 5.4918 \times 10^{-5} \text{ mm/kg},$$

- Lid's lower part:

$$\lambda_0 = \frac{1-\nu_0}{E_0 h_0} = \frac{1-0.33}{12200 \times 0.5} = 10.9836 \times 10^{-5} \text{ mm/kg},$$

- Determinant:

$$D = \lambda_0\lambda_1 + \lambda_1\lambda_2 + \lambda_2\lambda_0 = (343.8341 + 171.9170 + 60.3197) \times 10^{-10}$$
$$= 576.0708 \times 10^{-10} \text{ mm}^2/\text{kg}^2$$

- Ratio:

$$\frac{\Delta t}{D} = \frac{180}{576.0708 \times 10^{-10}} = 0.31246 \times 10^{10} \frac{°\text{C} \times \text{kg}^2}{\text{mm}^2}$$

Forces acting in mid-cross section of a long assembly:

- In the chip:

$$T_0^* = [(\alpha_0 - \alpha_1)\lambda_2 + (\alpha_0 - \alpha_2)\lambda_1]\frac{\Delta t}{D} = 1.5 \times 5.4918 \times 0.031246 = 0.2574 \text{ kg/mm}$$

- In the substrate:

$$T_1^* = [(\alpha_1 - \alpha_2)\lambda_0 + (\alpha_1 - \alpha_0)\lambda_2]\frac{\Delta t}{D} = (-1.5 \times 10.9836 - 1.5 \times 5.4918) \times 0.031246$$
$$= -0.7722 \text{ kg/mm}$$

- In the lid:

$$T_2^* = [(\alpha_2 - \alpha_1)\lambda_0 + (\alpha_2 - \alpha_0)\lambda_1]\frac{\Delta t}{D} = 1.5 \times 10.9836 \times 0.031246 = 0.5148 \text{ kg/mm}$$

Thus, the substrate is in compression, and the lid and the chip are in tension, as it is expected.

Interfacial compliances:

- The lid's lower part (with the bonding layer):

$$\kappa_0 + \kappa_{01} = \frac{h_0}{G_0} + \frac{h_{01}}{G_{01}} = \frac{0.5}{4586} + \frac{0.05}{70} = (10.9027 + 71.4286) \times 10^{-5}$$
$$= 82.3313 \times 10^{-5} \text{ mm}^3/\text{kg}$$

- The substrate:

$$\kappa_1 = \frac{h_1}{3G_1} = \frac{1.0}{3 \times 900} = 37.0370 \times 10^{-5} \text{ mm}^3/\text{kg},$$

- The lid's upper part:

$$\kappa_2 = \frac{h_2}{3G_2} = \frac{1.0}{3 \times 4586} = 7.2685 \times 10^{-5} \text{ mm}^3/\text{kg},$$

- Partial parameters of the interfacial shearing stress:

$$k_1 = \sqrt{\frac{\lambda_0 + \lambda_1}{\kappa_0 + \kappa_1}} = \sqrt{\frac{10.9836 \times 10^{-5} + 31.3043 \times 10^{-5}}{82.3313 \times 10^{-5} + 37.0370 \times 10^{-5}}} = \sqrt{\frac{42.2879}{119.36833}} = 0.5952 \text{ mm}^{-1}$$

$$k_2 = \sqrt{\frac{\lambda_0 + \lambda_2}{\kappa_0 + \kappa_2}} = \sqrt{\frac{10.9836 \times 10^{-5} + 5.4918 \times 10^{-5}}{82.3313 \times 10^{-5} + 7.2685 \times 10^{-5}}} = \sqrt{\frac{16.4754}{89.5998}} = 0.4288 \text{ mm}^{-1}$$

- Parameter of the trimaterial assembly:

$$\delta = \frac{\lambda_0^2}{(\lambda_0 + \lambda_1)(\lambda_0 + \lambda_2)} = \frac{(10.9836 \times 10^{-5})^2}{(42.2879 \times 10^{-5})(16.4754 \times 10^{-5})} = 0.1732.$$

- Parameter of the interfacial shearing stress:

$$k = \sqrt{\frac{k_1^2 + k_2^2}{2}\left[1 + \sqrt{1 + \delta\left(\frac{2k_1k_2}{k_1^2 + k_2^2}\right)^2}\right]} =$$

$$= \sqrt{\frac{0.3543 + 0.1839}{2}\left[1 + \sqrt{1 + 0.5560\left(\frac{2 \times 0.5952 \times 0.4288}{0.3543 + 0.1839}\right)^2}\right]} = 0.7738\ \text{mm}^{-1}$$

- "External" force

The length of the substrate's portion between the chip and the lid is $\Delta = 7.5\,\text{mm}$. Then, with

$$\kappa_{0m} = 83.8218 \times 10^{-5}\,\text{mm}^3/\text{kg}; \kappa_{1m} = 37.0370 \times 10^{-5}\,\text{mm}^3/\text{kg};$$

$$k_m = 0.7525\,\text{mm}^{-1}; T_{1m}^* = 1.5442\,\text{kg/mm};$$

$$\kappa_{0p} = 82.3313 \times 10^{-5}\,\text{mm}^3/\text{kg}; \kappa_{1p} = 37.0370 \times 10^{-5}\,\text{mm}^3/\text{kg};$$

$$k_p = 0.7738\,\text{mm}^{-1}; T_{1p}^* = -0.7722\,\text{kg/mm}; \lambda_1 = 31.3043 \times 10^{-5}\,\text{mm/kg},$$

we obtain the following value for the "external" force:

$$\hat{T} = \frac{(\kappa_{0m} + \kappa_{1m})k_m T_{1m}^* - (\kappa_{0p} + \kappa_{1p})k_p T_{1p}^*}{(\kappa_{0p} + \kappa_{1p})k_p - (\kappa_{0m} + \kappa_{1m})k_m + \lambda_1\Delta} =$$

$$= \frac{140.4392 \times 10^{-5} + 71.3259 \times 10^{-5}}{92.3672 \times 10^{-5} - 90.9462 \times 10^{-5} + 234.7823 \times 10^{-5}} = \frac{211.7651}{236.2033} = 0.8965\,\text{kg/mm}$$

This force is somewhat larger than the thermal forces acting in the lid and in the substrate, but the difference is not as significant as in the case of an organic lid.

Stresses:

Normal stresses in the assembly components:

- Maximum normal stress in the chip

$$\sigma_0 = \frac{T_0^*}{h_0} = \frac{-15.2630}{0.5} = -30.5260\,\text{kg/mm}^2$$

- Maximum normal stress in the substrate in its midportion

$$\sigma_1 = \frac{T_1^* - \hat{T}}{h_1} = \frac{1.5442 - 0.8965}{1.0} = 0.6477\,\text{kg/mm}^2$$

- Maximum normal stress in the lid in its midportion

$$\sigma_2 = \frac{T_2^* + \hat{T}}{h_1} = \frac{13.7188 + 0.8965}{1.0} = 14.6153\,\text{kg/mm}^2$$

Interfacial shearing stresses:

- Maximum shearing stresses in the design's midportion at the chip–substrate interface

$$\tau_1(\pm l_m) = \mp k(T_1^* + \hat{T}) = \mp 0.7525(1.5442 + 0.8965) = \mp 1.8366\,\text{kg/mm}^2$$

- Maximum shearing stresses in the design midportion at the chip–lid interface

$$\tau_1(\pm l_m) = \mp k(T_2^* + \hat{T}) = \mp 0.7525(13.7188 + 0.8965) = \mp 10.9980\,\text{kg/mm}^2$$

- Maximum shearing stresses in the design's peripheral portion at the chip–substrate interface at its boundary with the midportion

$$\tau_1(l) = -k(T_1^* + \hat{T}) = -0.7738(-0.7722 + 0.8965) = -0.0962\,\text{kg/mm}^2$$

- Maximum shearing stresses in the design's peripheral portion at the chip–substrate interface at its free boundary

$$\tau_1(-l) = kT_1^* = 0.7738 \times (-0.7722) = -0.5975\,\text{kg/mm}^2.$$

- Maximum shearing stresses in the design peripheral portion at the chip–lid interface at its boundary with the midportion

$$\tau_2(l) = -k(T_2^* + \hat{T}) = -0.7738 \times (0.5148 + 0.8965) = -0.0921\,\text{kg/mm}^2,$$

- Maximum shearing stresses in the design's peripheral portion at the chip–substrate interface at its free boundary

$$\tau_2(-l) = kT_2^* = 0.7738 \times 0.5148 = 0.3984\,\text{kg/mm}^2.$$

The computed data are summarized here:

	Stresses, kg/mm^2	With Plastic Lid	With Copper Lid
1	Maximum normal stress in the chip	−14.1042	−30.5260
2	Maximum normal stress in the substrate in its midportion	3.0588	0.6477
3	Maximum normal stress in the lid in its midportion	3.9933	14.6153
4	Maximum shearing stresses in the design's midportion at the chip–substrate interface	∓5.6468	∓1.8366
5	Maximum shearing stresses in the design midportion at the chip–lid interface	∓3.9933	∓10.9980
6	Maximum shearing stresses in the design's peripheral portion at the chip–substrate interface at its boundary with the midportion	−1.1558	−0.0962
7	Maximum shearing stresses in the design's peripheral portion at the chip–substrate interface at its free boundary	−0.7934	−0.5975
8	Maximum shearing stresses in the design peripheral portion at the chip–lid interface at its boundary with the midportion	−2.4782	−1.0921
9	Maximum shearing stresses in the design's peripheral portion at the chip–substrate interface at its free boundary	0.5289	0.3984

The computed data indicate that the highest shearing stress occurs at the chip–bond interface and is significantly, by the factor of about 2.45, higher than the stress at the substrate–bond interface, but even the latter stress is about twice as high as the maximum shearing stress predicted on the basis of the bimaterial model. As to the normal stresses acting in the cross sections of the assembly components, the trimaterial model predicts that the highest stresses occur in the chip, the lowest in the substrate, and that the stresses in the bond are rather high, about 59% of the stresses in the chip. The bimaterial model, however, simply assumes that the normal stresses in the bond are zero. The normal stresses in the chip predicted on the basis of the bimaterial model are only about 78% of the stress predicted by the trimaterial model. The normal stresses in the substrate evaluated on the basis of the bimaterial model are almost twice as high as the trimaterial model predicts, but these stresses are low anyway: it is the state of stress in the chip and in the bonding layer, and the interfacial stress at the chip–bond interface that should be of concern to the device designer. It is concluded that while a simple bimaterial model can be successfully used for adhesively bonded assemblies, a trimaterial model should be employed for FC assemblies, especially when high-modulus solders are used. In addition, we conclude that the application of a copper lid, although might be desirable from the standpoint of thermal management of the design, leads to significantly higher maximum stresses, both the normal compressive stresses in the chip and the maximum interfacial shearing stresses; that the maximum normal stress in the chip is about twice as high in the design with a copper lid; that the maximum interfacial shearing stresses occur in the design midportion at the chip–substrate interface in the case of a plastic lid and at the chip–lid interface in the case of a copper lid; that the maximum interfacial shearing stress is also about twice as high in the design with a copper lid; and that the maximum shearing stresses are about a quarter of the maximum normal stresses. The developed model can be used in the analysis and design of the FC package of the type in question. The model can be particularly helpful when developing an FEA preprocessing simulation model. Future work should include FEA verifications of this model.

7.4 IS IT REALLY IMPORTANT THAT THE ENTIRE UNDERCHIP AREA IS ENCAPSULATED ("UNDERFILLED")?

From the induced stresses standpoint, only the end portions of a bonded assembly should be securely bonded [40] and the lengths of these portions should not be below $L = \frac{5}{k}$, where k is the parameter of the interfacial shearing stress. This parameter should be determined, of course, differently for a bi- and a trimaterial assembly, as the FC assembly in accordance with the aforementioned calculation procedures. The rationale behind this requirement for the length of the bonded/encapsulated region is that the maximum interfacial stress $\tau_{\max} = kT \tanh kl$ can be evaluated, in an approximate analysis, as the product of the parameter k of the interfacial shearing stress, the maximum thermally induced force (per unit assembly width) T acting in the midportion of a sufficiently long chip, when the interfacial stress at its interface with the solder system is of interest, or the maximum force in the midportion of the substrate, when the shearing stress and the substrate–"bond" interface is sought, and l is half the bonded length. Applying the formula to the maximum interfacial shearing stress at the bonded peripheral portions of a FC assembly bonded at the ends, it could be concluded that this stress will not change, if the assembly is already large enough and therefore the hyperbolic tangent in the aforementioned formula is close to 1.0 ($\tanh 2.5 = 0.9867$), and this requirement takes place, if the condition $L = 2l \geq \frac{5}{k}$ is fulfilled.

7.5 STRESS RELIEF IN AN FC DESIGN DUE TO THE APPLICATION OF AN INHOMOGENEOUS SOLDER JOINT SYSTEM

Suggestions made for the second level of interconnections (Chapter 6) are, to a great extent, applicable also to the FC designs. Significant stress relief can be achieved, if necessary, by using solder joints with elevated stand-offs [30]. Such joints are characterized also by an elevated interfacial compliance and owing to that lead to lower interfacial stresses. One should have in mind, however, that solder joint systems with elevated stand-offs add appreciable weight to the structure, and if drop tests are considered, such systems might be much more vulnerable than designs with conventional joints [41–43] (see also Chapter 9). Inhomogeneous solder joint system designs, in which the solder material at their peripheral portions have considerably lower Young's moduli and/or are applied at lower melting/fabrication temperatures, or when epoxies are employed at the assembly ends, are as advisable in the case of the encapsulated FC structures as in the case of chip-to-substrate second-level interconnections. The interfacial shearing stresses in the midportion of an assembly with an inhomogeneous solder system increase in its midportion from zero at the assembly's mid-cross section to its maximum values at the boundaries with the peripheral portions, then drop to low values at the inner boundaries of the peripheral portions, then increase again and reach their maximum values at the free ends of the assembly. Optimized stress relief (see Chapter 6) can be achieved if the lengths of the low-modulus peripheral portions are established in such a way that the two stress maxima, one at the boundary of the midportion with the peripheral portions and another one at the assembly ends, are equal. In such a case, each of these maxima turns out to be considerably lower than the maximum stresses at the ends of the assembly with a homogeneous bond. Strange as it may sound, the lowest interfacial stresses can be achieved with the stiffer midportions [13]. This is because such midportions result in lower interfacial displacements at the ends of these portions and, hence, in the lower peripheral displacements and, as a result, in lower interfacial stresses at the peripheral portions of the assembly. It is this phenomenon that explains the stress relief in quad-flat no-leads (QFN) assemblies [23]. This approach could also be applied if there is a need for that, to FC designs. If there is no way to avoid the inelastic strains in the peripheral solder joints of an FC design, then the lengths of the peripheral portions experiencing inelastic strains can be found from the equation, in which the left part is the maximum thermally induced force at the end of the elastic midportion of the assembly and the right part is the product of the yield stress in shear of the solder material and the length of the sought inelastic peripheral portion of the assembly. The number of peripheral joints that experience inelastic strain is important, because the fatigue lifetime of the interconnection, whose peripheral joints experience low cycle fatigue condition, is inversely proportional to the number of joints that are simultaneously subjected to inelastic deformations.

7.6 EFFECT OF THE UNDERFILL GLASS TRANSITION TEMPERATURE

7.6.1 Background

One of the critical problems associated with using underfills is the glass-transition temperature (T_g) of the underfill material [2–7, 44]. Some existing experiments show that packages with high-T_g underfills are less sensitive to the 85°C/85% relative humidity temperature-humidity bias, that high T_g underfills lead to somewhat higher induced curvatures than low T_g underfills, that packages with low-T_g underfills exhibit during cooling processes appreciable stress relaxation; that the final deformations of the packages (this takes place at low temperature conditions, when the difference in the thermal strains between the chip and its substrate is the highest) depend, in low T_g underfills, also on the cooling rate, and that the underfill behavior depends not only on the T_g level, but also on the width of the T_g region and its slope [45]. The authors of [45] conclude that "as long as the accurate temperature-dependent properties of underfill are not used, the predicted solder fatigue life based on a single value of T_g and a sharp transition is subject to skepticism" and that the "lifetime variation can be 80% less for steeper slope and doubled for a shallower slope." This particularly means that the "steep slope" assumption is conservative: it results in a somewhat shorter predicted fatigue life.

Below T_g, the epoxy is in a hard (glassy) state. Above the T_g, it transitions to a rubbery state and, as a result, the material's Young's modulus decreases and its CTE increases. There is also an indication that, for many epoxies, their cohesive and adhesive strength is lower at elevated temperatures, such as at temperatures above the T_g. The latter drawback might be, however, a lesser problem, considering that the applications of a low T_g epoxy might lead to considerably lower stresses. In the subsequent analysis, it is shown that this might indeed be the case. The role of the underfill-solder composite bond (USCB) thickness is also assessed. The stresses considered in this analysis include normal stresses acting in the cross sections of the chip, the substrate, and the USCB, and the shearing stresses at the USCB interfaces with the chip and the substrate. Since the USCB bonding layer is comprised of a relatively high-modulus solder and low-modulus epoxy underfill (even when the underfill epoxy is loaded with fillers), and is characterized, therefore, unlike in adhesively bonded joints, by a relatively high effective Young's modulus, a tri- and not a bimaterial predictive model is employed.

7.6.2 Assumptions

The following major assumptions were used in this analysis:

- A structural analysis (strength-of-materials) approach can be applied to evaluate the induced stresses so that no singular stresses can possibly occur at the assembly edges. From the theory-of-elasticity standpoint, the predicted stresses can be viewed as useful design-for-reliability parameters that characterize the state of stress in the bonded assembly of interest, including its edge portions.

- The assembly and its constituents (components) can be treated as thin elongated plates, experiencing small deflections. The engineering theory of such plates can therefore be employed to determine the stresses and the deflections.
- The interfacial shearing stresses and the total assembly curvature can be evaluated without considering the effect of the peeling stresses—normal interfacial stresses acting in the through-thickness direction of the assembly. The peeling stresses, which are proportional to the longitudinal gradient of the interfacial shearing stresses, can then be determined, if necessary, from the evaluated shearing stresses.
- The shearing stresses can be found, based on the concept of the interfacial compliance, assuming that the interfacial displacements of the assembly components can be represented as a sum of unrestricted (stress-free) thermal displacements; displacements, predicted using Hooke's law and calculated under an assumption that these displacements are the same for the entire cross section of the given assembly component; and corrections considering that, in reality, the interfacial displacements are somewhat larger than the displacements of the inner points of the given cross section. In addition, it is assumed that these corrections are proportional to the interfacial shearing stress in the given cross section and are not affected by the stresses and strains in the adjacent cross sections.
- The effect of the assembly bow is not considered in this analysis.
- The dependence of the mechanical properties (Young's modulus and CTE) of the underfill material of temperature is a step-wise function; in other words, the glass-transition temperature T_g is characterized by a single number; based on information from [45], it is a conservative assumption.

7.6.3 Thermally Induced Forces and Interfacial Stresses

The following formulas can be used for the evaluation of the thermally induced forces and the interfacial shearing stresses.

7.6.3.1 Thermally Induced Forces in the Midportion of a Long Flip-Chip/Substrate Assembly

$$T_0 = \frac{(\alpha_0 - \alpha_1)\lambda_2 + (\alpha_0 - \alpha_2)\lambda_1}{\lambda_0\lambda_1 + \lambda_1\lambda_2 + \lambda_2\lambda_0}\Delta t,$$

$$T_1 = \frac{(\alpha_1 - \alpha_2)\lambda_0 + (\alpha_1 - \alpha_0)\lambda_2}{\lambda_0\lambda_1 + \lambda_1\lambda_2 + \lambda_2\lambda_0}\Delta t,$$

$$T_2 = \frac{(\alpha_2 - \alpha_1)\lambda_0 + (\alpha_2 - \alpha_0)\lambda_1}{\lambda_0\lambda_1 + \lambda_1\lambda_2 + \lambda_2\lambda_0}\Delta t. \quad (7.46)$$

Here, Δt is the change in temperature from the elevated (manufacturing) temperature to the low (room, testing, operation, T_g) temperature; T_0, T_1, and T_2 are the thermally induced forces (per unite assembly width) acting in the midportion of the (underfilled) solder layer (zero component), the chip (component #1), and in the substrate (component #2), respectively; α_0, α_1, and α_2 are the effective CTEs of the materials; $\lambda_i = \frac{1-v_i}{E_i h_i}, i = 0,1,2$ are the effective axial compliances of the assembly components; $E_i, i = 0,1,2$, are the effective Young's moduli of the materials; $v_i, i = 0,1,2$ are their Poisson's ratios; and $h_i, i = 0,1,2$ are the assembly component thicknesses.

7.6.3.2 Distributed Thermally Induced Forces

$$T_0(x) = T_0\left(1 - \frac{\cosh kx}{\cosh kl}\right),$$

$$T_1(x) = T_1\left(1 - \frac{\cosh kx}{\cosh kl}\right),$$

$$T_2(x) = T_2\left(1 - \frac{\cosh kx}{\cosh kl}\right). \tag{7.47}$$

Here, l is half the assembly length, and k is the thus far unknown parameter of the assembly. Here,

$$k = \sqrt{\frac{k_1^2 + k_2^2}{2}\left[1 + \sqrt{1 + \delta\left(\frac{2k_1 k_2}{k_1^2 + k_2^2}\right)^2}\right]}. \tag{7.48}$$

is the parameter of the interfacial shearing stress;

$$k_1 = \sqrt{\frac{\lambda_0 + \lambda_1}{\kappa_0 + \kappa_1}}, \quad k_2 = \sqrt{\frac{\lambda_0 + \lambda_2}{\kappa_0 + \kappa_2}}, \tag{7.49}$$

are the "partial "parameters of the interfacial shearing stresses;

$$\delta = \frac{\lambda_0\lambda_1 + \lambda_1\lambda_2 + \lambda_2\lambda_0}{(\lambda_0 + \lambda_1)(\lambda_0 + \lambda_2)} \tag{7.50}$$

is the factor of the axial compliances of the assembly components; $\kappa_0 = \frac{h_0}{G_0}$ is the longitudinal interfacial compliance of the solder layer; G_0 is the effective shear modulus of this layer; $\kappa_i = \frac{h_i}{3G_i}, i = 1,2$ are the longitudinal interfacial compliances of the chip and the substrate, respectively; $G_i = \frac{E_i}{2(1+v_1)}, i = 0,1,2$ are the shear moduli of the components' materials.

7.6.3.3 Interfacial Shearing Stresses

$$\tau_1(x) = \frac{dT_1(x)}{dx} = -kT_1\frac{\sinh kx}{\cosh kl},$$
$$\tau_2(x) = \frac{dT_2(x)}{dx} = -kT_2\frac{\sinh kx}{\cosh kl} \quad (7.51)$$

The signs of these stresses depend on the signs of the forces T_1 and T_2, and the signs of these forces, whether tensile (+) or compressive (–), are determined by formulas (7.46).

The highest shearing stresses take place, in accordance with this model, at the end cross sections:

$$\tau_1(l) = -kT_1 \tanh kl, \quad \tau_2(l) = -kT_2 \tanh kl. \quad (7.52)$$

For $kl \succ 2.5$, these stresses become assembly size independent

$$\tau_1(l) = -kT_1, \quad \tau_2(l) = -kT_2. \quad (7.53)$$

7.6.4 Numerical Example

A case of high T_g (above the curing temperature), high-modulus, low-expansion, and 0.05-mm thick underfill

Component # **Properties**	**0 (Underfilled Solder)**	**1 (Chip)**	**2 (Substrate)**
Effective CTE, α, 1/°C	60×10^{-6}	2.2×10^{-6}	13.2×10^{-6}
Effective Young's modulus, E, kg/mm^2	2000	12300	2000
Poisson's ratio, ν	0.40	0.24	0.30
Shear modulus $G = \frac{E}{2(1+\nu)}$, kg/mm^2	714.3	4959.7	769.2
Thickness, h, mm	0.05	0.5	1.5
Axial compliance $\lambda = \frac{1-\nu}{Eh}$, mm/kg	60.0000×10^{-4}	1.2358×10^{-4}	2.3333×10^{-4}
Axial compliance factor (determinant) $F = \lambda_0\lambda_1 + \lambda_1\lambda_2 + \lambda_2\lambda_0$, mm^2/kg^2		217.0295×10^{-8}	
Thermal forces, kg/mm			
$T_0 = \frac{\Delta t}{F}[(\alpha_0-\alpha_1)\lambda_2 + (\alpha_0-\alpha_2)\lambda_1]$,			
$T_1 = \frac{\Delta t}{F}[(\alpha_1-\alpha_2)\lambda_0 + (\alpha_1-\alpha_0)\lambda_2]$,	150.3194	–620.0493	469.7299
$T_2 = \frac{\Delta t}{F}[(\alpha_2-\alpha_1)\lambda_0 + (\alpha_2-\alpha_0)\lambda_1]$.			
Normal stresses $\sigma = \frac{T}{h}$, kg/mm^2	3006.3880	–1240.0986	313.1533
Interfacial compliance, κ, $\kappa_0 = \frac{h_0}{G_0}$, $\kappa_1 = \frac{h_1}{3G_1}$, $\kappa_2 = \frac{h_2}{3G_2}$. mm^3/kg	0.7000×10^{-4}	0.3360×10^{-4}	6.5003×10^{-4}

Factor of the relative axial compliance $\delta = \frac{F}{(\lambda_0 + \lambda_1)(\lambda_0 + \lambda_2)}$		0.056858	
Chip-to-solder parameter of the interfacial shearing stress $k_1 = \sqrt{\frac{\lambda_0 + \lambda_1}{\kappa_0 + \kappa_1}}$, mm^{-1}		7.6882	
Substrate-to-solder parameter of the interfacial shearing stress $k_2 = \sqrt{\frac{\lambda_0 + \lambda_2}{\kappa_0 + \kappa_2}}$, mm^{-1}			2.9423
Parameter of the interfacial shearing stress $k = \sqrt{\frac{k_1^2 + k_2^2}{2}\left[1 + \sqrt{1 - \delta\left(\frac{2k_1k_2}{k_1^2 + k_2^2}\right)^2}\right]}$, mm^{-1}		8.2057	
Shearing stress at chip–solder interface $\tau_1(l) = -kT_1$, kg/mm^2		5087.9305	
Shearing stress at solder-PCB interface $\tau_2(l) = -kT_2$, kg/mm^2			−3854.4626

Curing temperature 150°C; low temperature −20°C; Change in temperature 170°C;

A case of Low T_g (below the lowest temperature possible), low-modulus, high-expansion, and 0.05-mm thick underfill

Component # Properties	0 (Underfilled Solder)	1 (Chip)	2 (Substrate)
Effective CTE, α, 1/°C	120×10^{-6}	2.2×10^{-6}	13.2×10^{-6}
Effective Young's modulus, E, kg/mm^2	400	12300	2000
Effective Poisson's ratio, v	0.45	0.24	0.30
Shear modulus $G = \frac{E}{2(1+v)}$, kg/mm^2	137.9310	4959.7	769.3
Thickness, h, mm	0.05	0.5	1.5
Axial compliance $\lambda = \frac{1-v}{Eh}$, mm/kg	275×10^{-4}	1.2358×10^{-4}	2.3334×10^{-4}
Axial compliance factor $F = \lambda_0\lambda_1 + \lambda_1\lambda_2 + \lambda_2\lambda_0$, mm^2/kg^2		984.4136×10^{-8}	
Thermal forces, kg/mm $T_0 = \frac{\Delta t}{F}[(\alpha_0 - \alpha_1)\lambda_2 + (\alpha_0 - \alpha_2)\lambda_1]$, $T_1 = \frac{\Delta t}{F}[(\alpha_1 - \alpha_2)\lambda_0 + (\alpha_1 - \alpha_0)\lambda_2]$, $T_2 = \frac{\Delta t}{F}[(\alpha_2 - \alpha_1)\lambda_0 + (\alpha_2 - \alpha_0)\lambda_1]$.	70.2611	−569.8619	499.6008
Normal stress $\sigma = \frac{T}{h}$, kg/mm^2	1405.2220	−1139.7238	333.0672

Interfacial compliance of the components, κ, $\kappa_0 = \frac{h_0}{G_0}, \kappa_1 = \frac{h_1}{3G_1}, \kappa_2 = \frac{h_2}{3G_{20}}$. mm^3/kg	3.6250×10^{-4}	0.3360×10^{-4}	6.4994×10^{-4}
Factor of the axial compliance $\delta = \frac{F}{(\lambda_0 + \lambda_1)(\lambda_0 + \lambda_2)}$		0.012850	
Partial (chip-to-solder) parameter of the interfacial shearing stress $k_1 = \sqrt{\frac{\lambda_0 + \lambda_1}{\kappa_0 + \kappa_1}}$, mm^{-1}		5.5931	
Partial (substrate-to-solder) parameter of the interfacial shearing stress $k_2 = \sqrt{\frac{\lambda_0 + \lambda_2}{\kappa_0 + \kappa_2}}$, mm^{-1}			3.2995
Parameter of the interfacial shearing stress $k = \sqrt{\frac{k_1^2 + k_2^2}{2}\left[1 + \sqrt{1 - \delta\left(\frac{2k_1k_2}{k_1^2 + k_2^2}\right)^2}\right]}$, mm^{-1}.		9.8427	
Shearing stress at chip–solder interface $\tau_1(l) = -kT_1, kg/mm^2$		4061.7032	
Shearing stress at solder-PCB interface $\tau_2(l) = -kT_2, kg/mm^2$			−3035.8392
Curing temperature is 150°C; low temperature is −20°C; Change in temperature from the curing temperature to the low temperature is therefore 170°C;			

A case of high T_g (above the curing temperature), high-modulus, low-expansion, and 0.1-mm thick underfill

Component # **Properties**	**0 (Underfilled Solder)**	**1 (Chip)**	**2 (Substrate)**
Effective CTE, α, 1/°C	60×10^{-6}	2.2×10^{-6}	13.2×10^{-6}
Effective Young's modulus, $E, kg/mm^2$	2000	12300	2000
Poisson's ratio, v	0.40	0.24	0.30
Shear modulus $G = \frac{E}{2(1+v)}, kg/mm^2$	714.3	4959.7	769.2
Thickness, h, mm	0.1	0.5	1.5
Axial compliance $\lambda = \frac{1-v}{Eh}$, mm/kg	30.0000×10^{-4}	1.2358×10^{-4}	2.3333×10^{-4}
Axial compliance factor $F = \lambda_0\lambda_1 + \lambda_1\lambda_2 + \lambda_2\lambda_0, mm^2/kg^2$		109.9595×10^{-8}	
Thermal forces, kg/mm $T_0 = \frac{\Delta t}{F}[(\alpha_0 - \alpha_1)\lambda_2 + (\alpha_0 - \alpha_2)\lambda_1]$,			
$T_1 = \frac{\Delta t}{F}[(\alpha_1 - \alpha_2)\lambda_0 + (\alpha_1 - \alpha_0)\lambda_2]$,	297.9190	−718.6920	420.7730
$T_2 = \frac{\Delta t}{F}[(\alpha_2 - \alpha_1)\lambda_0 + (\alpha_2 - \alpha_0)\lambda_1]$.			

Normal stress $\sigma = \frac{T}{h}, \text{kg/mm}^2$	2979.1900	−1437.3840	280.5153
Interfacial compliance, $\kappa_0 = \frac{h_0}{G_0},\ \kappa_1 = \frac{h_1}{3G_1},\ \kappa_2 = \frac{h_2}{3G_2}.\ \text{mm}^3/\text{kg}$	1.4000×10^{-4}	0.3360×10^{-4}	6.5003×10^{-4}
Factor of the relative axial compliance $\delta = \frac{F}{(\lambda_0 + \lambda_1)(\lambda_0 + \lambda_2)}.$		0.108875	
Chip-to-solder parameter of the interfacial shearing stress $k_1 = \sqrt{\frac{\lambda_0 + \lambda_1}{\kappa_0 + \kappa_1}},\ \text{mm}^{-1}$		4.2418	
Substrate-to-solder parameter of the interfacial shearing stress $k_2 = \sqrt{\frac{\lambda_0 + \lambda_2}{\kappa_0 + \kappa_2}},\ \text{mm}^{-1}$			2.0230
Parameter of the interfacial shearing stress $k = \sqrt{\frac{k_1^2 + k_2^2}{2}\left[1 + \sqrt{1 - \delta\left(\frac{2k_1k_2}{k_1^2 + k_2^2}\right)^2}\right]},\ \text{mm}^{-1}.$		4.6601	
Shearing stress at chip–solder interface $\tau_1(l) = -kT_1, \text{kg/mm}^2$		3349.1683	
Shearing stress at solder-PCB interface $\tau_2(l) = -kT_2, \text{kg/mm}^2$			−1960.8443

Curing temperature 150°C; low temperature −20°C; Change in temperature 170°C;

A case of low T_g (below the lowest temperature possible), low-modulus, high-expansion, and 0.1-mm thick underfill

Component # **Properties**	**0 (Underfilled Solder)**	**1 (Chip)**	**2 (Substrate)**
Effective CTE, α, 1/°C	120×10^{-6}	2.2×10^{-6}	13.2×10^{-6}
Effective Young's modulus, $E, \text{kg/mm}^2$	400	12300	2000
Effective Poisson's ratio, v	0.45	0.24	0.30
Shear modulus $G = \frac{E}{2(1+v)}, \text{kg/mm}^2$	137.9310	4959.7	769.3
Thickness, h, mm	0.1	0.5	1.5
Axial compliance $\lambda = \frac{1-v}{Eh}, \text{mm/kg}$	137.5×10^{-4}	1.2358×10^{-4}	2.3334×10^{-4}
Axial compliance factor $F = \lambda_0\lambda_1 + \lambda_1\lambda_2 + \lambda_2\lambda_0, \text{mm}^2/\text{kg}^2$	493.6486×10^{-8}		

Thermal forces, kg/mm $T_0 = \frac{\Delta t}{F}[(\alpha_0 - \alpha_1)\lambda_2 + (\alpha_0 - \alpha_2)\lambda_1]$, $T_1 = \frac{\Delta t}{F}[(\alpha_1 - \alpha_2)\lambda_0 + (\alpha_1 - \alpha_0)\lambda_2]$, $T_2 = \frac{\Delta t}{F}[(\alpha_2 - \alpha_1)\lambda_0 + (\alpha_2 - \alpha_0)\lambda_1]$.	140.1369	–615.6379	475.5010
Normal stress $\sigma = \frac{T}{h}$, kg/mm^2	1401.3690	–1231.2758	317.0007
Interfacial compliance of the components, κ, $\kappa_0 = \frac{h_0}{G_0}$, $\kappa_1 = \frac{h_1}{3G_1}$, $\kappa_2 = \frac{h_2}{3G_{20}}$. mm^3/kg	7.2500×10^{-4}	0.3360×10^{-4}	6.4994×10^{-4}
Factor of the axial compliance $\delta = \frac{F}{(\lambda_0 + \lambda_1)(\lambda_0 + \lambda_2)}$	0.025446		
Partial (chip-to-solder) parameter of the interfacial shearing stress $k_1 = \sqrt{\frac{\lambda_0 + \lambda_1}{\kappa_0 + \kappa_1}}$, mm^{-1}		4.2765	
Partial (substrate-to-solder) parameter of the interfacial shearing stress $k_2 = \sqrt{\frac{\lambda_0 + \lambda_2}{\kappa_0 + \kappa_2}}$, mm^{-1}			3.1891
Parameter of the interfacial shearing stress $k = \sqrt{\frac{k_1^2 + k_2^2}{2}\left[1 + \sqrt{1 - \delta\left(\frac{2k_1k_2}{k_1^2 + k_2^2}\right)^2}\right]}$, mm^{-1}.	5.3190		
Shearing stress at chip–solder interface $\tau_1(l) = -kT_1$, kg/mm^2		3274.5560	
Shearing stress at solder-PCB interface $\tau_2(l) = -kT_2$, kg/mm^2			–2529.1898

Curing temperature is 150°C; low temperature is –20°C; Change in temperature from the curing temperature to the low temperature is therefore 170°C;

A case of high T_g (above the curing temperature), high-modulus, low-expansion, and 0.75-mm thick underfill

Component # **Properties**	**0 (Underfilled Solder)**	**1 (Chip)**	**2 (Substrate)**
Effective CTE, α, 1/°C	60×10^{-6}	2.2×10^{-6}	13.2×10^{-6}
Effective Young's modulus, E, kg/mm^2	2000	12300	2000
Poisson's ratio, ν	0.40	0.24	0.30
Shear modulus $G = \frac{E}{2(1+\nu)}$, kg/mm^2	714.3	4959.7	769.2
Thickness, h, mm	0.75	0.5	1.5

Axial compliance $\lambda = \dfrac{1-v}{Eh}$, mm/kg	4.0000×10^{-4}	1.2358×10^{-4}	2.3333×10^{-4}
Axial compliance factor (determinant) $F = \lambda_0\lambda_1 + \lambda_1\lambda_2 + \lambda_2\lambda_0$, mm^2/kg^2		17.1604×10^{-8}	
Thermal forces, kg/mm $T_0 = \dfrac{\Delta t}{F}[(\alpha_0 - \alpha_1)\lambda_2 + (\alpha_0 - \alpha_2)\lambda_1]$, $T_1 = \dfrac{\Delta t}{F}[(\alpha_1 - \alpha_2)\lambda_0 + (\alpha_1 - \alpha_0)\lambda_2]$, $T_2 = \dfrac{\Delta t}{F}[(\alpha_2 - \alpha_1)\lambda_0 + (\alpha_2 - \alpha_0)\lambda_1]$.	1909.0414	−1771.9806	−137.0608
Normal stress $\sigma = \dfrac{T}{h}$, kg/mm^2	2545.3885	−3543.9612	−91.3739
Interfacial compliance, κ, $\kappa_0 = \dfrac{h_0}{G_0}$, $\kappa_1 = \dfrac{h_1}{3G_1}$, $\kappa_2 = \dfrac{h_2}{3G_2}$. mm^3/kg	10.4998×10^{-4}	0.3360×10^{-4}	6.5003×10^{-4}
Factor of the relative axial compliance $\delta = \dfrac{F}{(\lambda_0 + \lambda_1)(\lambda_0 + \lambda_2)}$		0.517496	
Chip-to-solder parameter of the interfacial shearing stress $k_1 = \sqrt{\dfrac{\lambda_0 + \lambda_1}{\kappa_0 + \kappa_1}}$, mm^{-1}		0.6951	
Substrate-to-solder parameter of the interfacial shearing stress $k_2 = \sqrt{\dfrac{\lambda_0 + \lambda_2}{\kappa_0 + \kappa_2}}$, mm^{-1}			0.6104
Parameter of the interfacial shearing stress $k = \sqrt{\dfrac{k_1^2 + k_2^2}{2}\left[1 + \sqrt{1 - \delta\left(\dfrac{2k_1k_2}{k_1^2 + k_2^2}\right)^2}\right]}$, mm^{-1}.		0.8531	
Shearing stress at chip–solder interface $\tau_1(l) = -kT_1$, kg/mm^2		1511.6946	
Shearing stress at solder-PCB interface $\tau_2(l) = -kT_2$, kg/mm^2			−116.9266
Curing temperature 150°C; low temperature −20°C; Change in temperature 170°C;			

A case of low T_g (below the lowest temperature possible), low-modulus, high-expansion, and 0.75-mm thick underfill

Component # **Properties**	**0 (Underfilled Solder)**	**1 (Chip)**	**2 (Substrate)**
Effective CTE, α, 1/°C	120×10^{-6}	2.2×10^{-6}	13.2×10^{-6}
Effective Young's modulus, E, kg/mm^2	400	12300	2000
Effective Poisson's ratio, v	0.45	0.24	0.30
Shear modulus $G = \dfrac{E}{2(1+v)}$, kg/mm^2	137.9310	4959.7	769.3
Thickness, h, mm	0.75	0.5	1.5
Axial compliance $\lambda = \dfrac{1-v}{Eh}$, mm/kg	18.3333×10^{-4}	1.2358×10^{-4}	2.3334×10^{-4}

Axial compliance factor $F = \lambda_0\lambda_1 + \lambda_1\lambda_2 + \lambda_2\lambda_0$, mm^2/kg^2	68.3188×10^{-8}		
Thermal forces, kg/mm $T_0 = \frac{\Delta t}{F}[(\alpha_0 - \alpha_1)\lambda_2 + (\alpha_0 - \alpha_2)\lambda_1]$, $T_1 = \frac{\Delta t}{F}[(\alpha_1 - \alpha_2)\lambda_0 + (\alpha_1 - \alpha_0)\lambda_2]$, $T_2 = \frac{\Delta t}{F}[(\alpha_2 - \alpha_1)\lambda_0 + (\alpha_2 - \alpha_0)\lambda_1]$.	1012.3846	−1185.7792	173.3947
Normal stress $\sigma = \frac{T}{h}$, kg/mm^2	1349.8461	−2371.5584	115.5965
Interfacial compliance of the components, κ, $\kappa_0 = \frac{h_0}{G_0}$, $\kappa_1 = \frac{h_1}{3G_1}$, $\kappa_2 = \frac{h_2}{3G_{20}}$. mm^3/kg	54.3750×10^{-4}	0.3360×10^{-4}	6.4994×10^{-4}
Factor of the axial compliance $\delta = \frac{F}{(\lambda_0 + \lambda_1)(\lambda_0 + \lambda_2)}$	0.168927		
Partial (chip-to-solder) parameter of the interfacial shearing stress $k_1 = \sqrt{\frac{\lambda_0 + \lambda_1}{\kappa_0 + \kappa_1}}$, mm^{-1}		0.5981	
Partial (substrate-to-solder) parameter of the interfacial shearing stress $k_2 = \sqrt{\frac{\lambda_0 + \lambda_2}{\kappa_0 + \kappa_2}}$, mm^{-1}			0.5827
Parameter of the interfacial shearing stress $k = \sqrt{\frac{k_1^2 + k_2^2}{2}\left[1 + \sqrt{1 - \delta\left(\frac{2k_1k_2}{k_1^2 + k_2^2}\right)^2}\right]}$, mm^{-1}.	0.8163		
Shearing stress at chip–solder interface $\tau_1(l) = -kT_1$, kg/mm^2		967.9516	
Shearing stress at solder-PCB interface $\tau_2(l) = -kT_2$, kg/mm^2			−141.5421

Curing temperature is 150°C; low temperature is −20°C; Change in temperature from the curing temperature to the low temperature is therefore 170°C;

Calculated stresses (summary)

		0.05-mm Thick Underfill		0.10-mm Thick Underfill		0.75-mm Thick Underfill	
Stress, kg/mm^2		High T_g	Low T_g	High T_g	Low T_g	High T_g	Low T_g
Normal stress	in the underfill σ_0	3006	1405	2979	1401	2545	1350
	in the chip σ_1	−1240	−1140	−1437	−1231	−3544	−2372
	in the substrate σ_2	313	333	280	317	−91	116
Shearing stress	at the underfill-chip interface τ_1	5088	4062	3349	3275	1512	968
	at the underfill-substrate interface τ_2	−3854	−3036	−1961	−2529	−117	−141

In the carried out numerical example, the highest shearing stress occurs at the chip–"bond" interface and is by the factor of about 2.45 higher than the maximum shearing stress at the substrate–"bond" interface, but even the latter stress is about twice as high as the stress predicted using the bimaterial model. As to the normal stresses acting in the cross sections of the assembly components, the trimaterial model predicts that the highest stresses occur in the chip, the lowest in the substrate, and that the stresses in the cross sections of the "bond" are not low at all, only about 59% of the normal stresses in the chip (the bimaterial model simply assumes that the normal stresses in the" bond" are zero). The compressive normal stresses in the chip predicted on the basis of the bimaterial model are only about 78% of the stress obtained using the trimaterial model. The normal stresses in the substrate predicted using the bimaterial model are, however, almost twice as high as the trimaterial model predicts, but these stresses are low anyway.

Calculations indicated that the T_g level of the underfill material had a significant effect on the induced stresses: the normal stresses in the USCB with a low-modulus (low T_g) underfill were about half the stresses of the design with high-modulus (high T_g) underfill, and this was true for both thin (0.05-mm thick) and thick (0.1-mm thick) USCB layers. As to the role of the USCB thickness, thicker USCBs exhibit somewhat lower normal stresses, than thin layers, but the effect is insignificant. The maximum predicted shearing stresses occur at the USCB/chip, and not at the USCB/substrate interface. This result is in agreement with the observed, in a number of experiments, delaminations at the USCB/chip interface, and not at the USCB/substrate interface. The obtained data also indicate that (in a way, contrary to the current practice) there is an incentive for using low T_g underfills, provided, of course, that their adhesive strength is proven to be sufficient for the lower stress level. This is an important requirement, of course, and might explain why electronic product manufacturers employ mostly high T_g underfills. As to the incentive for using thicker USCBs, the increase in this thickness from 0.05 mm to 0.1 mm resulted in a minor relief in the normal stress in the USCB for both high and low T_g underfills, but led to an appreciable relief in the interfacial stresses at the USCB/chip interface, especially for high T_g underfills: the predicted stress relief in this case was as significant as 34%. For low Tg underfills, the stress relief was much lower, but still appreciable: about 19%. Thicker USCB layers could be more effective, because, as has been shown in our earlier publications and confirmed experimentally, elevated stand-off heights of solder joint interconnections are able to provide appreciable stress relief in the solder material by making the bonding system more compliant. Indeed, for the thickness of 0.75 mm (impossible for FC designs, but rather typical for FPBGA systems) the decrease in the normal stress acting in the USCB cross sections is appreciable, and the decrease in the shearing stress at the USCB/chip interface is as high as 70% in the case of high T_g underfill and even higher, 76%, in the case of low T_g underfill. The employed analytical stress model used in this analysis can be used for the selection of the adequate underfill material and establishing the appropriate USCB thickness at the design stage. It is noteworthy that, as long as the linear approach is used and the induced stresses are proportional to the change in temperature, the developed model can be used also in situations, when the underfill's T_g is between the temperature extremes that the assembly of interest experiences during its accelerated testing and in actual operation conditions.

The carried out numerical example shows how the model could be used in practical computations. It shows also that copper lid leads to higher thermally induced stresses than an organic lid. This is true for both the normal stresses in the chip and the maximum interfacial shearing stresses. The developed model can be employed in the analysis of a flip-chip package design of the type in question. Future work should include FEA verifications. The suggested analytical stress model can be particularly helpful when developing an FEA preprocessing model.

The developed analytical thermal stress model can be used for the assessment of the effect of the T_g of the underfill encapsulant and the thickness of the USCB on the induced stresses. The calculations were carried out for two T_g levels, above and below the operation and testing temperature range for the FC or FPBGA assembly with a USCB, and for two thicknesses, 0.05 mm and 0.1 mm, of the USCB. The maximum predicted shearing stresses do indeed occur at the USCB/chip, and not at the USCB/substrate interface, and this result is in agreement with the observed delaminations at the USCB/chip interface.

The calculated data indicate that there is an incentive for using low T_g underfills, provided, of course, that their adhesive strength is proven to be sufficient. This might me less of a challenge, even if the cohesive and the adhesive strength of the USCB layer with a low T_g is not very high, but the applied stresses are also relatively low.

As to the incentive for using thicker USCBs, the increase in this thickness from 0.05 mm to 0.1 mm resulted in rather minor relief in the normal stress in the bond for both high and low T_g underfills, but led to an appreciable relief in the interfacial stresses at the USCB/chip interface, especially for high T_g underfills. The stress relief for them was as high as 34%. For low T_g underfills, the relief was much lower, about 19%. However, for the thickness of 0.75 mm (impossible for FC designs, but quite typical for FPBGA systems), the decrease in the normal stress in the bond is appreciable, and the decrease in the shearing stress at the USCB with the chip is as high as 70% in the case of high T_g underfill and 76% in the case of low T_g underfill.

Thicker USCB layers could be more effective, because, as has been shown earlier, elevated stand-off heights of solder joint interconnections are able to provide appreciable stress relief in the solder joints by making the bond more compliant. This should be attributed in part to the use of solder joints with elevated stand-off heights.

Future work should include FEA evaluations and experimental verification of the obtained information; development of a predictive model for the case when the T_g of the underfill is between the curing and the low temperatures, and for the situation when the dependences of the Young's modulus and CTE of the underfill are characterized by a nonzero width of the transition region; in effect, the developed model can be used for that; development of a methodology for the evaluation of the role of the viscoelastic behavior of the underfill material; and development of a methodology for the evaluation of the effective modulus and CTE of the USCB layer (this could be done particularly using [26]).

REFERENCES

1. E. Suhir, J.M. Segelken, "Mechanical Behavior of Flip-Chip Encapsulants," *ASME Journal of Electrical Packaging (JEP)*, vol. 112, No. 4, 1990.
2. D. Suryanarayana, R. Hsiao, T.P. Gall, J.M. McCreary, "Enhancement of Flip-Chip Fatigue Life by Encapsulation," *IEEE Transactions of Component Hybrids Manufacturing Technology*, vol. 14, 1991.
3. D. Suryanarayana, T.Y. Wu, J.A. Varcoe, "Encapsulants Used in Flip-Chip Packages," 43rd ECTC, Orlando, Florida, June 1993.
4. J. Clementi, J. McCreary, T.M. Niu, J. Palomaki, J. Varcoe, G. Hill, "Flip-Chip Encapsulation on Ceramic Substrates," IEEE 43rd ECTC, Orlando, Florida, June 1993.
5. J.H. Lau, *Flip Chip Technologies*, McGraw-Hill, 1995.
6. K. Chai, E. Wu, R. Hsieh, J.Y. Tong, "Challenge of Flip Chip Encapsulation Technologies," Proceedings of SPIE, Denver, Colorado, 2002.
7. J. Lau, L.M. Powers-Maloney, J.R. Baker, D. Rice, B. Shaw, "Solder Joint Reliability of Fine Pitch Surface Mount Technology Assemblies," 7th IEEE/CHMT International Electronic Manufacturing Technology Symposium, San Francisco, California, Sept. 1989.
8. J.B. Nysaether, P. Lundstrom, J. Liu, "Measurements of Solder Bumps Lifetime as a Function of the Underfill Material Properties," 1st IEEE International Symposium on Polymeric Electronics Packaging, PEP '97, Norrkoping, Sweden, Oct. 1997.
9. S. Rzepka, M. Korhonen, E. Meusel, "The Effect of Underfill and Underfill Delamination on the Thermal Stress in Flip-Chip Solder Joints," *ASME Journal of Electronic Packaging*, vol. 120, No. 4, 1998.
10. J.E. Semmens, T. Adams, "Flip Chip Package Failure Mechanism," *Solid State Technology*, vol. 41, 1998.
11. J.H. Lau, C. Chang, "How to Select Underfill Materials for Solder Bumped Flip-Chips on Low Cost Substrates?" *International Journal of Microcircuits and Electronic Packaging*, vol. 22, 1999.
12. P. Su, S. Rzepka, M. Korhonen, "The Effects of Underfill on the Reliability of Flip Chip Solder Joints," *Symposium on Fatigue and Internal Friction in Miniature Structures and Components*, TMS Annual Meeting, vol. 28, No. 9, Mar. 1999.
13. C.P. Wong, S. Lou, Z. Zhang, "Flip-the-Chip," *Science*, vol. 290, 2000.
14. R. Dudek, A. Schubert, B. Michel, "Analysis of Flip Chip Attach Reliability," 4th International Conference on Adhesive Joining and Coating Technology in Electronics Manufacturing, Espoo, Finland, June 2000.
15. J. Lau, S. Lee, C. Chang, "Effects of Underfill Material Properties on the Reliability of Solder Bumped Flip Chip on Board with Imperfect Underfill Encapsulants," *IEEE CPMT Transactions*, vol. 23, No. 2, 2000.
16. X.J. Fan, H.B. Wang, T.B. Lim, "Investigation of the Underfill Delamination and Cracking in Flip-Chip Modules under Temperature Cyclic Loading," *IEEE Transactions on Components Packaging Technology*, vol. 24, 2001.
17. K. Hirohata, N. Kawamura, M. Mukai, T. Kawakami, H. Aoki, K. Takahashi, "Mechanical Fatigue Test Method for Chip/Underfill Delamination in Flip-Chip Packages," *IEEE Transactions on Electronics Packaging Manufacturing*, vol. 25, 2002.
18. I. Dutta, A. Gopinath, C. Marshall, "Underfill Constraint Effects During Thermo-Mechanical Cycling of Flip-Chip Solder Joints," *Journal of Electronic Materials*, vol. 31, No. 4, 2002.
19. J. Qu, C.P. Wong, "Effective Elastic Modulus of Underfill Material for Flip-Chip Applications," *IEEE CPMT Transactions*, vol. 25, 2002.

20. T. Wang, Y. Lai, J. Wu, "Effect of Underfill Thermo-Mechanical Properties on Thermal Cycling Fatigue Reliability of Flip-Chip Ball Grid Array," *Symposium on Thermal Management of Electronic Systems*, vol. 126, No. 4, Nov. 2003.
21. L. Mercado, V. Sarihan, "Evaluation of the Die Attach Cracking in Flip-Chip PBGA Packages," *IEEE CPMT Transactions*, vol. 26, 2003.
22. T. Chen, J. Wang, D. Lu, "Emerging Challenges of Underfill for Flip-Chip Applications," 54th ECTC, Las Vegas, Nevada, June 2004.
23. C.J. Zhai, R.C. Blish, R.N. Master, "Investigation and Minimization of Underfill Delamination in Flip-Chip Packages," *IEEE Transactions on Developing Material Reliability*, vol. 4, 2004.
24. C.-T. Kuo, M.-C. Yip, K.-N. Chiang, "Time and Temperature-Dependent Mechanical Behavior of Underfill Materials in Electronic Packaging Application," *Microelectronics Reliability*, vol. 44, 2004.
25. B.-I. Noh, B.-Y. Lee, S.-B. Jung, "Thermal Fatigue Performance of SnAgCu Chip-Scale Package with Underfill," *Materials Science and Engineering A*, No. 483–484, 2008.
26. Z. Zhang, S.B. Park, K. Darbha, R.N. Master, "Effect of Glass Transition Slope of Underfill on Solder Joint Fatigue Life," 11th International Conference on Electronic Packaging Technology and High Density Packaging, Aug. 2010.
27. J.B. Kwak, S. Chung, "The Effects of Underfill on the Thermal Fatigue Reliability of Solder Joints in Newly Developed Flip Chip on Module," IEEE I-Therm, San Diego, California, May–June 2012.
28. R. Ghaffarian, "Underfill Optimization for FPGA Package/Assembly," Jet Propulsion Lab., Pasadena, California, 2012.
29. J.-Y. Chang, S-Y. Huang, C.-C. Lee, T.-C. Chuang, "Influence of Glass Transition Temperature of Underfill on the Stress Behavior and Reliability of Microjoints Within a Chip Stacking Architecture," *ASME Journal of Electronic Packaging*, vol. 137, No. 3, Sept. 2015.
30. E. Suhir, R. Ghaffarian, "Flip-Chip (FC) and Fine-Pitch-Ball-Grid-Array (FPBGA) Underfills for Application in Aerospace Electronics Packages–Brief Review," *Aerospace*, vol. 5, No. 3, July 2018.
31. E. Suhir, R. Ghaffarian, "Predicted Effect of the Underfill Glass Transition Temperature on Thermal Stresses in a Flip-Chip or a Fine-Pitch BGA Design," *Journal of Electrical and Electronic Systems (JEES)*, vol. 7, No. 4, 2018.
32. E. Suhir, "Avoiding Low-Cycle Fatigue in Solder Material Using Inhomogeneous Column-Grid-Array (CGA) Design," *Chip Scale Reviews*, March–April 2016.
33. E. Suhir, "Relieving Stress in Flip-Chip Solder Joints," *Chip Scale Reviews*, vol. 21, No. 5, Sept.–Oct. 2017.
34. J. Schijve, "Fatigue of Structures and Materials in the 20th Century and the State of the Art," *International Journal of Fatigue*, vol. 25, No. 8, 2003.
35. S. Suresh, *Fatigue of Materials*, Cambridge University Press, 2004.
36. E. Suhir, "Stresses in Bi-metal Thermostats," *ASME Journal of Applied Mechanics*, vol. 53, No. 3, Sept. 1986.
37. E. Suhir, "Interfacial Stresses in Bi-Metal Thermostats," *ASME Journal of Applied Mechanics*, vol. 56, No. 3, Sept. 1989.
38. E. Suhir, "Analysis of Interfacial Thermal Stresses in a Tri-Material Assembly," *Journal of Applied Physics*, vol. 89, 2001.
39. E. Suhir, "Flip-Chip Assembly: Is a Bi-Material Model Acceptable?" *Journal of Materials Science: Materials in Electronics*, vol. 28, No. 21, 2017.
40. E. Suhir, "'Global' and 'Local' Thermal Mismatch Stresses in an Elongated Bi-Material Assembly Bonded at the Ends," in E. Suhir, (ed.), *Structural Analysis in Microelectronic and Fiber-Optic Systems*, Symposium Proceedings, ASME Press, 1995.

41. E. Suhir, R. Ghaffarian, "Column-Grid-Array (CGA) vs. Ball-Grid-Array (BGA): Board-Level Drop Test and the Expected Dynamic Stress in the Solder Material," *Journal of Materials Science: Materials in Electronics*, vol. 27, No. 11, 2016.
42. E. Suhir, R. Ghaffarian, "Board Level Drop Test: Exact Solution to the Problem of the Nonlinear Dynamic Response of a PCB to the Drop Impact," *Journal of Materials Science: Materials in Electronics*, vol. 27, No. 9, 2016.
43. E. Suhir, R. Ghaffarian, "Predictive Modeling of the Dynamic Response of Electronic Systems to Impact Loading: Review," *Zeitschrift für Angewandte Mathematik und Mechanik (ZAMM)*, vol. 97, No. 6, 2017.
44. M.-Y. Tsai, Y.-C. Lin, C.Y. Huang, J.D. Wu, "Thermal Deformations and Stresses of Flip-Chip BGA Packages with Low- and High-T_g Underfills," *IEEE Transactions on Electronics Packaging Manufacturing*, vol. 28, 2005.
45. Z. Zhang, S.-B. Park, K. Darbha, R.N. Master, "Effect of Glass Transition Slope of Underfill on Solder Joint Fatigue Life," 11th International Conference on Electronic Packaging Technology and High Density Packaging, Aug. 2010.
46. E. Suhir, "Relieving Stress in Flip-Chip Solder Joints," *Chip Scale Reviews*, vol. 21, No. 5, Sept.-Oct. 2017.
47. E. Suhir, "Analytical Thermal Stress Model for a Typical Flip-Chip Package Design," *Journal of Materials Science: Materials in Electronics*, vol. 29, No. 4, 2018.
48. E. Suhir, "Low-Cycle-Fatigue Failures of Solder Material in Electronics: Analytical Modeling Enables to Predict and Possibly Prevent Them-Review," *Journal of Aerospace Engineering and Mechanics (JAEM)*, vol. 2, No. 1, 2018.

8 Assessed Interfacial Strength and Elastic Moduli of the Bonding Material from Shear-Off Test Data

"By asking for the impossible, obtain the best possible."

—Italian saying

8.1 BACKGROUND/INCENTIVE

Lap shear testing is widely performed to evaluate the interfacial shear strength and stiffness of composite materials, assemblies, and structures. Outside the field of electronic materials, such testing is usually carried out on fasteners, such as bolts, machine screws, and rivets. In electronics materials science, shear testing is performed to determine the shear strength of a bonding material (adhesive, solder) and can also be used to compare between different bonding materials and/or their thicknesses and/or for different lots for the same bonding material. Here is a brief summary of the existing shear test methods and standards.

ASTM D1002 addresses shear strength of single-lap-joint adhesively bonded metal specimens by tension loading and could be viewed as the most common standard test involving shear. ASTM D732 is the standard test method for shear strength of plastics by using a punch tool. ASTM D2344 is the standard test method for short-beam strength of polymer composite materials and laminates. ASTM D3163 is a standard test method for determining strength of adhesively bonded plastic lap-shear joints in shear by applying tension loading. ASTM D3164 is another standard test method for strength properties of adhesively bonded plastic lap-shear "sandwich" in shear by tension loading. ASTM D5868 provides a test method for lap shear adhesion for fiber reinforced plastic bonding. ASTM D5379 suggests double notch off-axis test rail shear test, as well as torsion tests to check adhesion. Inter-laminar shear strength (ILSS) procedure is suggested by EN ISO 14130, which replaced, with input from, ASTM D2344 and other standards. The test only provides an "apparent" shear strength using an assumed isotropic material stress distribution. This excludes the effect of other stresses due to the flexural loading and contact points. In addition, as follows from the analysis carried out in this paper, the interfacial shearing stress is distributed highly nonuniformly along the interface when shear-off tests are conducted. This important circumstance is not considered by this standard.

The bonding strength of various attachment materials, both adhesives and solders, has always been of significant importance in electronics system engineering and, hence, of great interest to reliability engineers, including modeling effort; therefore, the published work in this field, mostly experimental, is enormous (see, for example, [1–29]). The main shearing strength–related reliability standards in electronics are ASTM F1269-89 "Test Methods for Destructive Shear Testing of Ball Bonds," 1991 Annual Book of ASTM Standards, vol. 10.04, 1991, ASTM, Philadelphia and JESD22-B117 "BGA Ball Shear," July 2000, JEDEC Solid State Technology Association.

During the last decade or so, silver sintering has become increasingly important as an attractive bonding technique in power and, particularly, automotive electronics [30–47]. There is an indication that silver sintered pastes have excellent adhesion strength, good performance in active power cycling and in temperature cycling, high thermal stability, and improved performance over tin-silver-copper (SAC) and other lead-free solders widely used today. The bottleneck of the modeling effort, as far as design-for-reliability (DfR) with silver sintering technology is concerned, is the availability of the trustworthy information about the ultimate strength and mechanical properties of the silver sintering material, especially having in mind that these properties might be dependent on the fabrication conditions and the two adherents between which the bonding layer is sandwiched. Accordingly, in the analysis that follows, a simple and physically meaningful analytical stress model is developed in application to shear-off testing of a bonded assembly with an objective to evaluate the magnitude and the distribution of the interfacial shearing stress in the bonding material from the measured shear-off force. The model can also be used for the evaluation of the shear modulus of the bonding material if the corresponding interfacial displacement is also concurrently measured. The experimental data obtained using the suggested model for interpreting shear-off experimental data will enable one to develop a silver sintering material with the best power cycle performance. The model can also be used, of course, for conventional solders and adhesives, including developing of a suitable standard. The model was also used to explain the physics of specially prepared shear-off testing specimens suggested by J. Kivilahti and T. Reinikainen [48–50] to obtain the most accurate information of the solder material properties.

8.2 BASIC EQUATION

Consider a bonded elongated assembly schematically shown in Figure 8.1. Its lower component #2 ("substrate") is rigidly attached to the tester, and the upper component #1 ("chip") is subjected to an external shearing force $\hat{T}$. Using the concept of the interfacial compliance [19] (see Chapter 2), the longitudinal interfacial displacements of these components can be sought in the form

$$u_1(x) = \lambda_1 \int_0^x T_1(\xi)d\xi + \kappa_1 \tau(x), \quad u_2(x) = \lambda_2 \int_0^x T_2(\xi)d\xi - \kappa_2 \tau(x), \tag{8.1}$$

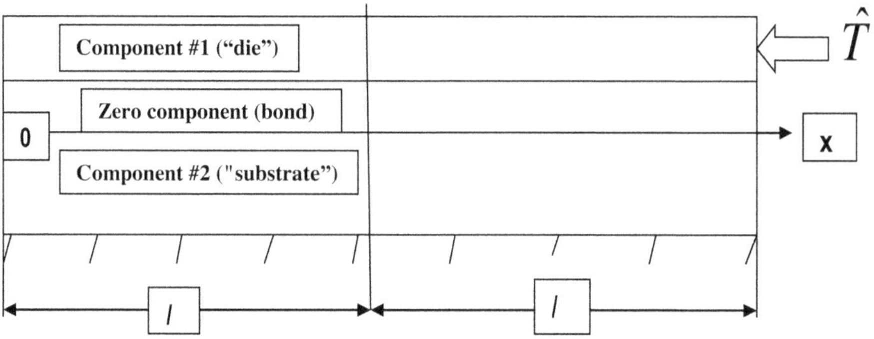

FIGURE 8.1 Bimaterial bonded assembly subjected to shear-off testing.

where the force

$$T_1(x) = \hat{T} - \int_0^x \tau(\xi)d\xi \tag{8.2}$$

acting in the cross sections of component #1 can be found by subtracting the force

$$T_2(x) = \int_0^x \tau(\xi)d\xi \tag{8.3}$$

acting in the cross sections of component #2 from the total external force $\hat{T}$; $\tau(x)$ is thus far unknown interfacial shearing stress;

$$\lambda_1 = \frac{1-\nu_1}{E_1 h_1}, \quad \lambda_2 = \frac{1-\nu_2}{E_2 h_2} \tag{8.4}$$

are the axial compliances of the assembly components; E_1 and E_2 are Young's moduli of the materials; ν_1 and ν_2 are their Poisson's ratios; h_1 and h_2 are the component thicknesses;

$$\kappa_1 = \frac{h_1}{3G_1}, \quad \kappa_2 = \frac{h_2}{3G_2} \tag{8.5}$$

are the interfacial compliances of the components [19]; and

$$G_1 = \frac{E_1}{2(1+\nu_1)}, \quad G_2 = \frac{E_2}{2(1+\nu_2)}, \tag{8.6}$$

are the shear moduli of the component materials. The origin of the coordinate x is at the left end of the assembly.

The first terms in expressions (8.1) are the displacements of the components' cross sections caused by the forces $T_1(x)$ and $T_2(x)$ and determined based on Hooke's law assuming that all the points of the given cross section of the given component have the same axial displacements. The second terms in (8.1) are, in effect, corrections to this assumption. They consider that the longitudinal interfacial displacements are somewhat greater than the displacements of the inner points of the cross section. The structure of these terms reflects an assumption that the corrections in question can be found as the product of the interfacial compliance of the component and the interfacial shearing stress acting in the given cross section. This is, actually, the substance of the concept of the interfacial compliance (8.19).

The condition of the compatibility of the displacements (8.1) can be written as

$$u_1(x) - u_2(x) = \kappa_0 \tau(x), \tag{8.7}$$

where

$$\kappa_0 = \frac{h_0}{G_0} \tag{8.8}$$

is the interfacial compliance of the bonding layer, h_0 is its thickness, and $G_0 = \dfrac{E_0}{2(1+\nu_0)}$ is its shear modulus. Introducing expressions (8.1) into the displacement compatibility condition (8.7), the following equation for the sought interfacial shearing stress function $\tau(x)$ can be obtained:

$$\kappa\tau(x) + \lambda \int_0^x T_1(\xi)d\xi = -\lambda_2 \hat{T}. \tag{8.9}$$

Here,

$$\kappa = \kappa_0 + \kappa_1 + \kappa_2 \tag{8.10}$$

is the total interfacial compliance of the assembly and

$$\lambda = \lambda_1 + \lambda_2 \tag{8.11}$$

is its total axial compliance. It is noteworthy that while the total longitudinal interfacial compliance considers the compliance of the bonding layer, the axial compliance is due to the two bonded components only. This assumption is valid as long as the bonding layer is thin, and its Young's modulus is significantly lower than the Young's moduli of the component materials. This is usually the case in bonded electronic assemblies.

Differentiating equation (8.9) with respect to the longitudinal coordinate x and considering that, in accordance with relationship (8.2),

$$\tau'(x) = -T_1''(x), \tag{8.12}$$

the following basic differential equation for the force $T_1(x)$ can be obtained:

$$\kappa T_1''(x) - \lambda T_1(x) = \lambda_2 \hat{T} \tag{8.13}$$

8.3 SOLUTION TO THE BASIC EQUATION

The solution to the differential equation (13) can be sought as

$$T_1(x) = C_0 + C_1 \sinh kx + C_2 \cosh kx. \tag{8.14}$$

Introducing this solution into equation (8.13), one could conclude that this equation is fulfilled, if the relationships

$$k = \sqrt{\frac{\lambda}{\kappa}}, \quad C_0 = -\frac{\lambda_2}{\lambda}\hat{T} \tag{8.15}$$

take place. The first formula determines the parameter of the interfacial shearing stress, and the second formula defines the particular solution to the differential equation (8.13). The constants C_1 and C_2 of integration can be found from the boundary conditions $T_1(0) = 0$ and $T_1(2l) = -\hat{T}$ and are as follows:

$$C_1 = -\frac{\hat{T}}{\sinh 2kl}\left(\frac{\lambda_1}{\lambda} + \frac{\lambda_2}{\lambda}\cosh 2kl\right), \quad C_2 = \frac{\lambda_2}{\lambda}\hat{T}. \tag{8.16}$$

Then, solution (8.14) yields

$$T_1(x) = -\hat{T}\left[\frac{\lambda_2}{\lambda}(1 - \cosh kx) + \left(\frac{\lambda_1}{\lambda} + \frac{\lambda_2}{\lambda}\cosh 2kl\right)\frac{\sinh kx}{\sinh 2kl}\right] \tag{8.17}$$

8.4 INTERFACIAL SHEARING STRESS

Formula (8.2) indicates that the interfacial shearing stress can be found from (8.17) by differentiation:

$$\tau(x) = -T_1'(x) = k\hat{T}\left[\left(\frac{\lambda_1}{\lambda} + \frac{\lambda_2}{\lambda}\cosh 2kl\right)\frac{\cosh kx}{\sinh 2kl} - \frac{\lambda_2}{\lambda}\sinh kx\right]. \tag{8.18}$$

This stress changes from

$$\tau(0) = k\hat{T}\left[\left(\frac{\lambda_1}{\lambda} + \frac{\lambda_2}{\lambda}\cosh 2kl\right)\frac{1}{\sinh 2kl}\right] \tag{8.19}$$

at the origin to its maximum value

$$\tau(2l) = k\hat{T}\left(\frac{\lambda_1}{\lambda}\coth 2kl + \frac{\lambda_2}{\lambda}\frac{1}{\sinh 2kl}\right) \tag{8.20}$$

at the end, where the external force is applied. For long enough assemblies with stiff interfaces ($kl \geq 2.5$), the shearing stress (8.19) becomes next-to-zero, and the stress (8.20) is

$$\tau(2l) = k\hat{T}\frac{\lambda_1}{\lambda}. \tag{8.21}$$

This formula indicates that the maximum interfacial shearing stress is proportional to the parameter k of the interfacial shearing stress and to the ratio $\frac{\lambda_1}{\lambda}$ of the axial compliance of component #1 that experiences the direct action of the applied force to the total axial compliance of the assembly. For very short assemblies with compliant interfaces, when $\cosh 2kl$ can be put equal to one, formulas (8.19) and (8.20) provide the same result. When the force $\hat{T}$ is measured, the induced interfacial shearing stress can be evaluated, for a long enough specimen, by using formula (8.21). Note that if the assembly is fabricated at an elevated temperature and is subsequently cooled down by the temperature Δt to a low (room, testing) temperature, it experiences already thermally induced stresses caused by the thermally induced force

$$T_1 = \frac{\Delta\alpha\Delta t}{\lambda}, \tag{8.22}$$

where $\Delta\alpha$ is the difference in the CTE of the bonded components. If this force is appreciable, formula (8.21) should be replaced with formula

$$\tau(2l) = k(\hat{T} + T_t)\frac{\lambda_1}{\lambda} \tag{8.23}$$

to consider for the fact that the bonding material is subjected to the combined action of the thermally induced and mechanical loading, and its actual ultimate bonding strength is, in effect, higher than that predicted by formula (8.21).

8.5 SHEAR MODULUS OF THE BONDING MATERIAL

The maximum interfacial displacement can be found as

$$u_{\max} = u(2l) = \kappa\tau(2l) = \kappa k\hat{T}\frac{\lambda_1}{\lambda} = \frac{\hat{T}}{k}\lambda_1. \tag{8.24}$$

Then, the parameter k of the interfacial shearing stress is

$$k = \lambda_1\frac{\hat{T}}{u_{\max}} \tag{8.25}$$

Using the first formula in (8.15), we have:

$$\kappa = \lambda \left(\frac{u_{max}}{\lambda_1 \hat{T}} \right)^2 \tag{8.26}$$

Then, using formula (8.10), one can determine the interfacial compliance of the bonding material as

$$\kappa_0 = \lambda \left(\frac{u_{max}}{\lambda_1 \hat{T}} \right)^2 - \kappa_1 - \kappa_2 \tag{8.27}$$

Then, the shear modulus of this material can be found as

$$G_0 = \frac{h_0}{\kappa_0}. \tag{8.28}$$

By assuming a suitable Poisson's ratio of the bonding material, its Young's modulus could be tentatively assessed as

$$E_0 = 2(1 + \nu_0)G_0 \tag{8.29}$$

8.6 NUMERICAL EXAMPLE

Input data

Structural element	#1 (Package)	#2 (Substrate)	0 (Bonding Material)
Young's modulus, kg/mm^2	8775	2320	TBD
Poisson's ratio	0.28	0.32	0.35
Shear modulus, kg/mm^2	3428	879	TBD
CTE, 1/°C	6.5E-6	15.0E-6	x
Thickness, mm	2.0	1.5	0.05
Half package (assembly) length, $l = 7.5$ mm			
Measured force (within the elastic region) $\hat{T} = 0.325\,\text{kg/mm}$			
Corresponding (measured) displacement, $u_{max} = 2.6 \times 10^{-5}$ mm			
Force at failure $\hat{T}_f = 36.0\,\text{kg/mm}$			

Calculated data

Axial compliances:

$$\lambda_1 = \frac{1-\nu_1}{E_1 h_1} = \frac{1-0.28}{8775 \times 2.0} = 4.1026 \times 10^{-5}\ \text{mm/kg},$$

$$\lambda_2 = \frac{1-\nu_2}{E_2 h_2} = \frac{1-0.32}{2320 \times 1.5} = 19.5402 \times 10^{-5}\ \text{mm/kg}$$

$$\lambda = \lambda_1 + \lambda_2 = 4.1026 \times 10^{-5} + 19.5402 \times 10^{-5} = 23.6428 \times 10^{-5}\ \text{mm/kg}$$

Total interfacial compliance of the assembly

$$\kappa = \lambda\left(\frac{u_{\max}}{\lambda_1 \hat{T}}\right)^2 = 23.6428\times10^{-5}\left(\frac{2.6\times10^{-5}}{4.1026\times10^{-5}\times0.325}\right)^2 = 89.9002\times10^{-5}\ \text{mm}^3/\text{kg}$$

Parameter of the interfacial shearing stress

$$k = \sqrt{\frac{\lambda}{\kappa}} = \sqrt{\frac{23.6428\times10^{-5}}{89.9002\times10^{-5}}} = 0.5128\,\text{mm}^{-1}$$

Shearing stress at the assembly end

$$\tau(2l) = k\hat{T}\frac{\lambda_1}{\lambda} = 0.5128\times0.325\times\frac{4.1026\times10^{-5}}{23.6428\times10^{-5}} = 0.0289\,\text{kg/mm}^2$$

Stress at failure

$$\tau_f = k\hat{T}_f\frac{\lambda_1}{\lambda} = 0.5128\times36.0\times\frac{4.1026\times10^{-5}}{23.6428\times10^{-5}} = 3.2034\,\text{kg/mm}^2$$

Interfacial compliances of the assembly components

$$\kappa_1 = \frac{h_1}{3G_1} = \frac{2}{3\times3428} = 19.4477\times10^{-5}\ \text{mm}^3/\text{kg}$$

$$\kappa_2 = \frac{h_2}{3G_2} = \frac{1.5}{3\times879} = 56.8828\times10^{-5}\ \text{mm}^3/\text{kg}$$

Interfacial compliance of the bonding layer

$$\kappa_0 = \kappa - \kappa_1 - \kappa_2 = (89.9002 - 19.4477 - 56.8828)\times10^{-5} = 13.5697\times10^{-5}\ \text{mm}^3/\text{kg}$$

Shear modulus of the bonding material

$$G_0 = \frac{h_0}{\kappa_0} = \frac{0.05}{13.5697\times10^{-5}} = 368.5\,\text{kg/mm}^2$$

Tentative Young's modulus

$$E_0 = 2(1+\nu_0)G_0 = 2\times1.35\times368.5 = 994.9\,\text{kg/mm}^2$$

8.7 POSSIBLE CHARACTERIZATION OF THE SOLDER MATERIAL PROPERTIES

The developed model was also used to explain a paradoxical situation that was originally detected on the basis of finite element analysis (FEA) by Jorma Kivilahti and Tommi Reinikainen [48]: deep enough transverse grooves in the adherends ("pins")

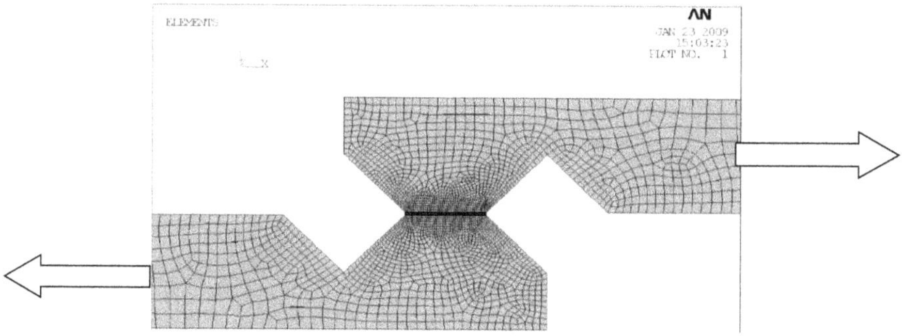

FIGURE 8.2 Transverse grooves in the adherends lead to lower and more uniformly distributed interfacial stresses.

(Figure 8.2) resulted in an appreciable reduction in, and to the more uniform distribution of, the interfacial shearing and peeling stresses. This phenomenon, important from the standpoint of testing solder materials, was explained based on analytical modeling by Suhir and Reinikainen [49, 50]. The model indicates that the observed phenomenon is due primarily to the increase in the interfacial compliance of the bonding structure: the grooves "convert" the adherend ("pin") portions located between the inner portions of the grooves and the bonding (adhesive, solder) layer into parts of the bonding structure, thereby increasing the compliance of this structure with respect to the shearing load. This positive effect overwhelms, as far as the magnitude and the distribution of the interfacial stresses are concerned, the negative effect of the increased axial compliance of the loaded portions of the adherends ("pins"), because of the grooves.

The analytical predictions agree well with the FEA data, despite the FEA overestimation of the increase in the interfacial stresses in the proximity of the joint edges (as is known, FEA method, which is based on one of the numerical methods of the elasticity theory, leads to singular stresses at the edges of assemblies comprised of dissimilar materials).

The obtained information explains the physics of the phenomenon in question, and the developed analytical models can be used in the analysis and physical design of soldered lap shear joints. It is also concluded that analytical modeling is able not only to come up with relationships that clearly indicate "what affects what and what is responsible for what," but, more importantly, can explain the physics of phenomena that neither the FEA modeling ("simulation"), nor even actual experimentation is able to. It is recommended that such tests are considered to obtain the mechanical properties of the solder materials for further analytical or FEA modeling.

8.8 CONCLUSION

An analytical stress model is developed in application to shear-off testing with an objective to evaluate the interfacial shearing stress in the bonding material from the measured shear-off force. The model can also be used for the evaluation of the elastic moduli of the bonding material if the interfacial displacement is also measured. In the author's opinion, the suggested methodology, based on the concept of the interfacial compliance, could become a basis for a new effective experimental method for assessing the interfacial shearing strength and elastic moduli of the bonding material in electronics. The methodology can be used particularly in application to the recently suggested sintered silver bonding materials to evaluate their bonding strength from the measured force-at-failure and shear modulus from the measured shearing force and displacement.

REFERENCES

1. M. Klein, B. Wiens, M. Hutter, H. Oppermann, R. Aschenbrenner, H. Reichl, "Behaviour of Platinum as UBM in Flip Chip Solder Joints," Proceedings of the 50th ECTC, Nevada, May 2000.
2. R.H. Uang, K.C. Chen, S.W. Lu, H.T. Hu, S.H. Huang, "The Reliability Performance of Low Cost Bumping on Aluminum and Copper Wafer," Proceedings of the 3rd Electronics Packaging Technology Conference (EPTC 2000), Singapore, Dec. 2000.
3. R.J. Coyle, P.P. Solan, A.J. Serafino, S.A. Gahr, "The Influence of Room Temperature Aging on Ball Shear Strength and Microstructure of Area Array Solder Balls," Proceedings of the 50th ECTC, Nevada, May 2000.
4. K.M. Levis, A. Mawer, "Assembly and Solder Joint Reliability of Plastic Ball Grid Array with Lead-free versus Lead-Tin Interconnect," Proceedings of the 50th ECTC, Nevada, May 2000.
5. Y. Tomita, Q. Wu, A. Maeda, S. Baba, N. Ueda, "Advanced Surface Plating on the Organic FC-BGA Package," Proceedings of the 50th ECTC, Nevada, May 2000.
6. I.S. Kang, et al., "The Solder Joint and Runner Metal Reliability of Wafer-Level CSP (Omega-CSP)," Proceedings of the 50th ECTC, Nevada, May 2000.
7. S.-W.R. Lee, C.C. Yan, Z. Karim, X. Huang, "Assessment on the Effect of Electroless Nickel Plating on the Reliability of Solder Ball Attachment to the Bond Pads of PBGA Substrate," Proceedings of the 50th ECTC, Nevada, May 2000.
8. C.K. Shin, J.Y. Huh, "Effect of Cu-containing Solders on the Critical IMC Thickness for the Shear Strength of BGA Solder Joints," Proceedings of the 3rd Electronics Packaging Technology Conference (EPTC2000), Singapore, Dec. 2000.
9. S.Y. Jang, K.W. Paik, "Comparison of Electroplated Eutectic Sn/Bi and Pb/Sn Solder Bumps on Various UBM Systems," Proceedings of the 50th ECTC, Nevada, May 2000.
10. S.J. Cho, J.Y. Kim, M.G. Park, I.S. Park, H.S. Chun, "Under Bump Metallurgies for a Wafer Level CSP with Eutectic Pb-Sn Solder Ball," Proceedings of the 50th ECTC, Nevada, May 2000.
11. S.-W.R. Lee, K. Newman, L. Hu, "Thermal Fatigue Analysis of PBGA Solder Joints with the Consideration of Damage Evolution," Packaging of Electronic and Photonic Devices, EEP, vol. 28, Florida, Nov. 2000.
12. D. Vogel, R. Kiihnert, M. Dost, B. Michel, "Determination of Packaging Material Properties Utilizing Image Correlation Techniques," *Journal of Electronic Packaging*, vol. 124, 2002.
13. A. Schubert, R. Dudek, E. Auerswald, A. Gollhardt, B. Michel, H. Reichl, "Fatigue Life Models for SnAgCu and SnPb Solder Joints Evaluated by Experiments and Simulation," 53rd ECTC, 2003.
14. R. Dudek, W. Faust, J. Vogel, B. Michel, "In-situ Solder Fatigue Studies Using a Thermal Lap Shear Test," Proceedings of the International Conference on Electronics Packaging Technology, 2004.

15. R. Dudek, H. Walter, R. Doring, B. Michel, "Thermal Fatigue Modelling for SnAgCu and SnPb Solder Joints," Proceedings of EuroSimE 2004, Brussels, Belgium, 2004.
16. A. Wymysłowski, et al., "Shearing tests for solder joints reliability assessment," XXXI International Conference of IMAPS Poland Chapter, Rzeszów-Krasiczyn, Poland, Sept. 2007.
17. J. Lau (ed.), *Solder Joint Reliability: Theory and Applications*, Van Nostrand Reinhold, 1990.
18. W. Engelmaier, "Reliability for Surface Mount Solder Joints: Physics and Statistics of Failure," Proceedings of Surface Mount International, vol. 1, San Jose, California, Aug. 1992.
19. E. Suhir, "Stresses in Bi-Metal Thermostats," *ASME Journal of Applied Mechanics*, vol. 53, No. 3, Sept. 1986.
20. E. Suhir, "Interfacial Stresses in Bi-Metal Thermostats," *ASME Journal of Applied Mechanics*, vol. 56, No. 3, September 1989.
21. A.Y. Kuo, "Thermal Stress at the Edge of a Bi-Metallic Thermostat," *ASME Journal of Applied Mechanics*, vol. 57, 1990.
22. C. Caswell, "Manufacturing and Reliability Challenges with QFN", IMAPS Conference, Atlantic City, New Jersey, June 2010.
23. E. Suhir, "Thermal Stress in a Bi-Material Assembly Adhesively Bonded at the Ends," *Journal of Applied Physics*, vol. 89, No. 1, 2001.
24. E. Suhir, "Thermal Stress in an Adhesively Bonded Joint with a Low Modulus Adhesive Layer at the Ends," *Journal of Applied Physics*, April 2003.
25. E. Suhir, "Interfacial Thermal Stresses in a Bi-Material Assembly with a Low-Yield-Stress Bonding Layer," *Modeling and Simulation in Materials Science and Engineering*, vol. 14, 2006.
26. E. Suhir, "On a Paradoxical Situation Related to Bonded Joints: Could Stiffer Mid-Portions of a Compliant Attachment Result in Lower Thermal Stress?" *Journal of Solid Mechanics and Materials Engineering (JSMME)*, vol. 3, No. 7, 2009.
27. E. Suhir, "Thermal Stress in a Bi-Material Assembly with a 'Piecewise-Continuous' Bonding Layer: Theorem of Three Axial Forces," *Journal of Physics D: Applied Physics*, vol. 42, 2009.
28. E. Suhir, L. Bechou, B. Levrier, "Predicted Size of an Inelastic Zone in a Ball-Grid-Array Assembly," *ASME Journal of Applied Mechanics*, vol. 80, March 2013.
29. E. Suhir, "Avoiding Low-Cycle Fatigue in Solder Material Using Inhomogeneous Column-Grid-Array (CGA) Design," ChipScale Reviews, March–April 2016.
30. H. Schwarzbauer, "Method of Securing Electronic Components to a Substrate," U.S Patent # 4 810 672, March 7, 1989.
31. H. Schwarzbauer, R. Kuhnert, "Novel Large Area Joining Technique for Improved Power Device Performance," *IEEE Transactions on Industry Applications*, vol. 27, No. 1, Jan. 1991.
32. C. Mertens, R. Sittig, "Low Temperature Joining Technique for Improved Reliability," International Conference on Integrated Power Electronic Systems, Bremen, Germany, 2002.
33. G. Bai, "Low-Temperature Sintering of Nanoscale Silver Paste for Semiconductor Device Interconnection," Ph.D. dissertation, Virginia Polytechnic Institute and State University, Blacksburg, Virginia, Oct. 2005.
34. C. Göbl, P. Beckedahl, H. Braml, "Low Temperature Sinter Technology: Die Attachment for Automotive Power Electronic Applications," Automotive Power Electronics, Paris, France, June 2006.
35. T. Wang, X. Chen, G.-Q. Lu, G.-Y. Lei, "Low-Temperature Sintering with Nano-Silver Paste in Die-Attached Interconnection," *Journal of Electronic Materials*, vol. 36, No. 10, 1333–1340, 2007.

36. N. Lubick, "Nanosilver Toxicity: Ions, Nanoparticles or Both?" *Environmental Science and Technology*, vol. 42, No. 23, 2008.
37. P.O. Quintero, T. Oberc, F. P. McCluskey, "High Temperature Die Attach by Transient Liquid Phase Sintering," HiTEC 2008, IMAPS, Albuquerque, New Mexico, May 2008.
38. M.H. Poech, M. Weiss, K. Gruber, "Chip Drop after Silver Sintering Process," COMSOL Proceedings, Milan, Italy, 2009.
39. B. McPherson, et al., "Packaging of High-Temperature 50kW SiC Motor Drive Module for Hybrid Electric Vehicles," *Advanced Microelectronics*, vol. 37, No. 1, Jan. 2010.
40. H.A. Mantooth, S. Ang, J.C. Balda, K. Okumura, T. Otsuka, "Packaging of High Temperature 50kW SiC Motor Drive Module for Hybrid-Electric Vehicles," *Advancing Microelectronics*, vol. 37, No. 1, January 2010.
41. D. Wakuda, K.-S. Kim, K. Suganuma, "Ag Nanoparticle Paste Synthesis for Room Temperature Bonding," *IEEE CPMT Transactions*, vol. 33, No. 1, March 2010.
42. T.G. Lei, J.N. Calata, G.-Q. Lu, X. Chen, S. Luo, "Low-Temperature Sintering of Nanoscale Silver Paste for Attaching Large-Area (>100 mm^2) Chips," *IEEE CPMT Transactions*, vol. 33, No. 1, March 2010.
43. M. Wrosch, A. Soriano, "Sintered Conductive Adhesives for High Temperature Packaging," ECTC, 2010.
44. V. Manikam, K.Y. Cheong, "Die Attach Materials for High Temperature Applications: A Review," *CPMT Transactions*, vol. 1, No. 4, April 2011.
45. C. Buttay, et al., "Die Attach of Power Devices Using Silver Sintering–Bonding Process Optimization and Characterization," VDE Verlag GMBH, Berlin, Offenbach, Germany, Feb. 2012.
46. S. Kraft, A. Schletz, M. Maertz, "Reliability of Silver Sintering on DBC and DBA Substrates for Power Electronic Applications," *CIPS 2012*, Nuremberg, Germany, Mar. 2012.
47. H. Jin, S. Kanagavel, W.-F. Chin, "Novel Conductive Paste Using Hybrid Silver Sintering Technology for High Reliability Power Semiconductor Packaging," ECTC, 2014.
48. E. Suhir, T. Reinikainen, "On a Paradoxical Situation Related to Lap Shear Joints: Could Transverse Grooves in the Adherends Lead to Lower Interfacial Stresses?" *Journal of Applied Physics D*, vol. 41, 2008.
49. T. Reinikainen, E. Suhir, "Novel Shear Test Methodology for the Most Accurate Assessment of Solder Material Properties," IEEE ECTC, 2009.
50. E. Suhir, T. Reinikainen, "Interfacial Stresess and a Lap Shear Joint (LSJ): The 'Transverse Groove Effect' (TGE)," *Journal of Solid Mechanics and Materials Engineering (JSSME)*, vol. 3, No. 6, 2009.

9 Board-Level Dynamic Tests

"You can see a lot by observing."

—Yogi Berra, American baseball player

"It is easy to see. It is hard to foresee."

—Benjamin Franklin, American scientist and statesman

9.1 BACKGROUND

Dynamic response of materials and structures to shocks and vibrations has always been an important topic of applied science and engineering [1], including the field of electronics [2–6]. In military, avionic, space, automotive, and maritime electronics, dynamic loading occurs during the normal operation of the system, and is considered by many MIL-SPECs and other qualification requirements (see, for example, [7–9]). In commercial electronics, dynamic loading takes place during mishandling or transportation of electronic equipment and instrumentation. In addition, random vibrations are often applied as a time- and cost-effective means to detect and weed out infant mortalities, even though a particular product might not be intended for a dynamic environment. In recent decades, the necessity to protect portable electronics (cellular phones, personal digital assistants, digital cameras, notebook computers, and so on) from shock loading (primarily because of an accidental drop) triggered an increased interest to the field of dynamic loading. The effort is on both modeling and experimental effort, mostly accelerated testing and particularly failure-oriented accelerated testing (FOAT). Conventional engineering practice is a "design-failure-redesign" process: the initial reliability testing is carried out on a fabricated prototype, the design shortcomings are detected, and then new specimens are produced and tested. The process is repeated until all parts pass the environmental tests. The product development time and cost could be minimized considerably if effective and physically meaningful design-for-reliability (DfR) predictive modeling is conducted.

In the review that follows, we consider publications addressing the dynamic response of electronic materials to shocks and vibrations [10–22]; role, significance, and attributes of predictive modeling [23–31]; the situations when it is sufficient to consider the deterministic (nondeterministic) linear response [32–34]; and when nonlinear [35–41] and probabilistic [42–45] effects are significant and cannot be ignored. The emphasis of the review is on the reliability of solder joint interconnections [46–92] and especially on the board-level drop testing [93–107] of solder materials and systems used as the second level of interconnections (package-to-substrate): printed circuit boards (PCBs) carrying surface-mounted devices (SMDs).

It is the board-level drop testing that has become the most popular way to establish if, based on the existing qualification standards, the reliability of the second level of interconnections is adequate. We address ball-grid arrays (BGAs), the bottleneck of current technologies, and column-grid arrays (CGAs), which are viewed as a promising improvement of the state of the art in the second level of interconnections. The boards could also carry, in accelerated test conditions, heavy concentrated masses that are sometimes attached to the opposite side of the board to enhance its dynamic response, especially when there is an intent to conduct FOAT or highly accelerated life testing (HALT). The loading applied to the PCB support contour during board-level testing is transmitted to the SMDs and, first of all, to solder joint interconnections through the PCB deformations. As has been established [35–41], these deformations are due to both bending and in-plane (membrane) tension.

9.2 DROP TESTING

The published work on drop testing of electronic equipment is enormous. Let us indicate the most significant publications in this field. Goyal and Buratynski [10] have found that due to the multiple impacts that result during accidental drop of a portable electronic product, the propensity for its damage could be significantly higher than in a single drop impact. This observation agrees with the author's (unpublished) observations of the behavior of Nokia cell phones tested in 2008 at the Instituto Nokia de Tecnologia (INdT) in Manaus, Brazil. In the author's observations, it was found that the bounce height and the number of hits (impacts) depend on the initial "angle-of-attack" of the portable device, and that the lowest bounce heights (leading to the largest damage of the device's PCB and solder joints) and the largest number of post-drop hits occur in the region of the initial "angles-of-attack" between 15 and 25 degrees. The tests with strain cable attached have confirmed the intuitively obvious fact (based on the energy considerations) that the strains in the PCB inside the device indeed increase with a decrease in the bounce height. Goyal and Buratynski suggested an experimental method of drop testing that combines, in their opinion, the advantages of the constrained and free drop methods by suspending the test specimen into a guided drop-table. Wong et al. [11] simulated drop tests using finite element analysis (FEA) techniques and carrying out parametric studies varying drop height, PCB size and thickness, solder bump stand-off, size and number, and so on. Lim et al. [12] and Luan and Tee [13] addressed the effect of the impact pulse (its duration, shape, peak acceleration and width, etc.), on the dynamic response of the PCB and solder material. Their findings have shown that there has been large variation of test results even though the impact pulse used by different companies are all within the JEDEC test specification, which is believed to be too loose. The authors conclude that the impulse has to be controlled, because otherwise the test results from different tests or different manufacturers could not be compared. They also conclude that (as long as the response is linear) it is possible using numerical modeling to convert the drop test result from one test condition to another without performing additional testing, and that a life prediction model can be built based on a carefully and adequately designed experiment. Tee et al. [14] investigated board-level solder joint reliability performance during a drop test in the light of a new

JEDEC standard. Comprehensive dynamic responses of PCB and solder joints, such as acceleration, strains, and resistance, were measured and analyzed with a multi-channel real-time electrical monitoring system, and simulated using a novel input acceleration (Input-G) method. It has been found, both experimentally and numerically, that the mechanical shock causes multiple PCB bending or vibration, which lead to solder joint fatigue failure. It has also been found that it is the peeling stress in the critical solder joint that is the dominant failure indicator by simulation. This correlates with the observations and assumptions of the experiment, as well as with the earlier findings of the author of this review [23]: the maximum stresses and strains in the solder joints occur in the through-thickness (axial) direction ("peeling effect"), although the thermal expansion/contraction mismatch of the soldered components is associated with mismatched shear deformations.

The reliability of WL-CSP component boards was studied by Mattila et al. [15] by executing mechanical shock tests at different temperatures in accordance with the JESD22-B111 [9] standard. The tests were carried out at four different temperatures. It has been observed that the number of drops-to-failure increased significantly with an increase in temperature. While the authors of this publication associate this observation with metallurgical and crack propagation phenomena, the author of the current review is inclined to attribute the observed circumstance mostly to the decrease, at the elevated temperatures in the induced thermal stress, caused by the dissimilar materials of the package and the PCB.

Suhir [16] analyzed the response of a heavy electronic (power) component to harmonic vibrations applied to its external electric leads soldered into the PCB. The vibrations are caused by the lateral vibrations of the PCB. It has been found that the natural frequencies of vibration of actual (elongated) components can be significantly lower than in a system idealized as a cantilever beam with a lump mass at the end; that if the ratio of the frequency of the external excitations to the natural fundamental frequency of vibrations of the PCB exceeds a certain threshold, no induced PCB vibrations could possibly occur; that if the frequency ratio is small, the amplitudes of the induced vibrations increase rapidly with an increase in this ratio; and that a dynamically stable design should be characterized by a high frequency ratio.

Zhou et al. [17] designed a shock table test vehicle to mimic real-life drop conditions for components and systems adopted in portable electronics. Experiments were carried out to determine the dynamic characteristics of typical portable electronic components under drop impact. Then, a series of shock table tests with different constraint conditions were designed to mimic the real-life impact state. By comparing the typical results from shock table tests and those from drop tests, the correlation of shock table test parameters and drop tests conditions were investigated. The results revealed that the conventional fully constrained shock table test cannot adequately mimic the real-life drop impact conditions: the appropriate shock table test method should allow the sample to rotate freely. Theoretical analysis is developed to explain the mechanics of the impact scenarios. It has particularly been found that due to the Hertz contact spring effect and the rotational acceleration during the impact, the acceleration of the test sample could differ significantly from that of the table. It has also been found that the acceleration estimated by the traditional force-divided-by-mass method may appreciably underestimate the actual acceleration of components located inside the product.

Gu et al. [18] applied health monitoring and prognostics methodology for assessing the reliability of electronic components mounted on a PCB using strain gages and an accelerometer to monitor the vibration loads. The computed stresses were then used in a vibration failure fatigue model for damage assessment. Damage estimates were accumulated using Miner's rule and then used to predict the life consumed and remaining useful life.

Yu and Zhu [19] designed a drop test system to conduct repetitive drop tests with arbitrarily specified orientation of the tested device. The drop impact was generated by a pendulum, and the impact pulse was measured by strain gages attached to a Hopkinson bar. A simplified beam-rod model is proposed to provide a guideline to the drop/impact protection design for a typical portable electronic device. It is concluded that the acceptable shock level can be evaluated before the prototype is fabricated and tested. Zhou et al. [20] conducted shock table tests with different constraint conditions to mimic actual drop tests using the suggestions contained in [10]. The test results for different shock durations have confirmed the estimations provided in [36] on the basis of analytical modeling. It is concluded that to fully mimic the actual drop conditions, an appropriate shock table test method should allow the sample to rotate freely and the shock duration should be much shorter than the duration of the actual drop.

The late Steinberg [21], the recognized guru in the field of dynamics of electronic products, addressed electronic components failures produced by thermal and vibration cycling, how manufacturing methods and material properties can produce failures in electronic equipment, how viscoelastic damping materials might cause problems, if these materials are not used properly and carefully, the appropriate types of test equipment, vibration test fixtures and adapters, design considerations concerning vibration fixtures, and other relevant considerations associated with drop testing and shock protection.

Recently, an approximate method has been suggested for the tentative assessment of the dynamic strength of a bonding material in an electronic device from the static shear-off test data [22]. The method is based on an assumption that the initial potential energy of the device converts completely into the strain energy of the bonding material at its failure during shear-off testing. It is important, of course, that this finding is confirmed experimentally.

9.3 ROLE OF MODELING

Predictive modeling, both simulation-based and analytical, are equally important and complement each other in any important design and testing effort when there is a need to understand the physics of a phenomenon or the behavior and performance of a particular electronic product. Broad application of powerful computer programs and particular those using FEA have by no means made analytical solutions less important. Simple analytical relationships have invaluable advantages, because of the clarity and compactness of the obtained information and clear indication on interactions and dependencies of different material and design factors. FEA simulations and analytical solutions are typically based on different assumptions and use different methods. It is always advisable to obtain a solution to the problem of interest on the basis of different approaches. If the results are in a satisfactory agreement, one could be confident that they are most likely correct. It is always advisable to analytically investigate the problem before carrying out computer-aided analyses. Such a preliminary investigation helps to reduce computer time and expense, develop the most feasible and effective preprocessing model and, in many cases, avoid fundamental errors. Analytical ("mathematical") modeling [25] enabled one to obtain closed form solutions for numerous electronic packaging problems, including stresses in solder joints configured as short cylinders [23], twist-off testing of solder joint interconnections [24], and various paradoxical situations [31] encountered in electronic packaging materials science. Analytical modeling was also effectively used to predict, quantify and, through that, to assure microelectronics reliability [28]. This type of modeling was also used in a prognostics-and-health monitoring (PHM) three-step-concept (TSC) [30], when the recently suggested physically meaningful, powerful, and flexible probabilistic Boltzmann–Arrhenius–Zhurkov (BAZ) model [29, 48] was "sandwiched" between two statistical models: Bayes' equation, aimed at identification, on the probabilistic basis, the malfunctioning device(s), and beta-distribution (see, for example, [47]) aimed at updating reliability if a failure has been detected despite the anticipated low probability of its possible occurrence. Zhu and Marcienkiewicz [26] used advanced modeling technique based on computer simulations to model drop tests of chip scale package (CSP) and fine pitch BGA packages. Three drop tests ("board drop, board with fixture drop or shock, and system level phone drop") were conducted and "explicit-implicit sequential modeling techniques" were used to characterize the dynamic responses of CSP/BGA packages in different board designs. Failure criteria and effects of strain rate and edge support on BGA in multicomponent boards were also investigated. A validation test with data acquisition was used to correlate the test results and computed data. Marjamaki et al. [27] and many others used FEA software to model various aspects of drop tests, and particularly for lead-free solders.

9.4 LINEAR RESPONSE

Whenever possible, a linear approach should be applied first to investigate/model the behavior of a dynamic system. In many cases such an approach is sufficient, unless the loading is significant and/or new phenomena that could not be captured by a linear analysis are addressed. But even when nonlinear phenomena are addressed, the linear behavior of the system should be analyzed first to capture its most important aspects that will take place at low levels of loading. Let us indicate several critical problems, in which the application of a linear approach enabled one to obtain practically important solutions in electronics packaging materials science.

1. There is an obvious incentive for using viscous damping to reduce the vibration amplitudes in structural elements subjected to drop testing. Application of viscous damping has proven to be effective in many engineering systems subjected to vibrations, especially when resonant or close-to-resonant

conditions cannot be avoided. It is less obvious to what extent extensive damping could be advantageous in systems experiencing impact loading. The question is to what extent elevated viscous damping might be able to effectively minimize the maximum "breaking acceleration" (deceleration) during "landing" of an electronic device, as well as the "breaking distance", such as the distance that the falling device covers from the moment when its enclosure touches the floor to the moment when the velocity of the vulnerable structural element in the device becomes zero. It has been shown [32] that the "breaking distance" always decreases with an increase in the level of the viscous damping, and that the effectiveness of such damping is higher for a low-level damping. As to the maximum acceleration (deceleration), the situation is different: there is a certain, a relatively low, level of damping that results in decelerations that are indeed somewhat lower than those in a system with no damping. However, if damping is significant, the maximum decelerations can significantly exceed the decelerations in a system with no damping at all. For this reason, viscous damping should be introduced in a system subjected to drop impact with caution, and the level of damping should be established beforehand, based on the carried out theoretical analysis [32], depending on the mass of the body to be protected and the spring constant of the protective cushion (restoring force).

2. The level of dynamic loading is typically expressed in electronics engineering by the magnitude of the induced accelerations (decelerations). Although such an approach is commonly accepted and is widely employed, as far as functional (electrical, optical, thermal) performance is concerned, it might be misleading when the mechanical and materials reliability is of interest [33]. In many cases, high accelerations indeed go together with high dynamic stresses, but there are also situations when these stresses are much higher in systems that experience relatively low accelerations. For instance, structural elements of the type of a low weight beam supported at the ends are characterized by high natural frequencies and, as a result of that, experience high drop induced accelerations (these are proportional to the vibration frequencies squared). On the other hand, structural elements that could be idealized as cantilever beams carrying heavy lumped masses at the end might experience high dynamic stresses despite relatively low drop induced accelerations (Figure 9.1).
3. In accelerated tests in electronics, there is always a temptation to replace relatively complicated, time-consuming, and expensive testing with simpler, faster, and cost-effective tests. Shock testers are widely available in the electronic industry, and so are drop testers. Shock testers are much cheaper and much simpler to operate than drop testers, and the dynamic response of specimens tested on shock testers is much easier to measure and to interpret than the drop test response. A natural question arises of whether drop test conditions could be adequately mimicked by shock conditions, and if they could, how the shock tester should be tuned so that the results of shock testing would adequately mimic drop test conditions.

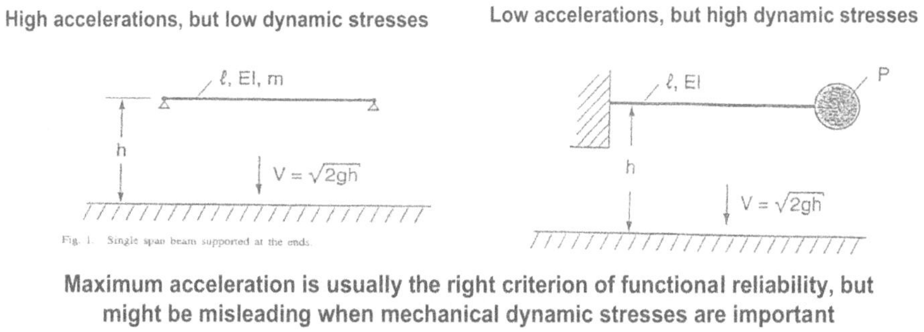
High accelerations, but low dynamic stresses
Low accelerations, but high dynamic stresses
ℓ, EI, m
h
V = √2gh
Fig. 1 Single span beam supported at the ends.
ℓ, EI
P
h
V = √2gh
Maximum acceleration is usually the right criterion of functional reliability, but might be misleading when mechanical dynamic stresses are important

FIGURE 9.1 The structural element idealized as a simply supported beam (left sketch) experiences high accelerations, but low stresses in its mid-cross section. The structural element idealized as a cantilever beam with a lump mass at the end (right sketch) experiences low accelerations, but high stress at the clamped end. (From E. Suhir, "Is the Maximum Acceleration an Adequate Criterion of the Dynamic Strength of a Structural Element in an Electronic Product?" IEEE CPMT Transactions, Part A, vol. 20, No. 4, Dec. 1997.)

It has been shown [36] that substitution of drop testing with shock testing is indeed possible, provided that the magnitude and the duration of the drop impact is preestablished by modeling, and that, based on the predicted drop test response, the shock tester is tuned accordingly. It has been particularly suggested that, assuming that the dynamic response remains linear, the shock tester should be able to produce very short impact loads (preferably shorter than one eighth of the period of vibrations) in order to adequately reflect the drop test conditions. Based on the results of the theoretical analysis [36], a shock table has been designed and built [17] to mimic real-life drop conditions. The obtained experimental data are in good agreement with the theoretical predictions. It has also been an attempt to show [22] that if the ultimate drop height, such as the height that leads to the failure, is selected as an adequate criterion of its dynamic strength, then the shear-off data for the ultimate shear-off force can be effectively used to assess the ultimate drop height. Such an assessment is expected to be conservative, because the dynamic stress–strain diagrams for a wide variety of materials indicate that that the same material is characterized by a somewhat higher level of the loading-at-failure than what could be concluded from the static stress–strain diagram.

4. As is known (see, for example, [42]), short-term loadings generate higher harmonics. An instantaneous impulse, in the absence of dissipation, generates a vibration spectrum with an infinitely large number of harmonics. How important is it that the role of these harmonics is considered? This question can be answered, using linear approach, using the reasoning in Figure 9.2 (see, for example, [5]). Consider an elongated simply supported thin plate mimicking a PCB. The board is dropped from a height H, hits the floor with its supports, and, as a result of that, experiences shock-exited

Energy Balance in Drop Tests

$$w(x,t) = \sum_{i=1}^{n} f_i(t)\cos\frac{i\pi x}{2a}, i = 1,3,5,\ldots$$

$$\dot{w}(x,t) = \sum_{i=1}^{n} \dot{f}_i(t)\cos\frac{i\pi x}{2a}, i = 1,3,5,\ldots \qquad \dot{f}_i(0) = \frac{4}{i\pi}\sqrt{2gH}$$

$$K = \frac{1}{2}m\sum_{i=1}^{n}\dot{f}_i^2(0)\int_{-a}^{a}\cos^2\frac{\pi x}{2a}dx = \frac{16}{\pi^2}magH\sum_{i=1}^{n}\frac{1}{i^2} = \left(\frac{8}{\pi^2}\right)2magH\left(\frac{\pi^2}{8}\right) = \frac{\pi^2}{8}K_1 = 1.2337K_1 = 2magH$$

Initial kinetic energy   Initial kinetic energy of the first tone of vibrations   Factor that considers the role of the higher modes of vibrations on the kinetic energy of the PCB   Initial potential energy

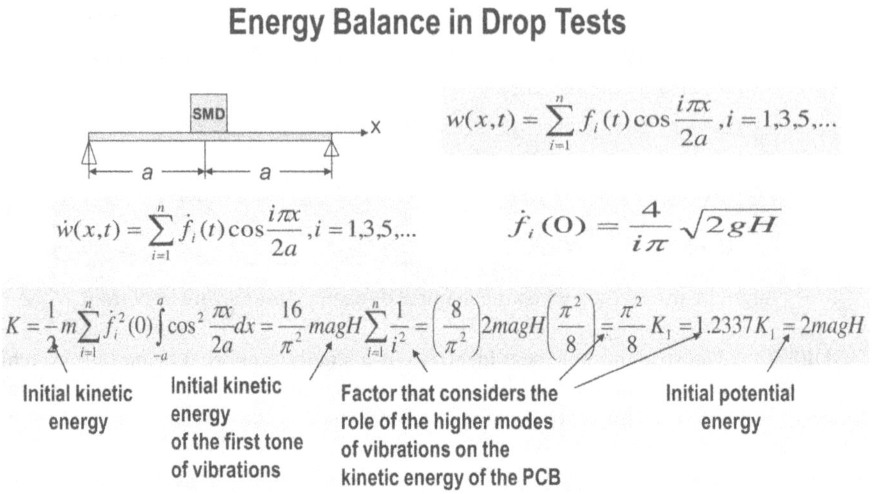

FIGURE 9.2   Role of the higher harmonics in the linear dynamic response of an elongated PCB to the drop impact applied to its ends. The formula for the kinetic energy, based on the energy balance for the case of undamped vibrations, indicates that the higher harmonics "consume" only about 23% of the initial potential energy. In the cases of damped vibrations and nonlinear vibrations, the role of the higher harmonics is even smaller. This circumstance justifies consideration of the fundamental mode only for drop impact-induced vibrations.

vibrations. Using the method of principal coordinates (see, for example, [5] or [42]), the induced deflections can be sought in the form of the series

$$w(x,t) = \sum_{i=1,3,5,\ldots}^{n} f_i(t)\cos\frac{i\pi x}{2a}.$$

Here, $f_i(t)$ is the principle coordinate of the ith mode, and $\cos\dfrac{i\pi x}{2a}$ is the coordinate function. This function is chosen in such a way that the zero boundary conditions for the deflection and the curvature are fulfilled at the plate's supports. The lateral velocities can be determined by differentiation

$$\dot{w}(x,t) = \sum_{i=1,3,5,\ldots}^{n} \dot{f}_i(t)\cos\frac{i\pi x}{2a}. \tag{9.1}$$

At the initial moment of time, all the points of the board have the same velocities $\dot{w}(x,0) = \sqrt{2gH}$. On the other hand, these velocities can be found from (9.1) as

$$\dot{w}(x,0) = \sum_{i=1,3,5,\ldots}^{n} \dot{f}_i(0)\cos\frac{i\pi x}{2a}. \tag{9.2}$$

Multiplying both parts of equation

$$\sqrt{2gH} = \sum_{i=1,3,5,...}^{n} \dot{f}_i(0)\cos\frac{i\pi x}{2a} \tag{9.3}$$

by the coordinate function $\cos\frac{i\pi x}{2a}$ and integrating along the plate, the following formula for the initial velocity of the principal coordinate can be obtained:

$$\dot{f}_i(0) = \frac{4}{i\pi}\sqrt{2gH}, \quad i = 1,3,5,... \tag{9.4}$$

Thus, while the initial velocities of all the points on the board are the same, the initial velocities of the principal coordinates are not: higher modes are characterized by lower initial velocities and, hence, by lower energies than the lower modes. The lower formula in Figure 9.2 reflects the change in the initial kinetic energy of the vibration modes. The first term is the total kinetic energy of the board. Introducing formula (9.4) into this term, the second term could be obtained. The series in the second term are converging: $\sum_{i=1,3,5,...}^{\infty} \frac{1}{i^2} = \frac{8}{\pi^2}$. This enables one to obtain the total energy in the form of the third term. The fourth term can be obtain considering that the structure of the second term and that the kinetic energy of the fundamental mode of vibrations is $K_1 = \frac{16}{\pi^2} magH$. The fourth and the fifth terms indicate that the total kinetic energy of the plate exceeds the energy of the fundamental mode by the factor of $\frac{\pi^2}{8} = 1.2337$. The sanity check indicates that the total kinetic energy of the board is indeed equal to its initial potential energy. Thus, the higher modes of vibrations are "responsible" for only 23.4% of the total energy, and therefore, in an approximate and practical analysis, one can get away with considering the fundamental mode of the induced vibrations only. This is even more true for the nonlinear vibrations subsequently addressed. This is because the energy associated with the in-plane (membrane) stresses is also included into the fundamental mode.

5. When there is intent to understand the physics underlying the behavior and performance of solder joint interconnections during board-level testing, one has to be able to model and analyze, first of all, the response of the bare board itself. Figure 9.3 [6] indicates that the vibration characteristics of a PCB can be predicted with high accuracy (the FEA has been carried out by M. Vujosevic, Intel).

	Linear Theory	Linear FEM
Linear frequency	$\omega = \frac{\pi^2}{2a}\sqrt{\frac{D}{M}} = \frac{\pi^2}{300}\sqrt{\frac{717.48}{7.4025x10^{-6}}} = 323.51\sec^{-1}$	$\omega = 323.04\frac{rad}{s}$
Period	T=0.0194	T=0.0194
Peak amplitude	$A_0 = \frac{16}{\pi^2}\frac{\sqrt{2gH}}{\omega} = \frac{16}{\pi^2}\frac{4996}{321} = 25.13mm$	A_0=24.70mm
Maximum acceleration	$\ddot{f}_{max} = -\omega^2 A_0 = -323.51^2 x25.13 = -2.6\times10^6 mm/\sec^2$	a = 2.578 e6 mm/s^2

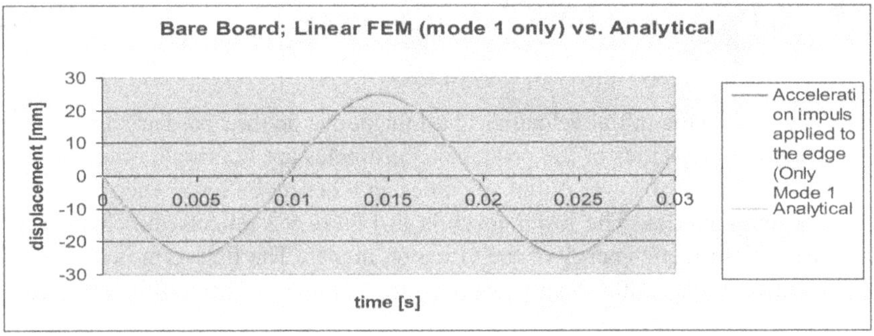

FIGURE 9.3 When the vibration amplitudes and the in-plane forces (stresses) are low (small) the linear theory well predicts the induced amplitudes and accelerations. But how small is "small"? This question can only be answered based on a nonlinear analysis.

9.5 NONLINEAR RESPONSE

In a typical situation, when the PCB contour remains nondeformable during the board's deformations caused by the postimpact vibrations, and when the impact is significant, as it should be in FOAT, HALT or even in qualification tests (QT), appreciable reactive in-plane (membrane) stresses arise in the board. Because of that, the board vibrations become (geometrically) nonlinear: the deflections are not proportional anymore to the applied load. It has been determined [35–41] that appreciable nonlinearity manifests itself even for small drop heights. It has also been determined that it is sufficient to consider the fundamental mode only, and that the nonlinear response of a PCB with a nondeformable support contour could be described by the Duffing equation

$$\ddot{f}(t) + \omega^2 f(t) + \mu f^3(t) = 0 \tag{9.5}$$

for the principal coordinate $f(t)$. In this equation, ω is the frequency of the linear vibrations,

$$\mu = 3(1+\nu)\left(\frac{\pi}{2a}\right)^6 \frac{D}{mh^2} \tag{9.6}$$

is the parameter of nonlinearity (for the case of an elongated board), a is half the board's length, D is its flexural rigidity, m is board's mass per unit area, and h is its thickness. The Duffing equation addresses nonlinear vibrations with the rigid cubic characteristic of the restoring force (nonlinearity) and lends itself to an exact solution [35–39] (see also the Appendix to this chapter), no matter how significant the loading and, hence, the nonlinearity level might be, provided, of course, that the elasticity limit of the PCB material is not exceeded.

Here is how the amplitude A of the induced vibrations could be obtained in a rather elementary way, without even solving nonlinear equation (9.5). This equation can be written as

$$\frac{d}{dt}\left[\dot{f}^2(t)+\omega^2 f^2(t)+\frac{1}{2}\mu f^4(t)\right]=0 \tag{9.7}$$

Hence, the expression in the brackets should be constant:

$$\dot{f}^2(t)+\omega^2 f^2(t)+\frac{1}{2}\mu f^4(t)=C=\text{const.} \tag{9.8}$$

It has to be this way, because the sum of the kinetic energy (first term) and the strain energy (the second and the third terms), whether linear or nonlinear, should be constant, as long as there no dissipation in the system. At the initial moment of time, the principal coordinate $f(t)$ is zero and the initial velocity is $\dot{f}(t)=\dot{f}(0)=\sqrt{2gH}$. Hence, the constant C is $C=2gH$. The velocity should be zero when the displacement reaches its amplitude value A. Then, the biquadratic equation

$$A^4+2\frac{\omega^2}{\mu}A^2-\frac{4gH}{\mu}=0 \tag{9.9}$$

that can be obtained from (9.8) yields

$$A=\eta_A A_0, \tag{9.10}$$

where

$$A_0=\frac{\sqrt{2gH}}{\omega} \tag{9.11}$$

is the linear amplitude,

$$\eta_A=\sqrt{\frac{\sqrt{1+2\bar{\mu}}-1}{\bar{\mu}}} \tag{9.12}$$

is the factor of the nonlinear amplitude, and

$$\bar{\mu}=\mu\left(\frac{A_0}{\omega}\right)^2 \tag{9.13}$$

is the dimensionless parameter of nonlinearity. The factor (12) changes from one to zero, when the linear amplitude A_0 changes from zero to infinity. The nonlinear acceleration (deceleration) can be found from (9.5). The obtained relationships indicate that the nonlinear amplitude can be substantially lower than the linear amplitude, while the induced acceleration (deceleration) can be significantly higher.

It is natural to assume that the mechanical behavior of the solder joint interconnections has a minor effect on the PCB vibrations, as long as the size of the SMDs is small in comparison with the size of the board and, because of that, these devices do not change the flexural rigidity of the board. It has been suggested, therefore [39], that the nonlinear drop impact excited vibrations of a "flexible-and-heavy" PCB are evaluated assuming that the mounted devices do not affect its flexural rigidity. The SMD masses could be considered in an approximate way, by spreading their total mass over the PCB surface, thereby making the board both "flexible" and "heavy." The attributes of the particular SMD (package) and, more importantly, its solder joint interconnections could be considered at the next step from the known (calculated or measured) tensile forces and bending moments acting in the PCB cross sections located in the close proximity of the particular SMD of interest. Figures 9.4 and 9.5 illustrate the significance of the in-plane stresses and how the induced bending moment and the membrane force can be experimentally determined. Of course, these force and moment can also be evaluated theoretically, and the results can and should be compared.

Combined Action of Tension and Bending: Dynamic Response

The developed stress model was used, as an appropriate illustration of the possible use of this model, for the evaluation of the dynamic response of a PCB/SMD assembly to impact loading. The loading was applied to the PCB support contour and is transmitted to the SMD through an attachment of finite compliance.

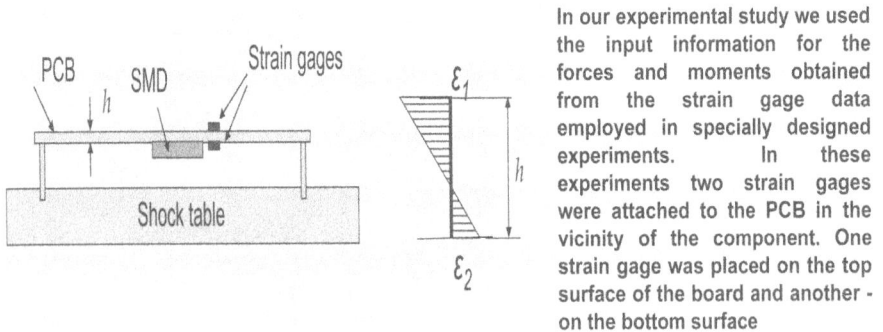

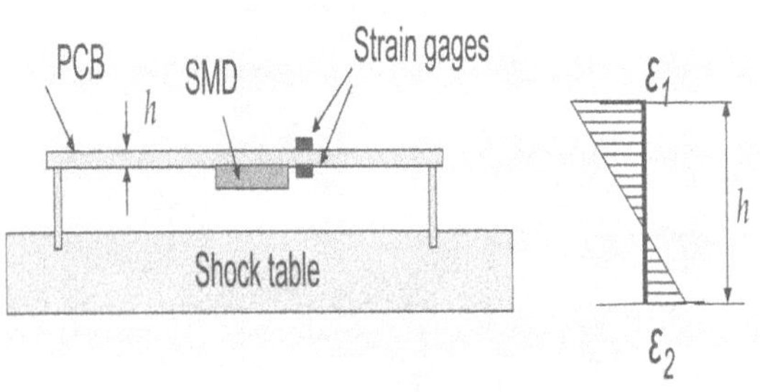

In our experimental study we used the input information for the forces and moments obtained from the strain gage data employed in specially designed experiments. In these experiments two strain gages were attached to the PCB in the vicinity of the component. One strain gage was placed on the top surface of the board and another - on the bottom surface

FIGURE 9.4 Experimental setup for drop tests. The PCB experiences shock excited vibrations that are transmitted to the SMD and particularly to the solder joint interconnections that attach the SMD to the PCB. The gages placed at the upper and the lower surfaces of the PCB show different maximum strains. This is an indication on the occurrence of in-plane (membrane) stresses.

Top: computed bending and membrane component of the strain Bottom: Forces and moments are computed at the time t=0.012sec (gray line on strain plots).

The external tensile forces

$$\hat{T} = \frac{Eh}{1-\nu} \frac{\varepsilon_1 + \varepsilon_2}{2},$$

and bending moments

$$\hat{M} = \frac{Eh^2}{6(1-\nu^2)} \frac{\varepsilon_1 - \varepsilon_2}{2}$$

calculated from the strains measured on the PCB surfaces

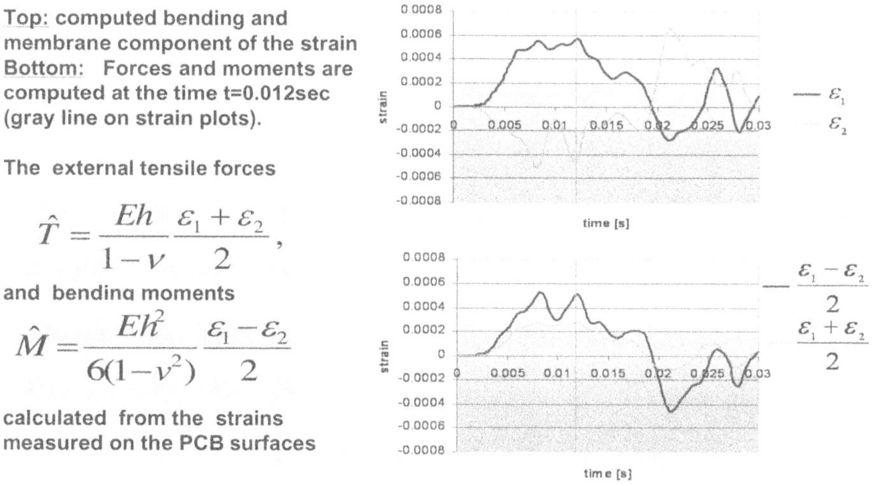

FIGURE 9.5 The in-plane tensile forces and bending moments acting in the PCB cross sections can be determined from the computed or measured strains at the upper and the lower surfaces of the PCB experiencing drop impact-induced vibrations.

Nonlinear dynamic response of flexible structural elements experiencing periodic shock loads distributed over their length (surface) and idealized as a train of instantaneous impacts (impulses) was addressed, apparently for the first time, in connection with the observed stochastically unstable ("chaotic") motions of such elements [1, 35, 42]. Distributed dynamic loading might occur, for example, as a result of acoustic (sonic) pulses acting on the plate (membrane). Linear and nonlinear, steady state and transitional, and undamped and damped vibrations were addressed and analyzed. The approach and the obtained solutions were later on extended and applied to PCBs employed in electronics [36–39].

It has been found [1, 35] that

1. the linear transient response of a single-degree-of-freedom (SDOF) linear system with viscous damping can lead, for particular combinations of the frequency ratio and the level of damping, to higher amplitudes and accelerations than in the subsequent steady-state vibrations; that

2. in linear SDOF systems, low level damping might be responsible for the very fact that the vibrations stabilize, such as become steady-state, but might not affect the nonresonant response, and this response can therefore be determined by simply imposing the conditions of periodicity on the system (these conditions state that the sequential impulse does not change the position of the system, but changes its velocity in a step-wise fashion, with the step defined as the ratio of the magnitude of the impulse to the system's mass); that

3. the nonlinear response of a PCB can be reduced if the fundamental mode of vibrations is considered to the Duffing type of equation (9.5) for the principal coordinate, that is, to the equation for a SDOF oscillator with the rigid cubic characteristic of the restoring force; and that

4. an exact closed-form solution to equation (9.5) can be obtained for time-independent practically important types of excitations (instantaneous impulses, or suddenly applied and then suddenly removed constant loading), no matter how significant these excitations and the corresponding nonlinear effects might be.

The loading on a PCB in electronic equipment of an aircraft, spacecraft, or a missile during the take-off of the flying apparatus can be idealized as a suddenly applied (in a step-wise fashion) constant load whose duration considerably exceeds the linear period of vibrations of the PCB, not to mention its nonlinear period. Figures 9.6 and 9.7 illustrate the response of a large and flexible PCB to such a load. It has been shown [36] that the actual (nonlinear) accelerations in the midportion of the PCB can be significantly larger than those predicted on the basis of the linear theory, even if the combined effect of the "external" acceleration (experienced by the PCB support contour) and the induced accelerations caused by the PCB vibrations with respect to the contour is considered. While the external acceleration applied in the carried out example to the PCB support contour was only 25 g, the predicted linear acceleration at the center of the PCB, because of the PCB (linear) vibrations, was about 65.5 g, while the maximum nonlinear acceleration turned out to be as high as 118.5 g. If a certain device was designed to the highest allowable acceleration of 100 g, this device could not be placed inside the shaded rectangular in Figure 9.7. Thus, the application of a linear approach would be misleading in the situation in question.

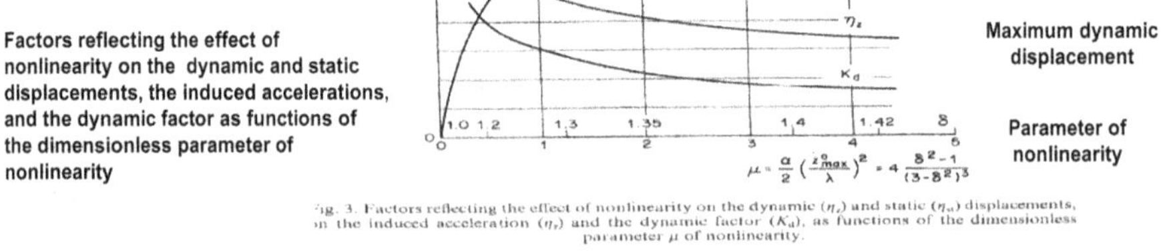
Factors reflecting the effect of nonlinearity on the dynamic and static displacements, the induced accelerations, and the dynamic factor as functions of the dimensionless parameter of nonlinearity
μ 0 0.2 0.5 1.0 2.0 3.0 4.0
η_z 1 0.7117 0.5897 0.5000 0.4176 0.3737 0.3447
η_st 1 0.9217 0.8477 0.7709 0.6823 0.6282 0.5898
K_d 2 1.5493 1.3913 1.2972 1.2240 1.1898 1.1689
η_z̈ 1 1.263 1.458 1.636 1.821 1.927 2.000
Factors of
Maximum acceleration
Maximum static displacement
Maximum dynamic displacement
Parameter of nonlinearity
Fig. 3. Factors reflecting the effect of nonlinearity on the dynamic (η_z) and static (η_{st}) displacements, on the induced acceleration ($\eta_{\ddot{z}}$) and the dynamic factor (K_d), as functions of the dimensionless parameter μ of nonlinearity.

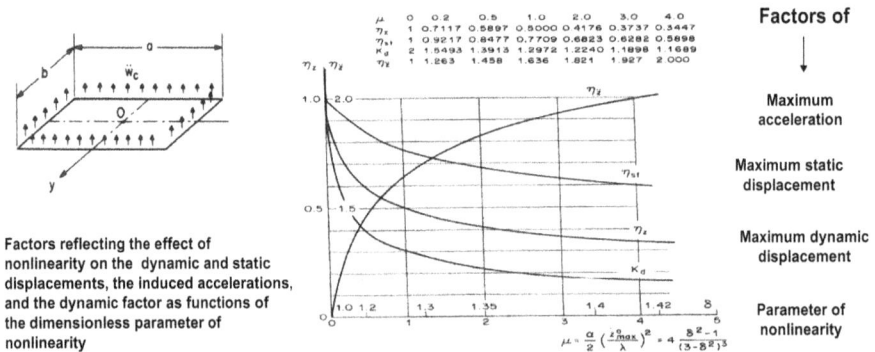

FIGURE 9.6 Response of a flexible PCB to a constant instantaneously applied load acting on the PCB's support contour. All the response characteristics, but the induced accelerations, decrease with an increase in the level of the nonlinearity (i.e., with the level of the induced in-plane stresses). The maximum nonlinear acceleration could be by a factor of three higher than the linear acceleration. (From E. Suhir, "Nonlinear Dynamic Response of a Flexible Thin Plate to Constant Acceleration Applied to its Support Contour, with Application to Printed Circuit Boards Used in Avionic Packaging," International Journal of Solids and Structures, vol. 29, No. 1, 1992.)

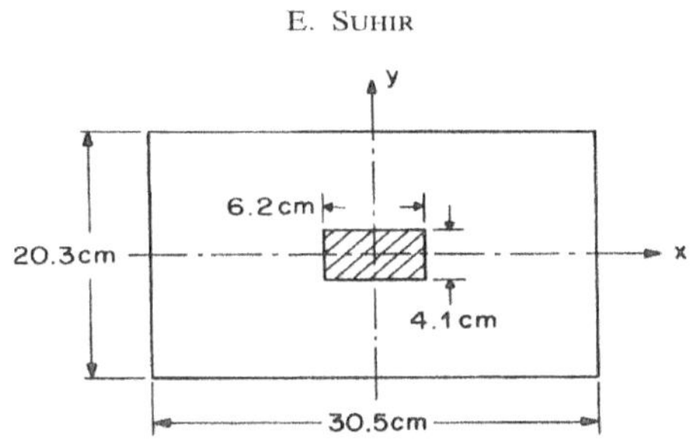

➢ASTM/NEMA Class G-10 fiber-glass PCB simply supported on its contour is subjected to constant suddenly-applied acceleration of 25g. The device can withstand acceleration as high as 100g.

➢The calculations indicate that 100g is the acceleration at the boundaries of the shaded rectangular contour.
➢The acceleration at the center of this contour is as high as 118.45g.
➢A linear approach would result in the maximum acceleration of 65.5g.
➢The linear and non-linear dynamic factors are 2.62 and 4.74, respectively.

FIGURE 9.7 If the maximum allowable acceleration of an electronic device of interest is restricted to the acceleration of the 100 g level, it has to be placed on the PCB outside the shaded rectangular, because the predicted induced nonlinear acceleration at the PCB center is as high as 118.45 g and is by the factor of 4.74 larger than the acceleration of the contour. The maximum linear acceleration at the PCB center is "only" 65.5 g. Although this value is by the factor of 2.52 greater than the acceleration of the PCB contour, it is still below the allowable maximum acceleration of the device, and, if the design decision is based on the linear approach, an erroneous decision will be made. The maximum nonlinear acceleration is in this example by the factor of 1.81 greater than the maximum linear acceleration. The theory based on the exact solution to the Duffing equation for the principal coordinate indicates that this factor could be as high as three.

The method of principal coordinates (known in the theory of differential equations of mathematical physics as Fourier method) was employed in the aforementioned solutions to separate in the equation of motion the (time-dependent) principal coordinate from the (location-dependent) mode (coordinate function), and it has been assumed that the nonlinear effects could be accounted for by considering their impact on the principal coordinate only, while the coordinate function was assumed to be the same as in a linear approach. The only crucial requirement was that this function should satisfy the physically justifiable boundary conditions at the PCB support contour. Suhir and Arruda [40] have shown that this assumption is valid for a simply supported PCB, while the "linear coordinate function" assumption for PCBs with other boundary conditions could lead to an appreciable error. This is especially true for response characteristics (angles of rotation or curvatures) that require differentiation of the coordinate function with respect to one of the coordinates in the x–y plane. It should be pointed out that although the edges of an actual PCB seem to be closer to the clamped ones, rather than to simply supported ones, the conclusion obtained in [41] is less restrictive than one might imagine. This is because the analysis is intended to be applied to an experimental setup rather than to a PCB of an actual device.

The analysis in [41] was applied for building a Nokia experimental setup with an objective to evaluate the response not so much of the PCB itself but rather of the BGA interconnections of the devices mounted onto the board. It is noteworthy that a test vehicle does not have to necessarily mimic the one in an actual structure, but has to be simple and lend itself to an easy and straightforward interpretation of the test data. Because of that, it is recommended that a simply supported PCB is used in an adequate experimental setup.

As has been indicated, the assumption that the analysis of the impact exited vibrations could be restricted to the consideration of the fundamental mode is even more applicable to the nonlinear response than to the linear one. The validity of this statement was confirmed by FEA [39]. In Figure 9.8, the effect of higher modes of vibration is manifested by just small "white caps" on the "crests" of the vibration "waves."

As has been indicated, it has been found [1, 35, 42] that a train of impact short-term loads can lead to the phenomenon of "stochastic instability" ("chaos") even in the case of a weak nonlinearity and in the simplest case of a SDOF system: the system behaves in a quasi-random fashion, although neither the excitation, nor the system itself, are characterized by a random loading or by random properties of

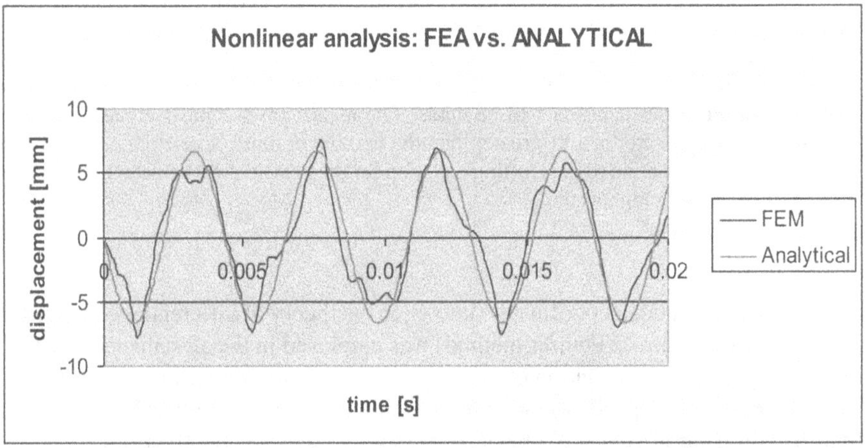

Analytical: A_0 **= 6.78mm @ time t=0.0011**

FEM: A_0 **=7.78mm @ time t=0.00119s**

FIGURE 9.8 Analytical versus finite element method (FEM) analysis of the shock-exited nonlinear vibrations. While analytical solution considers the fundamental mode of vibrations only, the FEM accounts for all the modes. As one could see from the diagram, the role of the higher modes has manifested itself in the little "wavelets" near the first mode maxima. Not only these maxima, but even the phase angles were predicted correctly by the analytical solution.

the system. Stochasticity occurs in situations when the period of the train of impulses substantially exceeds the period of natural linear vibrations (so that the stochastic phase approximation is applicable), but, on the other hand, is low enough, so that the responses caused by the sequential impulses still overlap. Certainly, the damping in the system does not have to be high either, otherwise the vibrations caused by sequential impacts will not overlap. The random phase approximation (stating that the phase angles at the moments of impact are evenly distributed between zero and 2π) and the Smoluchowski equation [42] can be used to obtain the condition of the "stochastic instability" of the vibrations and to characterize the quasi-random behavior of the "chaotic" vibrations. He and Fulton [37] and He and Stallybrass [38] have extended Suhir's work in application to PCBs used in electronic packaging.

An additional way to enhance the loading on the solder material and solder joint interconnections when conducting accelerated reliability testing on the board level, is to attach heavy concentrated masses to the PCB. This measure increases the inertia forces and, hence, the total accumulated strain energy in the board, the induced lateral accelerations, and the PCB in-plane deformations. Suhir, Vujosevic, and Reinikainen [39] have carried out an analysis of the dynamic response of a "flexible-and-heavy" PCB experiencing an impact load applied during drop tests to the board's support contour. Only a simply supported board was considered, for the sake of making the analysis as simple as possible and to be suitable for an experimental setup and for a straightforward interpretation of the results. It was assumed in the analysis that the attached heavy concentrated masses, if any, were sufficiently small in size, so that their effect on the flexural rigidity of the board was insignificant and did not have to be considered, while their total mass did have an appreciable effect on the induced inertia forces, the accelerations and the stresses, and should be accounted for. Elliptic functions were used to find the exact solution to the basic equation for the principal coordinate and to establish, on the basis of this solution, the amplitudes, frequencies, accelerations (decelerations), and stresses experienced by the board. Unlike in linear systems, the dynamic factors for different response characteristics (amplitudes, accelerations, frequencies, stresses) are different for different characteristics and are highly dependent on the level of the dynamic loading.

As is known, it is considerably easier to carry out the analysis of the dynamic response of a system when the actual short-term impact loading can be substituted by an instantaneous impulse. Having in mind a nonlinear system with the rigid cubic characteristic of the restoring force, Suhir and Arruda carried out the corresponding analysis for a nonlinear system [40]. The important conclusion is that, as far as the amplitudes of vibrations are concerned, the nonlinear system is much less sensitive to the dynamic nature of the applied load than a linear system, but, when the induced accelerations are important, the situation is just the opposite: the induced accelerations in the nonlinear system might be substantially larger than in a linear one. It is concluded that modeling based on the analysis carried out in [41] should always precede the construction of the experimental setup, the design of the experiments, and the analysis (interpretation) of the experimental data.

Numerous problems associated with the random loading on PCBs idealized as thin plates were addressed in the monograph [42] and in [44] and [45].

9.6 SOLDER JOINTS

Solder joint interconnections in flip-chip, BGA (Figure 9.9) and land-grid array (LGA) designs provide both electrical connections and mechanical support. When the stresses and strains in the solder material exceed, during temperature excursions or during drop tests, the yield point of the material, low-cycle fatigue conditions occur. The long-term reliability of this material and the device as a whole could be compromised if the inelastic strains are significant and the fatigue damage accumulates, with an increase in the number of temperature cycles and/or the number of drops impacts, to an unaccepted level. The situation has recently become more important and more complicated, because of the introduction of lead-free solders and the use of BGA and LGA systems. The published work that addresses the mechanical behavior and performance of tin-lead and lead-free BGA solder materials and interconnections, when subjected to thermal or dynamic loading, is enormous. In this section, we will indicate just a few major publications [46–87] that do not necessarily address board-level testing per se (the next section will be devoted to this important field) and also employ, in addition to experimental evaluations, modeling techniques.

Dynamic properties of solder materials were addressed by Deshpande et al. [51], Syed and Pang [52], Siviour et al. [59], Ong et al. [63], Field et al. [66], Ghaffarian [67, 72],

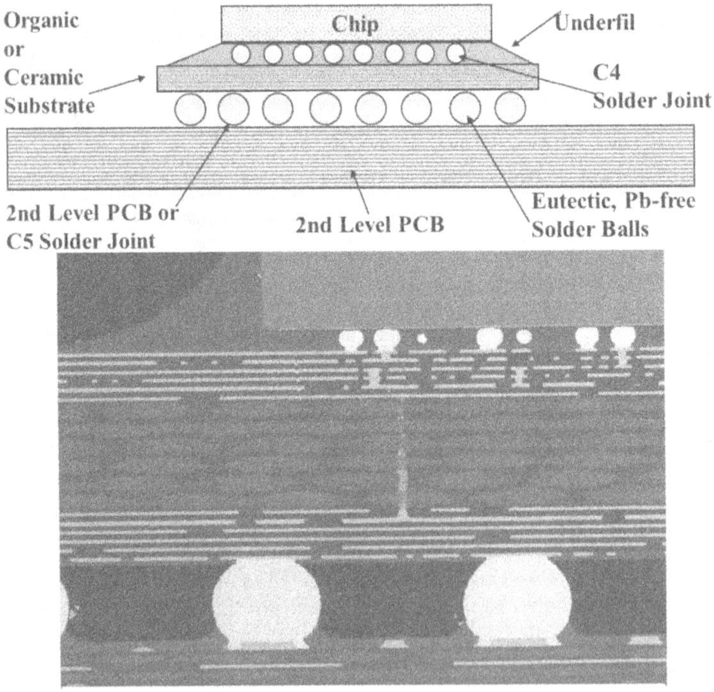

FIGURE 9.9 Flip-chip (first level of interconnections) and ball-grid array (second level of interconnections) technologies.

Siviour et al. [71], Williamson at al [74], Tan et al. [80], and Steinberg [81]. The results have shown a clear need for consistent and trustworthy experimental data for high rates of strain, which are relevant to the loading environment for portable electronic devices experiencing impact loading.

Drop impact conditions were studied by Sogo and Hara [53], Zhu [54], Yu et al. [55], Mahiro et al. [56], Xu et al. [60], Date et al. [62], Yeh and Lai [64, 65], Syed et al. [68], Britzer et al. [69], and Wong et al. [70]. The studies addressed the role of the geometry of the solder balls under drop loading, ductile-to-brittle transition in solder joints measured by impact tests, effect of drop impact on BGA/CSP package reliability and other aspects of the drop impact conditions affecting solder material behavior and reliability.

Shock and thermal cycling synergism effects were investigated by Ghaffarian [50, 76, 79, 82], Tee et al. [58], Chiu et al. [61], and Mattila et al. [73]. The authors considered the combined action of shock and thermal loading on solder materials in array and 3D packages, crucial accelerated reliability tests, meaningful predictive models, possible role of additives, and so on. Shock and thermal loading are important for many vehicular applications when high reliability level is crucial.

Vibration and thermal loading synergism was addressed mostly by UMD investigators: Vodzak et al. [46], Barker et al. [49], Qi et al. [75, 77], and Sakthievelan et al. [78]. The authors have focused on the ability to accurately model and estimate life expectancy of BGA solder interconnections under the combined action of vibration and temperature cycling conditions. Such loading is typical in many automotive, aircraft, airspace, and maritime systems.

The BGA and LGA systems are now commonly viewed both as an opportunity and as the major bottleneck of the short- and long-term reliability of surface-mount electronic technologies and package designs. To overcome the shortcomings of BGA and LGA technologies employing short-and-rigid solder joints, it has been recently suggested that short-cylinder-like solder joints are employed [83] and that CGA technologies [84–87] (Figures 9.10 and 9.11) are used for lower stresses in the solder material. There is a reason to believe that these stresses could be brought down to an extent that inelastic strains in the solder material could be avoided. If this happens, the fatigue life of the solder material will be improved dramatically (Figure 9.12).

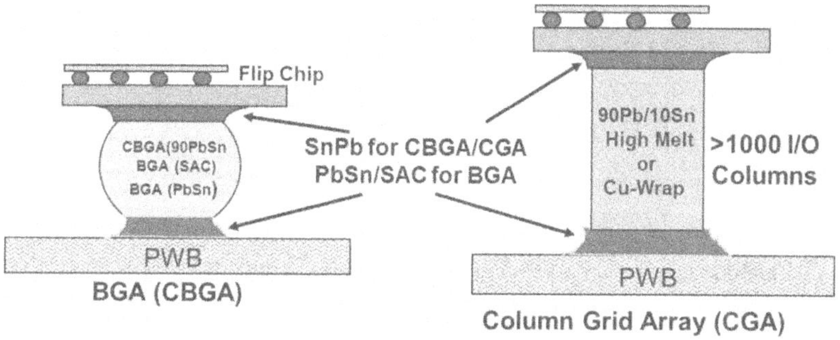

FIGURE 9.10 Ceramic ball-grid array (CBGA)/BGA and CGA technologies

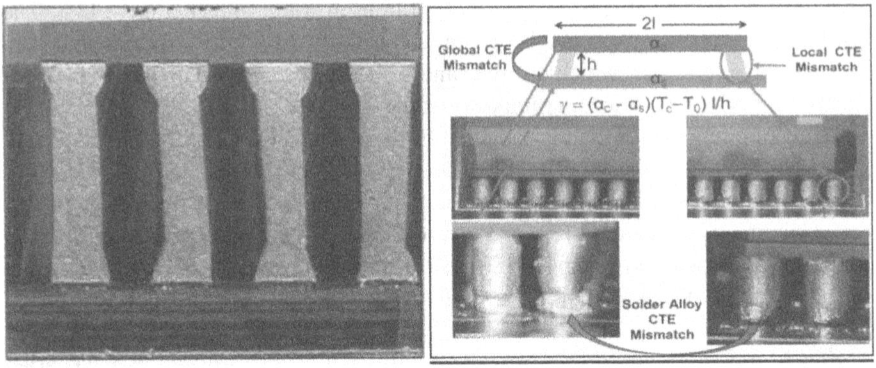

FIGURE 9.11   CGA technology.

$\beta1=6.08, \eta1=92.68, \rho=0.98$
$\beta2=6.54, \eta2=1715.35, \rho=0.97$
$\beta3=7.89, \eta3=3612.64, \rho=0.98$

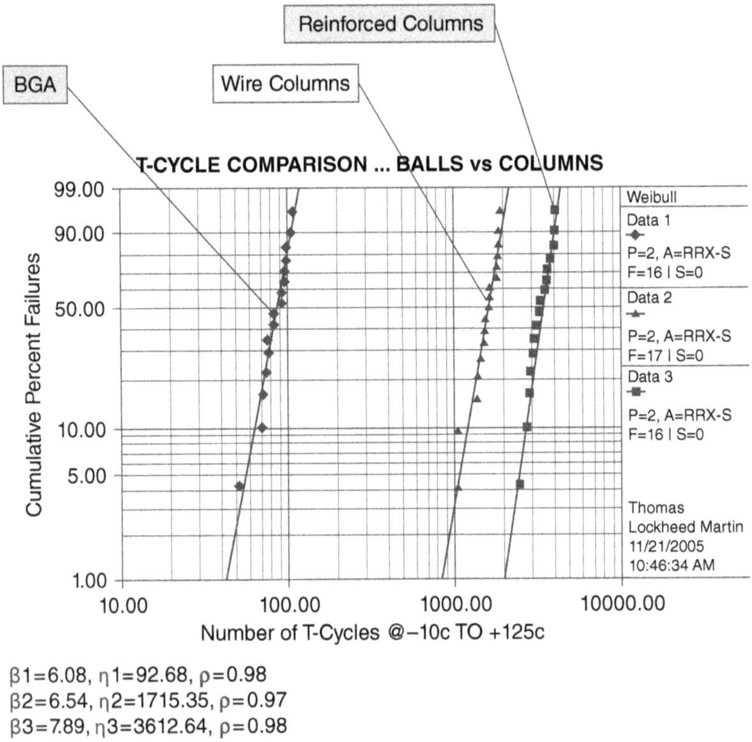

FIGURE 9.12   Thermal fatigue life comparison between BGA and CGA (Six Sigma data).

9.7   BOARD-LEVEL TESTING

Board-level testing [88–107] has become the test vehicle of choice when evaluating and assuring solder joint reliability. To understand the reliability physics underlying the behavior and performance of solder materials and structures, one has to understand, first of all, the behavior of the PCB itself, whether bare or carrying SMDs. PCBs, as far as their mechanical (physical) behavior is concerned, can be idealized and modeled as thin isotropic plates whose contours experience shock or vibration loadings (excitations). In approximate, but still practically useful, analyses, the PCB's anisotropy and the composite structure is not considered, nor the possible viscoelastic or inelastic effects in the PCB material are.

Response to the drop impact during board-level testing was addressed by Seah et al. [88], Goyal et al. [89, 90], Ong et al. [95], Luan and Tee [97], Chai et al. [100], Luan et al. [102], and recently by Ghaffarian [105]. Mechanical response, methods for realistic drop testing, shock-protection suspension design, comparison of the PCBs dynamic response when subjected to product level and board-level drop impact tests, effect of the attributes of the impact pulse, and the reliability of the PCB itself were addressed, analyzed, and discussed.

Modeling and simulation of PCB drop tests were carried out by Wang [91], Luan et al. [92], and Tee et al. [93, 98], including modal analysis, thin-and-fine-pitch BGA, compliance with JEDEC standards, and so on, with an objective to understand the test board's basic mechanical properties and variations of stress and strain on test board. ANSYS software was used to perform drop test simulations. The results between full model and quarter model were compared to verify the accuracy and efficiency of finite element analysis.

Tan [94] analyzed board-level solder joint failures under static and dynamic loading. Mattila et al. [103] considered board-level reliability of lead-free solder material under shock and vibration loads. Tee et al. [96] and Chiu et al. [99] considered the combined action of drop tests and temperature cycling tests or thermal aging. Correlations between board-level drop reliability and package level ball impact tests, as well as correlations between package level high speed solder ball shear/pull and board-level drop tests with brittle fracture failure modes were addressed by Luan et al. [101] and Song et al. [104]. It has been established that when the shear and pull test speeds increased, a continual increase in the brittle failure mode percentage was observed, and that lead-free solders showed more susceptibility to brittle fracture than tin-lead solders in both shear and pull tests.

9.8 CONCLUSIONS

The following major conclusions could be drawn from the carried out review.

- Recently published work on dynamic response of electronic systems to shock loading has been reviewed with an emphasis on solder joint materials, interconnections, and on the board-level reliability testing.
- Predictive modeling is an important DfR tool for electronic materials, packages, and modules subjected to dynamic loading. Analytical and computer-aided modeling should always be considered and conducted, in addition to accelerated testing, in the analysis, design, and manufacturing of electronic systems subjected to drop impacts, vibration, and thermal loading. It is always advisable to carry out both analytical and FEA-based modeling and to make sure that analytical and simulation data are in good agreement.

- Analytical modeling occupies a special place in predictive modeling, since it enables one to obtain simple and physically meaningful close form solutions that clearly indicate the role of various factors affecting the phenomenon, the material, and the structure of interest. It is always advisable that both analytical modeling and FEA simulations are carried out in all the important cases, and their results are compared. Then, there is confidence that the obtained predictions are sufficiently accurate.
- It is well known that it is the experiment that is the supreme and final judge of microelectronics reliability, including solder joint dynamics, and therefore the development of an appropriate test vehicle should be viewed as one of primary and particular importance. A crucial requirement for such a vehicle is its simplicity and the ability to correctly and straightforwardly interpret the obtained data and to extrapolate them on the actual operation conditions in the field.
- Accordingly, future work should include the development of analytical predictive stress models with an ultimate objective to suggest an appropriate experimental setup design for drop testing of surface-mounted devices at the board level, having primarily in mind the reliability evaluations of the second-level solder joint interconnections. It is suggested that a simply supported, rather than a clamped, board is considered in the test vehicle in question. This is because the experimental data for a simply supported board can be easily and reliably interpreted and effectively extrapolated for practical applications.
- The drop-induced vibrations are highly (geometrically) nonlinear because of the significant reactive in-plane (membrane) stresses. These take place because of the nondeformable support contour, even, as numerical data indicate, for not-very-large drop heights. Because of that it is particularly imperative that a drop impact experimental setup is developed with consideration of the role of the nonlinear response of the PCB to the drop impact applied to the board support contour.
- When the method of principal coordinates is used to solve the obtained nonlinear equation (of the Duffing type) for this time-dependent coordinate, an exact solution for the first mode of vibrations can be easily obtained, no matter how significant the nonlinearity level might be. For the simply supported board, the coordinate function remains the same whether linear or nonlinear shock-excited vibrations are addressed. This circumstance should be considered when building the stress vehicle and when interpreting the test data.
- Possible (and highly desirable) stress relief in solder joints configured as short-cylinders configuration should be assessed based on the developed predictive models. While the advantages of the BGA with elevated stand-offs and CGA designs has been demonstrated for thermally induced loading, it is imperative that merits, if any, of these technologies are evaluated as well, and the appropriate models are developed.
- The application of solder joints with elevated stand-offs and particularly CGA designs might enable one to avoid inelastic dynamic stresses in the

solder materials during the accelerated drop tests, and, hence, in the actual operation conditions. This has been proven for thermally induced loading on the basis of analytical models developed for static thermal loading and should be determined also for dynamic loading on the basis of a model that should consider board-level drop testing of package/PCB assemblies with BGA and CGA solder joint interconnections. It should also be proven for BGA and CGA designs when subjected to drop impacts.

APPENDIX A: EXACT SOLUTION TO THE PROBLEM OF THE NONLINEAR DYNAMIC RESPONSE OF A PCB TO THE DROP IMPACT DURING BOARD-LEVEL DROP TESTS

A.1 BACKGROUND/INITIATIVE

An analytical predictive model has been developed for the evaluation of the nonlinear dynamic response of a PCB to the drop impact during board-level testing. The hypothesis of a "heavy-and-flexible" PCB is used in the analysis: the SMDs are assumed to be small enough not to affect the PCB's flexural rigidity, but their masses have been considered and accounted for by "spreading out" the SMD total mass over the PCB surface. The analysis is restricted to the fundamental mode of vibrations, and the method of principal coordinates is used to evaluate the response. The exact solution to the nonlinear differential equation for the principal coordinate has been obtained. Another important finding is that the nonlinear amplitudes were determined even without solving the nonlinear differential equation of motion. The main objective of the analysis is to provide design guidelines for constructing a feasible experimental setup. A simply supported board is suggested as the most appropriate structure for an adequate test vehicle: the experimental data for such a board, as far as the behavior of the solder material in the second level of interconnections is concerned, can be easily and reliably interpreted and extrapolated for the practical use. The developed model enables one to predict the induced bending moments and the in-plane (membrane) forces that could be applied in the subsequent analyses to the PCB areas in the proximity of the package and its solder joint interconnections.

Board-level drop testing has become the test vehicle of choice when evaluating reliability of the second level of solder joint interconnections [A.1–A.23]. Response to the drop impact during board-level testing was addressed by Goyal and Buratynski [A.1], Seah et al. [A.2], Ong et al. [A.3], Luan and Tee [A.15], Chai et al. [A.18], Luan et al. [A.20], and, recently, by Ghaffarian [A.23]. Mechanical response, methods for realistic drop testing, shock-protection suspension design, comparison of the PCB's dynamic response when subjected to product level and board-level drop impact tests, effect of the attributes of the impact pulse, and the reliability of the PCB itself were addressed, analyzed, and discussed. Modeling and simulation of PCB drop tests were carried out by Wang [A.10], Luan et al. [A.11], Tee et al. [A.12, A.16], including modal analysis, thin-and-fine-pitch BGAs, compliance with JEDEC standards, and so on, with an objective to understand the test board's basic mechanical properties and variations of stresses and strains. FEA ANSYS software was used to perform drop test simulations. The results between full model and quarter model were compared to verify the FEA accuracy and efficiency. Tan [A.13] analyzed board-level solder joint failures under static and dynamic loading. Mattila et al. [A.21] considered board-level reliability of lead-free solder material under shock and vibration loads. Tee et al. [A.14] and Chiu et al. [A.17] considered the combined action of drop tests and temperature cycling tests or thermal aging. Correlations between board-level drop reliability and package level ball impact tests, as well as correlations between package level high-speed solder ball shear/pull and board-level drop tests with brittle fracture failure modes were addressed by Luan et al. [A.20] and Song et al. [A.22]. It has been established that when the shear and pull test speeds increased, a continual increase in the brittle failure mode percentage was observed, and that lead-free solders showed more susceptibility to brittle fracture than tin-lead solders in both shear and pull tests.

To understand the reliability physics underlying the behavior and performance of solder materials, one has to understand, first of all, the behavior of the PCB itself, whether bare or carrying SMDs. PCBs, as far as their mechanical (physical) behavior is concerned, can be idealized and modeled as thin isotropic plates whose contours experience shock or vibration loadings (excitations). In approximate, but still practically useful, analyses, the PCBs anisotropy and the composite structure is not considered, nor are the possible viscoelastic or inelastic effects in the PCB material.

In a typical situation, when the PCB contour remains nondeformable during the board's deformations caused by the postimpact vibrations, and when the impact is significant, as it should be in FOAT, HALT, or even QT, appreciable reactive in-plane (membrane) stresses arise in the board. Because of that, the board vibrations become geometrically nonlinear: the deflections are not proportional anymore to the applied load. It has been determined [A.23–A.28] that appreciable nonlinearity manifests itself even for small drop heights. The measured strains on the upper and the lower surfaces of the board (Figure 9.1) are quite different, and this is a clear evidence of the appreciable in-plain stresses and, hence, of the nonlinearity of the PCB response.

In the analysis that follows, an analytical predictive model is developed for the evaluation of the nonlinear dynamic response of the PCB to the drop impact during board-level testing. The analysis is restricted to the fundamental mode of vibrations, and the method of principal coordinates is used to evaluate the response. The nonlinear differential equation for the principal coordinate lends itself to an exact solution, using elliptic functions, no matter how significant the level of the nonlinearity might be. Another important finding is that the nonlinear amplitudes and the induced accelerations (decelerations) can be found even without solving the differential equation of motion. A simply supported board is considered. This is because the experimental data for a simply supported board can be easily and reliably interpreted and then effectively extrapolated for actual operation conditions. The present analysis is a modification and extension of an earlier study [A.28] carried out for an experimental set up used by Intel.

A.2 ASSUMPTIONS

The following major assumptions are used in this analysis:

- Basic equations of the engineering theory of thin plates and the engineering elasticity theory (see, for example, [A.24, A.25]) are applicable in the problem in question.
- The hypothesis of "heavy-and-flexible" PCB is used in the analysis. In accordance with this hypothesis, the SMD are assumed to be small enough not to affect the PCB's flexural rigidity, so that the flexural rigidity of the bare board is considered even though a plurality of SMDs are mounted on it. On the other hand, the SMD masses are accounted for by "spreading out" the total mass of all the SMDs over the PCB surface.
- The physical/mechanical behavior of the solder joint interconnections does not affect the PCB dynamic response. The solder joints behavior and performance are intended to be evaluated at the later step from the predicted or measured PCB response. Specifically, it is the membrane force and bending moments that have to be evaluated and applied to the package of interest to determine the stresses and strains in the solder material.
- Only the fundamental mode of the impact induced vibrations is considered.

Some additional, less important, assumptions are introduced, when necessary, in the text of the Appendix.

A.3 KINETIC AND STRAIN ENERGIES

Consider a square PCB whose sides are $2a$ and its thickness is h. The kinetic energy, K, and the strain energy, V, of the board experiencing the combined action of bending and in-plane (membrane) tension after PCB contour hits the floor are expressed as follows (see, for example, [A.1, A.2]):

$$K = \frac{1}{2} m \int_A \left(\frac{\partial w}{\partial t} \right)^2 dA, \tag{A.1}$$

$$V = \frac{1}{2} D \int_A \left[(\Delta w)^2 - 2(1-\nu)L(w,w) \right] dA + \frac{1}{2}\frac{h}{E} \int_A \left[(\Delta \varphi)^2 - 2(1+\nu)L(\varphi,\varphi) \right] dA. \tag{A.2}$$

Here, A is the PCB's area, m is its mass per unit area; $D = \dfrac{Eh^3}{12(1-\nu^2)}$ is the PCB's flexural rigidity; E and ν are the effective elastic constants of the PCB material;

$$w = w(x,y,t) = f(t)\cos\frac{\pi x}{2a}\cos\frac{\pi y}{2a} \tag{A.3}$$

are the sought PCB's drop impact induced deflections; $f(t)$ is the principal coordinate of the postimpact vibrations; $\Delta = \frac{\partial^2}{\partial x^2} + \frac{\partial^2}{\partial y^2}$ and $L = \frac{\partial^2}{\partial x^2}\frac{\partial^2}{\partial y^2} - \left(\frac{\partial^2}{\partial x \partial y}\right)^2$ are the Laplace operator and the operator of the in-plane stress, respectively; and φ is the stress (Airy) function. This function can be determined from the continuity equation

$$\nabla^4 \varphi = -EL(w,w), \tag{A.4}$$

where $\nabla^4 = \Delta^2 = \frac{\partial^4}{\partial x^4} + 2\frac{\partial^4}{\partial x^2 \partial y^2} + \frac{\partial^4}{\partial y^4}$ is the biharmonic operator. The origin of the coordinates x, y is in the PCB's center.

A.4 CONDITION OF NONDEFORMABILITY OF THE PCB SUPPORT CONTOUR

When the PCB's contour is nondeformable (and this is indeed the case in actual drop testing situation), the following conditions should be fulfilled:

$$\int_0^a \left[\frac{1}{E}\left(\frac{\partial^2 \varphi}{\partial y^2} - \nu\frac{\partial^2 \varphi}{\partial x^2}\right) - \frac{1}{2}\left(\frac{\partial w}{\partial x}\right)^2\right] dx = 0, \quad \int_0^a \left[\frac{1}{E}\left(\frac{\partial^2 \varphi}{\partial x^2} - \nu\frac{\partial^2 \varphi}{\partial y^2}\right) - \frac{1}{2}\left(\frac{\partial w}{\partial y}\right)^2\right] dy = 0. \tag{A.5}$$

These conditions reflect the requirement that the total in-plane displacement, which is due to both the in-plane stresses (first terms in the brackets) and the bending deformations (second terms), should compensate each other and their interaction should lead to the total zero in-plane displacements for a nondeformable contour.

A.5 STRESS (AIRY) FUNCTION

The stress (Airy) function $\varphi = \varphi(x,y,t)$ can be sought in the form

$$\varphi = f^2(t)\Phi(x,y). \tag{A.6}$$

Introducing formulas (A.3) and (A.6) into equation (A.4), the following biharmonic equation for the coordinate function $\Phi(x,y)$ can be obtained:

$$\nabla^4 \Phi = -\frac{\pi^4 E}{32a^4}\left(\cos\frac{\pi x}{a} + \cos\frac{\pi y}{a}\right). \tag{A.7}$$

The solution to this equation is

$$\Phi = C_0(x^2 + y^2) + C_2\left(\cos\frac{\pi x}{a} + \cos\frac{\pi y}{a}\right), \tag{A.8}$$

where C_0 and C_2 are constants of integration. The constant C_2 can be found by introducing solution (A.8) into equation (A.7):

$$C_2 = -\frac{E}{32}. \tag{A.9}$$

The constant C_0 can be determined by introducing expression (A.8) into (A.6), then the obtained expression into conditions (A.5) and considering (A.9):

$$C_0 = \frac{E}{1-\nu}\left(\frac{\pi}{16a}\right)^2. \tag{A.10}$$

Thus,

$$\Phi(x, y) = \frac{E}{32}\left[\frac{\pi^2}{2(1-\nu)}\frac{x^2+y^2}{a^2} - \left(\cos\frac{\pi x}{a} + \cos\frac{\pi y}{a}\right)\right]. \tag{A.11}$$

A.6 IN-PLANE (MEMBRANE) STRESSES AND STRAINS

The in-plane stresses acting in the cross sections of the PCB can be found from (A.6) by differentiation

$$\sigma_x^0 = \frac{\partial^2 \varphi}{\partial y^2} = f^2(t)\frac{\pi^2 E}{32a^2}\left(\frac{1}{1-\nu} + \cos\frac{\pi y}{a}\right), \quad \sigma_y^0 = \frac{\partial^2 \varphi}{\partial x^2} = f^2(t)\frac{\pi^2 E}{32a^2}\left(\frac{1}{1-\nu} + \cos\frac{\pi x}{a}\right). \tag{A.12}$$

These formulas indicate that the normal stresses acting in the PCB cross sections in the x direction are x-independent, and the stresses acting in the y direction are y-independent. The in-plane shearing stresses are zero. The normal stresses (A.12) change from $f^2(t)\frac{\pi^2 E}{32a^2}\frac{2-\nu}{1-\nu}$ in the middle of the PCB to $f^2(t)\frac{\pi^2 E}{32a^2}\frac{\nu}{1-\nu}$ at its edges, that is, decrease by the factor of $\frac{2-\nu}{\nu}$. For $\nu = 0.3$, this factor is as high as 5.6667.

The normal strains can be determined from Hooke's law equations

$$\begin{aligned}
\varepsilon_x^0 &= \frac{1}{E}\left(\sigma_x^0 - \nu\sigma_y^0\right) = \frac{\pi^2}{32}\left(\frac{f(t)}{a}\right)^2\left(1 + \cos\frac{\pi y}{a} - \nu\cos\frac{\pi x}{a}\right) \\
\varepsilon_y^0 &= \frac{1}{E}\left(\sigma_y^0 - \nu\sigma_x^0\right) = \frac{\pi^2}{32}\left(\frac{f(t)}{a}\right)^2\left(1 + \cos\frac{\pi x}{a} - \nu\cos\frac{\pi y}{a}\right)
\end{aligned} \tag{A.13}$$

and depend on both coordinates x and y. These strains change from $f^2(t)\frac{\pi^2}{32a^2}(2-\nu)$ in the middle of the PCB to $f^2(t)\frac{\pi^2}{32a^2}\nu$ at its edges, that is, get lower by the same factor of $\frac{2-\nu}{\nu}$, as the stresses do. This means particularly that the SMD under test should be placed, for the highest stresses, at the center of the board.

In an approximate and a conservative analysis, the in-plane forces (per unit PCB width) can be found as

$$\hat{T} = \frac{\pi^2}{32}\frac{2-\nu}{\nu}A^2\frac{Eh}{a^2}, \tag{A-14}$$

where A is the amplitude of the impact induced vibrations.

A.7 PARAMETER OF NONLINEARITY

Introducing formulas (A.3) and (A.6), with consideration of solution (A.11), into the expressions (A.1) and (A.2), the following formulas for the kinetic and strain energies can be obtained:

$$K = \frac{1}{2}M\dot{f}^2(t), \quad V = \frac{1}{2}M\left(\omega^2 f^2(t) + \frac{1}{2}\mu f^4(t)\right). \tag{A.15}$$

Here,

$$M = ma^2 \tag{A.16}$$

is the generalized mass of the PCB,

$$\omega = \frac{\pi^2}{2a}\sqrt{\frac{D}{M}} \tag{A.17}$$

is the frequency of its free linear vibrations, and

$$\mu = \frac{3\pi^4}{32}(3-\nu)(1+\nu)\frac{D}{Ma^2h^2} = \frac{\pi^4}{128}\frac{E}{M}\frac{3-\nu}{1-\nu}\frac{h}{a^2} \tag{A.18}$$

is the parameter of nonlinearity. This parameter increases with an increase in the Young's modulus of the PCB material and its thickness, and decreases with an increase in the PCB mass (and, hence, the masses of the SMDs) and its size. Note that the PCB's generalized mass M is only a quarter of its actual mass. This is because the expression (A.16) for the generalized mass considers that different points on the PCB respond differently to the same initial velocity of different vibration modes at the moment of impact.

A.8 BASIC EQUATION AND ITS SOLUTION

Introducing formulas (A.15) into the Lagrange equation (see, for example, [A.26],)

$$\frac{d}{dt}\frac{\partial K}{\partial \dot{f}} + \frac{\partial V}{\partial f} = 0, \tag{A.19}$$

the following basic differential equation (of Duffing type) can be obtained for the principal coordinate:

$$\ddot{f}(t)+\omega^2 f(t)+\mu f^3(t)=0. \tag{A.20}$$

This equation lends itself to an exact solution, no matter how significant the level of the nonlinearity is:

$$f(t)=Acn(\sigma t,\varepsilon)=Acnu. \tag{A.21}$$

Here, A is the vibration amplitude, *cnu* is the elliptic cosine (see, for example, [A.27]), σ is the parameter of the nonlinear frequency, t is time, and ε is the initial phase angle. Using formulas

$$(cnu)'=-snudnu,\quad (snu)'=cnudnu,\quad (dnu)'=-k^2 snucnu, \tag{A.22}$$

for differentiating the elliptic functions, and formulas

$$cn^2u+sn^2u=1,\quad dn^2u+k^2sn^2u=1 \tag{A.23}$$

of the "elliptic geometry," we obtain the following expressions for the velocity and the acceleration (deceleration):

$$\dot{f}(t)=-A\sigma snudnu,\quad \ddot{f}(t)=-A\sigma^2 cnu(1-2k^2sn^2u), \tag{A.24}$$

where k is the modulus of the elliptic function, *snu* is the elliptic sine, and *dnu* is the function of delta-amplitude. Introducing solution (A.21) for the displacement and the second formula in (A.24) for the acceleration into equation (A.20) of motion, one could conclude that equation (A.20) is fulfilled, if the relationships

$$\sigma=\sqrt{\omega^2+\mu A^2}=\frac{\omega}{\sqrt{1-2k^2}},\quad k=\frac{A}{\sigma}\sqrt{\frac{\mu}{2}}=\sqrt{\frac{1}{2}\left(1-\frac{\omega^2}{\sigma^2}\right)} \tag{A.25}$$

take place. These relationships define the parameter, σ, of the nonlinear frequency and the modulus, k, of the elliptic function, respectively.

A.9 VIBRATION AMPLITUDE

The amplitude, A, of the induced vibrations can be found, without even solving equation (A.20), based on the following simple reasoning. Equation (A.20) can be written as

$$\frac{d}{dt}\left[\dot{f}^2(t)+\omega^2 f^2(t)+\frac{1}{2}\mu f^4(t)\right]=0,$$

and therefore

$$\dot{f}^2(t)+\omega^2 f^2(t)+\frac{1}{2}\mu f^4(t)=C=Const.$$

The maximum deflection $f_{max}=A$ takes place, when the velocity $\dot{f}(t)$ is zero, so that

$$C=\omega^2 A^2+\frac{1}{2}\mu A^4. \tag{A.26}$$

On the other hand, at the initial moment of time, $t=0$, the principal coordinate, $f(0)$, is zero, while the initial velocity $\dot{f}(0)=V_0$ has its maximum value. This yields $C=V_0^2$. Hence, the amplitude, A, can be found from the biquadratic equation

$$A^4+2\frac{\omega^2}{\mu}A^2-2\frac{V_0^2}{\mu}=0. \tag{A.27}$$

as

$$A=\eta_A A_0, \tag{A.28}$$

where

$$A_0=\frac{V_0}{\omega} \tag{A.29}$$

is the amplitude of the corresponding linear vibrations ($\mu=0$), and the factor

$$\eta_A=\sqrt{\frac{\sqrt{1+2\bar{\mu}}-1}{\bar{\mu}}} \tag{A.30}$$

considers the effect of nonlinearity on the vibration amplitude. Here,

$$\bar{\mu}=\mu\left(\frac{A_0}{\omega}\right)^2 \tag{A.31}$$

is the dimensionless parameter of nonlinearity. The factor (A.30) changes from one to zero when the linear amplitude changes from zero to infinity.

A.10 EFFECTIVE INITIAL VELOCITY

From (A.3), we find:

$$\dot{w}=\dot{f}(t)\cos\frac{\pi x}{2a}\cos\frac{\pi y}{2a}. \tag{A.32}$$

If the PCB's vibrations are caused by a drop impact, the velocities of all the points of the PCB at the initial moment of time are the same and are equal to

$$\sqrt{2gH} = \dot{f}(0)\cos\frac{\pi x}{2a}\cos\frac{\pi y}{2a} = V_0\cos\frac{\pi x}{2a}\cos\frac{\pi y}{2a} \tag{A.33}$$

for any point x,y of the PCB. Here, H is the drop height. Using Fourier transform for the expression (A.33), we obtain the following formula for the effective initial velocity:

$$V_0 = \sqrt{2gH}\,\frac{\int_{-a}^{a}\cos\frac{\pi x}{2a}dx\int_{-a}^{a}\cos\frac{\pi y}{2a}dy}{\int_{-a}^{a}\cos^2\frac{\pi x}{2a}dx\int_{-a}^{a}\cos^2\frac{\pi y}{2a}dy} = \frac{16}{\pi^2}\sqrt{2gH}. \tag{A.34}$$

The factor $\frac{16}{\pi^2} = 1.6211$ in this formula reflects the effect of the coordinate function on the initial velocity.

A.11 NONLINEAR FREQUENCY

The period of the elliptic cosine $cn(t,\varepsilon)$ is $4K(k)$ (A.27), where $K(k) = \int_0^{\pi/2}\frac{d\xi}{\sqrt{1-k^2\sin^2\xi}}$ is the complete elliptic integral of the first kind. Hence, the period of the function $cn(\sigma t,\varepsilon)$, that is, the period of the induced vibrations, is $\frac{4}{\sigma}K(k)$, and the corresponding frequency is

$$p = \frac{\pi\sigma}{2K(k)}. \tag{A.35}$$

In the linear case, $K(k) = \frac{\pi}{2}$, and $p = \sigma = \omega$. In the nonlinear case, $\sigma \geq p \geq \omega$, however, even for highly nonlinear vibrations the parameter σ of the nonlinear frequency, is not very far away from the nonlinear frequency p itself (see numerical example).

A.12 BENDING MOMENTS

The bending moments can be found, using expression (A.3) for the deflection function, as (see, for example, [A.25])

$$\begin{aligned} M_x &= -D\left(\frac{\partial^2 w}{\partial x^2} + \nu\frac{\partial^2 w}{\partial y^2}\right) = -Df(t)\left(\frac{\pi}{2a}\right)^2(1+\nu)\cos\frac{\pi x}{2a}\cos\frac{\pi y}{2a} = \\ &= -\frac{E}{1-\nu}\frac{h^3}{12}f(t)\left(\frac{\pi}{2a}\right)^2\cos\frac{\pi x}{2a}\cos\frac{\pi y}{2a} \end{aligned} \tag{A.36}$$

and

$$M_y = -D\left(\frac{\partial^2 w}{\partial y^2} + \nu\frac{\partial^2 w}{\partial x^2}\right) = -\frac{E}{1-\nu}\frac{h^3}{12}f(t)\left(\frac{\pi}{2a}\right)^2 \cos\frac{\pi x}{2a}\cos\frac{\pi y}{2a} \tag{A.37}$$

The bending moments change from

$$M_b = \frac{E}{1-\nu}\frac{h^3}{12}A\left(\frac{\pi}{2a}\right)^2 \tag{A.38}$$

in the middle of the PCB to zero at its edges.

A.13 EQUIVALENT STATIC LOADING

The equation of bending in the case of static loading is

$$D\nabla^4 w(x,y) - hL(w,\varphi) = q, \tag{A.39}$$

where q is the static loading and

$$L(w,\varphi) = \frac{\partial^2 w}{\partial x^2}\frac{\partial^2 \varphi}{\partial y^2} - 2\frac{\partial^2 w}{\partial x \partial y}\frac{\partial^2 \varphi}{\partial x \partial y} + \frac{\partial^2 w}{\partial y^2}\frac{\partial^2 \varphi}{\partial x^2} \tag{A.40}$$

is the function of the in-plane tensile stresses.

The static displacement $w(x,y)$ and the static stress function $\varphi(x,y)$ can be sought in the form

$$w = w(x,y) = f_{st}\cos\frac{\pi x}{2a}\cos\frac{\pi y}{2a}, \quad \varphi = \varphi(x,y) = f_{st}^2\Phi(x,y), \tag{A.41}$$

where f_{st} is the static displacement at the PCB center and the function $\Phi(x,y)$ is expressed by formula (A.12). Then, equation (A.39) yields

$$\frac{f_{st}}{h}\cos\frac{\pi x}{2a}\cos\frac{\pi y}{2a} + \frac{3}{8}(1-\nu^2)\left(\frac{f_{st}}{h}\right)^3 \cos\frac{\pi x}{2a}\cos\frac{\pi y}{2a}\left(\frac{2}{1-\nu} + \cos\frac{\pi x}{a} + \cos\frac{\pi y}{a}\right)$$
$$= \frac{48}{\pi^4}\frac{1-\nu^2}{E}\left(\frac{a}{h}\right)^4 q$$

Using Galerkin's method, we obtain the following relationship between the static loading q and the induced displacement f_{st}:

$$q = \left(\frac{\pi}{2}\right)^6 \frac{Dh}{a^4}\left[\frac{f_{st}}{h} + \frac{3}{2}(1+\nu)\left(\frac{f_{st}}{h}\right)^3\right] \tag{A.42}$$

This relationship can be particularly helpful when one considers replacing drop or shock tests with static bending tests to generate similar curvatures and strains in the PCB.

In the case of small deflections,

$$q = q_0 = \left(\frac{\pi}{2}\right)^6 \frac{Dh}{a^4}\frac{f_{st}}{h} \tag{A-43}$$

Comparing expressions (A.42) and (A.43), we conclude that the factor

$$\chi = 1 + \frac{3}{2}(1+\nu)\left(\frac{A}{h}\right)^2 \tag{A-44}$$

considers the effect of the nonlinearity on the equivalent static loading. When $A = h$ and $\nu = \frac{1}{3}$, the factor χ is as high as $\chi = 3$.

A.14 NUMERICAL EXAMPLE

See also (A.28).

Input data:

Drop height $H = 1.273$ m; Board size $2a \times 2a = 300\ \text{mm} \times 300\ \text{mm}$; Board thickness $h = 1.5$ mm;

Young's modulus of the PCB material $E = 22.75\ \text{GPa} = 2321\ \text{kg/mm}^2$; Poisson's ratio $\nu = 0.3$;

Distributed mass of the PCB (including the mass of the SMDs and solder material) $m = 7.878 \times 10^{-10}\ \text{kg} \times \text{sec}^2/\text{mm}^3$.

Computed data:

Initial velocity:

$$\sqrt{2gH} = \sqrt{2 \times 9810 \times 1273} = 4996\ \text{mm/sec};$$

Flexural rigidity of the PCB:

$$D = \frac{Eh^3}{12(1-\nu^2)} = \frac{2321 \times 1.5^3}{12 \times 0.91} = 717.48\ \text{kg} \times \text{mm};$$

Actual mass of the PCB:

$$M_P = 4ma^2 = 4 \times 7.878 \times 10^{-10} \times 150^2 = 70.9 \times 10^{-6}\ \text{kg} \times \text{sec}^2/\text{mm}.$$

Generalized mass of the PCB:

$$M = ma^2 = 0.25 \times 70.9 \times 10^{-6} = 17.725 \times 10^{-6}\ \text{kg} \times \text{sec}^2/\text{mm};$$

Linear frequency

$$\omega = \frac{\pi^2}{2a}\sqrt{\frac{D}{M}} = \frac{\pi^2}{300}\sqrt{\frac{717.48}{17.725\times 10^{-6}}} = 209.31\,\text{sec}^{-1}$$

(Note that the FEA prediction [A.28] was $\omega = 209.48\,\text{sec}^{-1}$)

Linear amplitude of the induced vibrations

$$A_0 = \frac{16}{\pi^2}\frac{\sqrt{2gH}}{\omega} = \frac{16}{\pi^2}\frac{4996}{209.31} = 38.69\text{ mm}$$

(The FEA prediction [A.28] was $A_0 = 38.13$ mm).

Maximum linear acceleration (deceleration)

$$\ddot{f}_{\max} = -\omega^2 A_0 = -209.31^2 \times 38.69 = -1695035\text{ mm/sec}^2 = -173\text{ g}$$

Parameter of nonlinearity

$$\mu = \frac{3\pi^4}{32}(3-\nu)(1+\nu)\frac{D}{Ma^2h^2} = \frac{3\pi^4}{32}\times 2.7\times 1.3\frac{717.48}{17.725\times 10^{-6}\times 150^2\times 1.5^2}$$

$$= 25629.2\text{ mm}^{-2}\,\text{sec}^{-2}$$

Dimensionless parameter of nonlinearity

$$\bar{\mu} = \mu\left(\frac{A_0}{\omega}\right)^2 = 25629\left(\frac{38.69}{209.31}\right)^2 = 875.69$$

Factor of the nonlinear amplitude (the ratio of the linear amplitude, when the membrane forces are neglected, to the nonlinear amplitude, obtained with the consideration of these forces)

$$\eta_A = \sqrt{\frac{\sqrt{1+2\bar{\mu}}-1}{\bar{\mu}}} = \sqrt{\frac{\sqrt{1+2\times 875.69}-1}{875.69}} = 0.216$$

Nonlinear amplitude

$$A = \eta_A A_0 = 0.216\times 38.69 = 8.36\text{ mm (the FEA prediction [A.28] } A = 8.48\text{ mm).}$$

Parameter of the nonlinear frequency

$$\sigma = \sqrt{\omega^2 + \mu A^2} = \sqrt{209.31^2 + 25629\times 38.69^2} = 6197\text{ sec}^{-1}$$

Maximum nonlinear acceleration (deceleration)

$$\ddot{f}_{\max} = -\sigma^2 A = -6197^2\times 8.36 = -321047483\text{ mm/sec}^2 = -32726\text{ g}$$

Modulus of the elliptic function

$$k = \frac{A}{\sigma}\sqrt{\frac{\mu}{2}} = \frac{8.36}{6197}\sqrt{\frac{25629}{2}} = 0.153$$

Elliptic integral [A.27] $K(\mathrm{k}) = 1.58$.

Nonlinear frequency

$$p = \frac{\pi\sigma}{2K(k)} = \frac{\pi \times 6197}{2 \times 1.58} = 6161\ \mathrm{sec}^{-1}$$

Note that the nonlinear frequency p is not very much different of the parameter σ of the nonlinear frequency.

Nonlinear period of vibrations is $T = \frac{2\pi}{p} = 0.0010198$ sec.

The duration of the equivalent shock impact should be as low as $t_0 \approx \frac{T}{8} = 0.0001275$ sec, so that it could be replaced in the analyses by an instantaneous impulse. Note that in the case of linear vibrations, the duration of the equivalent shock impact load would be significantly (by a factor of 29.4) larger:

$$t_0 = \frac{\pi}{4\omega} = \frac{\pi}{4 \times 209.31} = 0.003752\ \mathrm{sec}$$

The magnitude of the impact pulse

$$S = M\frac{16}{\pi^2}\sqrt{2gH} = 17.725 \times 10^{-6} \times 1.6211 \times 4996 = 0.1436\ \mathrm{kg} \times \mathrm{sec}$$

The maximum force for a half-sine impulse

$$Q = \frac{\pi S}{2t_0} = \frac{\pi \times 0.1436}{2 \times 0.0001275} = 1768.6381\ \mathrm{kg}$$

The maximum value of the corresponding acceleration

$$\ddot{w}_{\max} = \frac{Q}{M} = \frac{8}{\pi}\frac{\sqrt{2gH}}{t_0} = 99782119.8196\ \mathrm{mm/sec}^2 = 10182\ \mathrm{g}$$

The maximum in-plane stress occurs at the PCB center

$$\sigma^0_{\max} = A^2\frac{\pi^2 E}{32a^2}\frac{2-\nu}{1-\nu} = 8.36^2\frac{\pi^2 \times 2321}{32 \times 150^2}\frac{1.7}{0.7} = 5.4001\ \mathrm{kg/mm}^2$$

The in-plane stress at the PCB contour:

$$\sigma^0_{\min} = A^2 \frac{\pi^2 E}{32a^2} \frac{\nu}{1-\nu} = 8.36^2 \frac{\pi^2 \times 2321}{32 \times 150^2} \frac{0.3}{0.7} = 0.9530 \text{ kg/mm}^2$$

The corresponding strains:

$$\varepsilon^0_{\max} = \frac{\pi^2}{32}\left(\frac{A}{a}\right)^2 (2-\nu) = \frac{\pi^2}{32}\left(\frac{8.36}{150}\right)^2 \times 1.7 = 0.001629,$$

$$\varepsilon^0_{\min} = \frac{\pi^2}{32}\left(\frac{A}{a}\right)^2 \nu = \frac{\pi^2}{32}\left(\frac{8.36}{150}\right)^2 \times 0.3 = 0.0002875.$$

The total tensile in-plane force at the PCB center

$$T^0_{\max} = 2ha\sigma^0_{\max} = 2 \times 1.5 \times 150 \times 5.4001 = 2430 \text{ kg.}$$

The total tensile in-plane force at the PCB contour

$$T^0_{\min} = 2ha\sigma^0_{\min} = 2 \times 1.5 \times 150 \times 0.9530 = 428.85 \text{ kg}$$

The maximum curvatures occur at the PCB center and, in an approximate analysis, can be found as

$$\kappa_{\max} = -A\left(\frac{\pi}{2a}\right)^2 (1+\nu) = -8.36^2\left(\frac{\pi}{300}\right)^2 1.3 = -0.0100 \text{ mm}^{-1}$$

The maximum bending moment (per unit PCB width)

$$M_b = -D\kappa_{\max} = 717.48 \times 0.0100 = 7.1486 \text{ kg}$$

The maximum bending stress

$$\sigma_b = \frac{6M_b}{h^2} = \frac{7.1486}{1.5^2} = 3.1772 \text{ kg/mm}^2,$$

Maximum bending strain

$$\varepsilon_b = \frac{\sigma_b}{E} = \frac{3.1772}{2321} = 0.001369$$

Total maximum stress on the convex side of the PCB

$$\sigma_{cv} = \sigma^0_{\max} + \sigma_b = 5.4001 + 3.1772 = 8.5773 \text{ kg/mm}^2$$

Total maximum stress on the concave side of the PCB

$$\sigma_{cc} = \sigma^0_{\max} - \sigma_b = 5.4001 - 3.1772 = 2.2229 \text{ kg/mm}^2,$$

that is, also tensile (because of the very high in-plane forces).
Total maximum strain on the convex side of the PCB

$$\varepsilon^t_{cv} = 0.001629 + 0.001369 = 0.002998.$$

Total maximum strain on the concave side of the PCB

$$\varepsilon^t_{cc} = 0.001629 - 0.001369 = 0.00026.$$

The equivalent distributed static loading resulting in the maximum deflection f_{st} that is equal to the nonlinear amplitude A of impact induced vibrations is

$$q = \left(\frac{\pi}{2}\right)^6 \frac{Dh}{a^4}\left[\frac{f_{st}}{h} + \frac{3}{2}(1+\nu)\left(\frac{f_{st}}{h}\right)^3\right] = \left(\frac{\pi}{2}\right)^6 \frac{717.48 \times 1.5}{150^4}\left[\frac{8.36}{1.5} + \frac{3}{2} \times 1.3 \times \left(\frac{8.36}{1.5}\right)^3\right]$$
$$= 0.0110 \text{ kg/mm}^2$$

This loading, high as it is, is considerably lower than the maximum distributed inertia load

$$q_{in} = -m\ddot{f}_{\max} = 7.878 \times 10^{-10} \times 321047483 = 0.2529 \text{ kg/mm}^2$$

Thus, a simple, easy-to-use, and physically meaningful predictive analytical (mathematical) stress model is suggested for the evaluation of the dynamic response of a PCB subjected to a drop impact during board-level drop testing. An exact solution is obtained for the Duffing-type equation for the principal coordinate of the induced vibrations. This means that this solution is applicable no matter how significant the level of the nonlinearity might be. Evaluation of the stresses and strains in the solder joints of the second level of interconnections (package to the PCB) is intended to be considered as the next step. Particularly, the advantage of CGA technology, as far as the possible stress relief in the solder material is concerned, is intended to be considered at this step. The induced membrane forces and the bending moments could be either predicted, or determined from the measured strains on the upper and the lower surfaces of the PCB. Experimental verification of the suggested model is certainly required. Particularly, the predicted and the measured strains on the concave and the convex surfaces of the PCB should be compared. Future work will include a methodology for evaluating the drop impact induced dynamic shearing and peeling stresses in the solder material of BGA and CGA designs.

REFERENCES

General Publications

1. E. Suhir, "Linear and Nonlinear Vibrations Caused by Periodic Impulses," AIAA/ASME/ASCE/AHS 26th Structures, Structural Dynamics and Materials Conference, Orlando, Florida, April 1985.
2. D. Steinberg, *Vibration Analysis for Electronic Equipment*, 2nd ed., John Wiley, 1988.
3. E. Suhir, "Dynamic Response of Micro-Electronic Systems to Shocks and Vibrations: Review and Extension," in E. Suhir, C.P. Wong, Y. C. Lee, (eds.), *Micro- and Opto-Electronic Materials and Structures: Physics, Mechanics, Design, Packaging, Reliability*, Springer, 2007.
4. E. Suhir, D. Steinberg, T. Yi, *Dynamic Response of Electronic and Photonic Systems to Shocks and Vibrations*, John Wiley, 2011.
5. E. Suhir, "Predictive Modeling of the Dynamic Response of Electronic Systems to Shocks and Vibrations," *ASME Applied Mechanics Reviews*, vol. 63, No. 5, March 2011.
6. E. Suhir, "Structural Dynamics of Electronics Systems," *Modern Physics Letters B (MPLB)*, vol. 27, No. 7, March 2013.

Industry Standards

7. ANSI/ASTM D3332-93, "Standard Test Methods for Mechanical Shock Fragility of Products Using Shock machines," ASTM, Philadelphia, Pennsylvania, 1996.
8. JEDEC Standard JESD22-B104-B, Mechanical Shock Test Method, 2001.
9. JEDEC Standard JESD22-B111, Board Level Drop Test Method of Components for Handheld Electronic Products, 2003.

Electronic Equipment

10. ASTM D4169.
11. E.H. Wong, K.M. Lim, N. Lee, S. Seah, C. Hoe, J. Wang, "Drop Impact Test–Mechanics and Physics of Failure," EPTC, Singapore, 2002.
12. C.T. Lim, C.W. Ang, L.B. Tan, S.K.W. Seah, E.H. Wong, "Drop Impact Survey of Portable Electronic Products," ECTC, 2003.
13. J.E. Luan, T.Y. Tee, "Analytical and Numerical Analysis of Impact Pulse Parameters on Consistency of Drop Impact Results," EPTC, IEEE Cat. No. 04EX971, Dec. 2004.
14. T.Y. Tee, J.E. Luan, E. Pek, C.T. Lim, Z.W. Zhong, "Advanced Experimental and Simulation Techniques for Analysis of Dynamic Responses During Drop Impact," ECTC, 2004.
15. T.T. Mattila, R. James, L. Nguyen, J.K. Kivilahti, "Effect of Temperature on the Drop Reliability of Electronic Assemblies," ECTC, 2007.
16. E. Suhir, "Response of a Heavy Electronic Component to Harmonic Excitations Applied to Its External Electric Leads," *Elektrotechnik & Informationstechnik (Austria)*, vol. 9, 2007.
17. C.Y. Zhou, T.X. Yu, E. Suhir, "Design of Shock Table Tests to Mimic Real-Life Drop Conditions," *IEEE CPMT Transactions*, vol. 32, No. 4, 2009.
18. J. Gu, D. Barker, M. Pecht, "Health Monitoring and Prognostics of Electronics Subject to Vibration Load Conditions," *IEEE Sensors Journal*, vol. 9, No. 11, 2009.
19. T.X. Yu, C.Y. Zhou, "Drop Impact of Typical Portable Electronic Devices: Experimentation and Modeling," in E. Suhir, D. Steinberg, T. Yi, (eds.), *Dynamic Response of Electronic and Photonic Systems to Shocks and Vibrations*, John Wiley, 2011.

20. C.Y. Zhou, T.X. Yu, S.W. Ricky Lee, E. Suhir, "Shock Test Methods and Test Standards for Portable Electronic Devices," in E. Suhir, D. Steinberg, T. Yi, (eds.), *Dynamic Response of Electronic and Photonic Systems to Shocks and Vibrations*, John Wiley, 2011.
21. D.S. Steinberg, "Test Equipment, Test Methods, Test Fixtures, and Test Sensors for Evaluating Electronic Equipment," in E. Suhir, D. Steinberg, T. Yi, (eds.), *Dynamic Response of Electronic and Photonic Systems to Shocks and Vibrations*, John Wiley, 2011.
22. E. Suhir, J.E. Morris, L. Wang, S. Yi, "Could Dynamic Strength of a Bonding Material in an Electronic Device be Assessed from Static Shear-Off Test Data?" *Journal of Materials Science, Electronic Materials*, 2016.

Modeling

23. E. Suhir, "Axisymmetric Elastic Deformations of a Finite Circular Cylinder with Application to Low Temperature Strains and Stresses in Solder Joints," *ASME Journal of Applied Mechanics*, vol. 56, No. 2, 1989.
24. E. Suhir, "Twist-Off Testing of Solder Joint Interconnections," *ASME Journal of Electronic Packaging (JEP)*, vol. 111, No. 3, Sept. 1989.
25. E. Suhir, "Analytical Modeling in Structural Analysis for Electronic Packaging: Its Merits, Shortcomings and Interaction with Experimental and Numerical Techniques," *ASME Journal of Electronic Packaging*, vol. 111, No. 2, June 1989.
26. L. Zhu, W. Marcinkiewicz, "Drop Impact Reliability Analysis of CSP Packages at Board and Product System Levels through Modeling Approaches," Intersociety Conference on Thermal and Thermo-mechanical Phenomena, 2004.
27. P. Marjamaki, T. Mattila, J. Kivilahti, "FEA of Lead-Free Drop Test Boards," ECTC Proceedings, 2005.
28. E. Suhir, "Could Electronics Reliability be Predicted, Quantified and Assured?" *Microelectronics Reliability*, vol. 53, April 2013.
29. E. Suhir, S. Kang, "Boltzmann-Arrhenius-Zhurkov (BAZ) Model in Physics-of-Materials Problems," *Modern Physics Letters B (MPLB)*, vol. 27, April 2013.
30. E. Suhir, "Three-Step Concept (TSC) in Modeling Microelectronics Reliability (MR): Boltzmann–Arrhenius–Zhurkov (BAZ) Probabilistic Physics-of-Failure Equation Sandwiched Between Two Statistical Models," *Microelectronics Reliability*, vol. 54, 2014.
31. E. Suhir, "Analytical Modeling Enables One to Explain Paradoxical Situations in the Behavior and Performance of Electronic Materials and Products (keynote presentation)," International Conference on Materials, Processing and Products Engineering (MPPE), Leoben, Austria, Nov. 2015.

Linear Response

32. E. Suhir, "Dynamic Response of a One-Degree-of-Freedom Linear System to a Shock Load During Drop Tests: Effect of Viscous Damping," *IEEE CPMT Transactions, Part A*, vol. 19, No. 3, 1996.
33. E. Suhir, "Is the Maximum Acceleration an Adequate Criterion of the Dynamic Strength of a Structural Element in an Electronic Product?" *IEEE CPMT Transactions, Part A*, vol. 20, No. 4, Dec. 1997.
34. E. Suhir, "Could Shock Tests Adequately Mimic Drop Test Conditions?" ECTC Proceedings, 2002.

Nonlinear and Probabilistic Responses

35. E. Suhir, "Response of a Flexible Printed Circuit Board to Periodic Shock Loads Applied to its Support Contour," *ASME Journal of Applied Mechanics*, vol. 59, No. 2, 1992.

36. E. Suhir, "Nonlinear Dynamic Response of a Flexible Thin Plate to Constant Acceleration Applied to its Support Contour, with Application to Printed Circuit Boards Used in Avionic Packaging," *International Journal of Solids and Structures*, vol. 29, No. 1, 1992.
37. X. He, R. Fulton, "Nonlinear Dynamic Analysis of a Printed Wiring Board," *ASME Journal of Electronic Packaging*, vol. 124, No. 2, 2002.
38. X. Heand, M. Stallybrass, "Impact Response of a Printed Wiring Board", *International Journal of Solids and Structures*, vol. 39, No. 24, 2002.
39. E. Suhir, M. Vujosevic, T. Reinikainen, "Nonlinear Dynamic Response of a "Flexible-and-Heavy" Printed Circuit Board (PCB) to an Impact Load Applied to its Support Contour," *Journal of Applied Physics, D*, vol. 42, No. 4, 2009.
40. E. Suhir, L. Arruda, "The Coordinate Function in the Problem of the Nonlinear Dynamic Response of an Elongated Printed Circuit Board (PCB) to a Drop Impact Applied to its Support Contour," *European Journal of Applied Physics*, vol. 48, No. 2, 2009.
41. E. Suhir, L. Arruda, "Could an Impact Load of Finite Duration Acting on a Duffing Oscillator be Substituted with an Instantaneous Impulse?" *Japan SME Journal of Solid Mechanics and Materials Engineering (JSMME)*, vol. 4, No. 9, 2010.
42. E. Suhir, *Applied Probability for Engineers and Scientists*, McGraw-Hill, 1997.
43. E. Suhir, "Probabilistic Design for Reliability," *Chip Scale Reviews*, vol. 14, No. 6, 2010.
44. E. Suhir, "Remaining Useful Lifetime (RUL): Probabilistic Predictive Model," *International Journal of Prognostics and Health Management*, vol. 2, No. 2, 2011.
45. E. Suhir, R. Mahajan, A. Vecero, "Do Electronic Industries Need New Approaches to Qualify Their Devices into Products?" Circuit Assembly, April 2011.

Solder Joints

46. J. Vodzak, D. Barker, A. Dasgupta, M. Pecht, "Combined Vibrational and Thermal Solder Joint Fatigue–A Generalized Strain Versus Life Approach," *ASME Journal of Electronic Packaging*, vol. 112, No. 2, 1990.
47. J.W. Evans, J.Y. Evans, R. Ghaffarian, A. Mawer, K. Lee, C. Shin, "Monte Carlo Simulation of BGA Failure Distributions for Virtual Qualification," ASME, Hawaii, 1991.
48. L. Condra, G. Johnson, M. Pecht, A. Christou, "Estimating the Vibration Fatigue Life of Quad Leaded Surface Mount Components," *IEEE Transactions on Components, Hybrids, and Manufacturing Technology*, vol. 15, No. 4, Aug. 1992.
49. D. Barker, A. Dasgupta, M. Pecht, "PWB Solder Joint Life Calculations Under Thermal and Vibrational Loading," *Journal of the Institute Environmental Sciences*, vol. 35, No. 1, 1992.
50. R. Ghaffarian, "Shock and Thermal Cycling Synergism Effects on Reliability of CBGA Assemblies," IEEE Aerospace Conference Proceedings, 2000.
51. A.M. Deshpande, G. Subbarayan, D. Rose, "A System for First Order Reliability Estimation of Solder Joint Area Array Packages," *Transactions of the ASME*, vol. 122, Mar. 2000.
52. A. Syed, J.H.L. Pang, "Solder Joint Reliability: Materials, Modeling and Testing," International Symposium on Advances in Packaging, Singapore, 2001.
53. T. Sogo, S. Hara, "Estimation of Fall Impact Strength for BGA Solder Joints," ICEP Conference, Tokyo, Japan, 2001.
54. L. Zhu, "Submodeling Technique for BGA Reliability Analysis of CSP Packaging Subjected to an Impact Loading," InterPACK Conference Proceedings, 2001.
55. Q. Yu, H. Kukuichi, S. Ikeda, M. Shiratori, M. Kakino, and N. Fujiwara, "Dynamic Behavior of Electronics Package and Impact Reliability of BGA Solder Joints," Intersociety Conference on Thermal Phenomena, 2002.

56. K. Mishiro, et al., "Effect of the Drop Impact on BGA/CSP Package Reliability," *Microelectronics Reliability*, vol. 42, 2002.
57. A.C. Shiah, X. Zhou, "A Low Cost Reliability Assessment for Double-Sided Mirror-Imaged Flip Chip BGA Assemblies," Proceedings of the Panpacific Conference, Surface Mount Technology Association, 2002.
58. K. Newman, "Board Level Solder Joint Reliability of High Performance Computers Under Mechanical Loading", EuroSimE, 2008.
59. C.R. Siviour, D.M. Williamson, S.J.P. Palmer, S.M. Walley, W.G. Proud, J.E. Field, "Dynamic Properties of Solders and Solder Joints," *Journal de Physique IV*, vol. 110, 2003.
60. L. Xu, et al., "Numerical Studies of the Mechanical Response of Solder Joints to Drop/Impact Load," Proceedings of the EPTC, 2003.
61. I.E. Anderson, J.L. Harringa, "Elevated Temperature Aging of Solder Joints Based on Sn-Ag-Cu: Effect of Joint Microstructure and Shears Strength", Journal of Electronic Materials, vol. 33, 2004.
62. M. Date, et al., "Ductile-to-Brittle Transition in Sn-Zn Solder Joints Measured by Impact Tests," *Scripta Materialia*, vol. 51, 2004.
63. K.C. Ong, V.B.C. Tan, C.T. Lim, E.H. Wong, X.W. Zhang, "Dynamic Materials Testing and Modeling of Solder Interconnects," ECTC, 2004.
64. C.-L. Yeh, Y.-S. Lai, "Effect of Solder Joint Shapes on Free Drop Reliability of Chip-Scale Packages," Proceedings of the IMAPS Taiwan Technology Symposium, 2004.
65. C.-L. Yeh, Y.-S. Lai, "Transient Simulation of Solder Joint Fracturing Under Impact Test," 6th EPTC, Singapore, Dec. 2004.
66. J.E. Field, S.M. Walley, W.G. Proud, H.T. Goldrein, C.R. Siviour, "Review of Experimental Techniques for High Rate Deformation and Shock Studies," *International Journal of Impact Engineering*, vol. 30, 2004.
67. R. Ghaffarian, "BGA Assembly Reliability," in K. Gilleo, (ed.), *Area Array Packaging Handbook*, McGraw-Hill, 2004.
68. A. Syed, et al., "A Methodology for Drop Performance Prediction and Application for Design Optimization of Chip Scale Packages," ECTC, 2005.
69. C. Britzer, et al., "Drop Test Reliability Improvement of Lead-Free Fine Pitch BGA Using Different Solder Ball Composition," ECTC, 2005.
70. E.H. Wong, et al., "Drop Impact: Fundamentals and Impact Characterization of Solder Joints," ECTC, 2005.
71. C.R. Siviour, S.M. Walley, W.G. Proud, J.E. Field, "Mechanical Properties of SnPb and Lead-Free Solders at High Rates of Strain," *Journal of Physics D: Applied Physics*, vol. 38, 4131–4139, 2005.
72. R. Ghaffarian, "Area Array Technology for High Reliability Applications", in E. Suhir, (ed.), *Micro-and Opto-Electronic Materials and Structures: Physics, Mechanics, Design, Reliability, Packaging*, Springer, 2006.
73. T.T. Mattila, E. Kaloinen, A. Syed, J. Kivilahti, "Reliability of SnAgCu Interconnections with Minor Additives of Ni or Bi under Mechanical Shock Loading at Elevated Temperatures," ECTC, 2007.
74. D.M. Williamson, J.E. Field, S.J.P. Palmer, C. R. Siviour, "Rate Dependent Strengths of Some Solder Joints," *Journal of Physics D: Applied Physics*, vol. 40, 2007.
75. H. Qi, M. Osterman, M. Pecht, "Modeling of Combined Temperature Cycling and Vibration Loading on PBGA Solder Joints Using an Incremental Damage Superposition Approach," *IEEE Transactions on Advanced Packaging*, vol. 31, No. 3, 2008.
76. R. Ghaffarian, "Thermal Cycle Reliability and Failure Mechanisms of CCGA and PBGA Assemblies with and Without Corner Staking," *IEEE Transactions on Components and Packaging Technologies*, vol. 31, No. 2, June 2008.

77. H. Qi, M. Osterman, M. Pecht, "A Rapid Life-Prediction Approach for PBGA Solder Joints Under Combined Thermal Cycling and Vibration Loading Conditions," *IEEE Transactions on Components and Packaging Technologies*, vol. 32, No. 2, 2009.
78. S. Sakthievelan, et al., "Thermal and Mechanical Behavior of Lead-Free Area Array Packages with Full/Corner/No-Underfill," InterPack, San Francisco, California, July, 2009.
79. R. Ghaffarian, "Thermal Cycle and Vibration/Drop Reliability of Area Array Package Assemblies," in E. Suhir, D. Steinberg, T. Yi (eds.), *Dynamic Response of Electronic and Photonic Systems to Shocks and Vibrations*, John Wiley, 2011.
80. V.B.C. Tan, K.C. Ong, C.T. Lim, J.E. Field, "Dynamic Mechanical Properties and Microstructural Studies of Lead-Free Solders in Electronic Packaging," in E. Suhir, D. Steinberg, T. Yi (eds.), *Dynamic Response of Electronic and Photonic Systems to Shocks and Vibrations*, John Wiley, 2011.
81. D.S. Steinberg, "Dynamic Response of Solder Joint Interconnections to Vibration and Shock," in E. Suhir, D. Steinberg, T. Yi (eds.), *Dynamic Response of Electronic and Photonic Systems to Shocks and Vibrations*, John Wiley, 2011.
82. R. Ghaffarian, "Damage and Failures of CGA/BGA Assemblies Under Thermal Cycling and Dynamic Loadings," ASME 2013 International Mechanical Engineering Congress and Engineering. (IMECE2013), San Diego, California, Nov. 2013.
83. E. Suhir, "Analysis of a Short Beam with Application to Solder Joints: Could Larger Stand-Off Heights Relieve Stress?" *European Physical Journal of Applied Physics (EPJAP)*, vol. 71, 2015.
84. E. Suhir, R. Ghaffarian, J. Nicolics, "Predicted Stresses in Ball-Grid-Array (BGA) and Column-Grid-Array (CGA) Interconnections in a Mirror-Like Package Design," *Journal of Materials Science: Materials in Electronics*, vol. 27, No. 3, 2016.
85. R. Ghaffarian, "Update on CGA Packages for Space Applications," *Microelectronics Reliability*, Jan. 2016.
86. E. Suhir, R. Ghaffarian, "Predicted Stresses in a Ball-Grid-Array (BGA)/Column-Grid-Array (CGA) Assembly with a Low Modulus Solder at its Ends," *Journal of Materials Science: Materials in Electronics*, 2016.
87. E. Suhir, R. Ghaffarian, "Predicted Stresses in a Ball-Grid-Array (BGA)/Column-Grid-Array (CGA) Assembly with Epoxy Adhesive at its Ends," *Journal of Materials Science: Materials in Electronics*, 2016.

Board-Level Testing

88. S.K.W. Seah, C.T. Lim, E.H. Wong, V.B.C. Tan, V.P.W. Shim, "Mechanical Response of PCBs in Portable Electronic Products during Drop Impact," EPTC, Singapore, Dec. 2002.
89. S. Goyal, E.K. Buratynski, "Methods for Realistic Drop Testing," *International Journal of Microcircuits and Electronic Packaging*, vol. 23, No. 1, 2000.
90. S. Goyal, E.K. Buratynski, G.W. Elko, "Shock-Protection Suspension Design for Printed Circuit Board," Proceedings of SPIE, vol. 4217, 2002.
91. Y.Q. Wang, "Modeling and Simulation of PCB Drop Test," Proceedings of the 5th EPTC, 2003.
92. J.E. Luan, T.Y. Tee, E. Pek, C.T. Lim, Z.W. Zhong, "Modal Analysis and Dynamic Responses of Board Level Drop Test," 5th EPTC Conference Proceedings, Singapore, 2003.
93. T.Y. Tee, H.S. Ng, C.T. Lim, E. Pek, Z.W. Zhong, "Board Level Drop Tests and Simulation of TFBGA Packages for Telecommunication Applications," ECTC Proceedings, May 2003.
94. L.B. Tan, "Board Level Solder Joint Failure by Static and Dynamic Loads," Proceedings of the 5th EPTC, 2003.

95. Y.C. Ong, V.P.W. Shim, T.C. Chai, C.T. Lim, "Comparison of Mechanical Response of PCBs Subjected to Product Level and Board-Level Drop Impact Tests," EPTC, Singapore, 2003.
96. T.Y. Tee, H.S. Ng, Z.W. Zhong, "Design for Enhanced Solder Joint Reliability of Integrated Passive Device Under Board Level Drop Test and Thermal Cycling Test," EPTC, Singapore, 2003.
97. J.E. Luan, T.Y. Tee, "Effect of Impact Pulse on Dynamic Responses and Solder Joint Reliability of TFBGA Packages During Board Level Drop Test," 6th EMAP Conference, Malaysia, Dec. 2004.
98. T.Y. Tee, J.E. Luan, E. Pek, C.T. Lim, Z.W. Zhong, "Novel Numerical and Experimental Analysis of Dynamic Responses Under Board Level Drop Test," EuroSime Conference Proceedings, 2004.
99. T.C. Chiu, et al., "Effect of Thermal Aging on Board Level Drop Reliability for Pb-Free BGA Packages," ECTC, 2004.
100. T.C. Chai, et al., "Board Level Drop Test Reliability of IC Packages," ECTC, 2005.
101. C.L. Yeh, Y.S. Lai, "Insights into Correlation between Board-Level Drop Reliability and Package-Level Ball Impact Test," ECTC, 2006.
102. J.-E. Luan, T.-Y. Tee, E. Pek, C.-T. Lim, Z. Zhong, "Dynamic Responses and Solder Joint Reliability Under Board Level Drop Test," *Microelectronics Reliability*, vol. 47, No. 2–3, Feb.-Mar. 2007.
103. H.S. Human, X. Fan, T. Zhou, "Modeling Techniques for Board Level Drop Test for a Wafer Level Package," ICEPT, China, 2008.
104. H. Dhiman, X. Fan, T. Zhou, JEDEC Board-Level Drop Test Simulation for Wafer Level Packages," ICEPT, China, 2009.
105. T.T. Matilla, P. Marjamali, J. Kivilahti, "Board-Level Reliability of Lead-Free Solder Under Mechanical Shock and Vibration Loads," in E. Suhir, D. Steinberg, T. Yi, (eds.), *Dynamic Response of Electronic and Photonic Systems to Shocks and Vibrations*, John Wiley, 2011.
106. F. Song, S.W. Ricky Lee, K. Newman, R. Sykes, S. Clark, "Correlation Between Package Level High-Speed Solder Ball Shear/Pull and Board-Level Mechanical Drop Tests with Brittle Fracture Failure Mode, Strength and Energy," in E. Suhir, D. Steinberg, T. Yi, (eds.), *Dynamic Response of Electronic and Photonic Systems to Shocks and Vibrations*, John Wiley, 2011.
107. R. Ghaffarian, "Reliability of Printed Circuit Boards," in C.F. Coombs (ed.), *Printed Circuit Handbook*, 7th ed., McGraw-Hill, 2016.

APPENDIX REFERENCES

A.1. S. Goyal, E.K. Buratynski, "Methods for Realistic Drop Testing," *International Journal of Microcircuits and Electronic Packaging*, vol. 23, No. 1, 2000.
A.2. S.R.W. Seah, et al., "Mechanical Response of PCBs in Portable Electronic Products during Drop Impact," EPTC, Singapore, Dec. 2002.
A.3. Y.C. Ong, et al., "Comparison of Mechanical Response of PCBs Subjected to Product Level and Board-Level Drop Impact Tests," EPTC, Dec. 2003.
A.4. T.Y. Tee, et al., "Novel Numerical and Experimental Analysis of Dynamic Responses under Board Level Drop Test," EuroSime Conference, 2004.
A.5. T.C. Chai, et al., "Board Level Drop Test Reliability of IC Packages," ECTC, 2005.
A.6. C.L. Yeh, Y.S. Lai, "Insights into Correlation Between Board-Level Drop Reliability and Package-Level Ball Impact Test," ECTC, 2006.
A.7. J.-E. Luan, et al., "Dynamic Responses and Solder Joint Reliability Under Board Level Drop Test," *Microelectronics Reliability*, vol. 47, No. 2–3, Feb.-Mar., 2007.

A.8. T.T. Matilla, et al., "Board-Level Reliability of Lead-Free Solder Under Mechanical Shock and Vibration Loads," in E. Suhir, D. Steinberg, T. Yi, (eds.), *Dynamic Response of Electronic and Photonic Systems to Shocks and Vibrations*, John Wiley, 2011.
A.9. R. Ghaffarian, "Reliability of Printed Circuit Boards," in C.F. Coombs (ed.), *Printed Circuit Handbook*, 7th ed., McGraw-Hill, 2016.
A.10. Y.Q. Wang, "Modeling and Simulation of PCB Drop Test," EPTC, 2003.
A.11. J.E. Luan, T.Y. Tee, E. Pek, C.T. Lim, Z.W. Zhong, "Modal Analysis and Dynamic Responses of Board Level Drop Test," EPTC, 2003.
A.12. T.Y. Tee, H.S. Ng, C.T. Lim, E. Pek, Z.W. Zhong, "Board Level Drop Tests and Simulation of TFBGA Packages for Telecommunication Applications," ECTC, 2003.
A.13. L.B. Tan, "Board Level Solder Joint Failure by Static and Dynamic Loads," EPTC, 2003.
A.14. T.Y. Tee, H.S. Ng, Z.W. Zhong, "Design for Enhanced Solder Joint Reliability of Integrated Passive Device Under Board Level Drop Test and Thermal Cycling Test," EPTC, 2003
A.15. J.E. Luan, T.Y. Tee, "Effect of Impact Pulse on Dynamic Responses and Solder Joint Reliability of TFBGA Packages During Board Level Drop Test," EMAP Conference, Malaysia, Dec. 2004.
A.16. T.Y. Tee, J.E. Luan, E. Pek, C.T. Lim, Z.W. Zhong, "Novel Numerical and Experimental Analysis of Dynamic Responses under Board Level Drop Test," EuroSime Conference, 2004.
A.17. T.C. Chiu, et al., "Effect of Thermal Aging on Board Level Drop Reliability for Pb-Free BGA Packages," ECTC, 2004.
A.18. T.C. Chai, et al., "Board Level Drop Test Reliability of IC Packages," ECTC, 2005.
A.19. C.L. Yeh, Y.S. Lai, "Insights into Correlation Between Board-Level Drop Reliability and Package-Level Ball Impact Test," ECTC, 2006.
A.20. J.-E. Luan, T.-Y. Tee, E. Pek, C.-T. Lim, Z. Zhong, "Dynamic Responses and Solder Joint Reliability Under Board Level Drop Test," *Microelectronics Reliability*, vol. 47, No. 2–3, Feb.-Mar., 2007.
A.21. T.T. Matilla, P. Marjamali, J. Kivilahti, "Board-Level Reliability of Lead-Free Solder Under Mechanical Shock and Vibration Loads," in E. Suhir, D. Steinberg, T. Yi, (eds.), *Dynamic Response of Electronic and Photonic Systems to Shocks and Vibrations*, John Wiley, 2011.
A.22. F. Song, S. Lee, W. Ricky, K. Newman, R. Sykes, S. Clark, "Correlation Between Package Level High-Speed Solder Ball Shear/Pull and Board-Level Mechanical Drop Tests with Brittle Fracture Failure Mode, Strength and Energy," in E. Suhir, D. Steinberg, T. Yi, (eds.), *Dynamic Response of Electronic and Photonic Systems to Shocks and Vibrations*, John Wiley, 2011.
A.23. R. Ghaffarian, "Reliability of Printed Circuit Boards," in C.F. Coombs, (ed.), *Printed Circuit Handbook*, 7th ed., McGraw-Hill,, 2016.
A.24. S.P. Timoshenko, J.M. Gere, *Theory of Elastic Stability*, 2nd ed., McGraw-Hill, 1988.
A.25. E. Suhir, *Structural Analysis in Microelectronics and Fiber Optics*, Van-Nostrand, 1997.
A.26. L.A. Pars, *A Treatise of Analytical Dynamics*, Heinemann, 1965.
A.27. I.N. Sneddon, *Special Functions of Mathematical Physics and Chemistry*, 3rd ed., Longman, 1980.
A.28. E. Suhir, M. Vujosevic, T. Reinikainen, "Nonlinear Dynamic Response of a "Flexible-and-Heavy" Printed Circuit Board (PCB) to an Impact Load Applied to Its Support Contour," *Journal of Applied Physics, D*, vol. 42, No. 4, 2009.

10 Failure-Oriented-Accelerated-Testing and Multiparametric Boltzmann–Arrhenius–Zhurkov Equation

"In the long run we are all dead."

—John Maynard Keynes, British economist

10.1 ACCELERATED TESTING

Thermal stresses have been addressed in many monographs and numerous papers (see, for example, [1–66]). It is impractical and uneconomical to wait for failures when the mean-time-to-failure (MTTF) for a typical electronic or a photonic device is hundreds of thousands of hours. Accelerated testing (see, for example, [16, 17, 25, 47, 57, 65, 66]) has become an inevitable and powerful means for improving reliability. This is true regardless of whether irreversible or reversible failures will or will not actually occur during the HALT (highly accelerated life testing, "discovery testing"), FOAT (failure-oriented accelerated testing, "testing to fail"), or QT (qualification testing, "testing to pass"). To accelerate the material's (device's) degradation and/or its failure, one has to deliberately "distort" ("skew") one or more parameters affecting the device's functional and mechanical performance and its environmental durability. Then, one has to bridge the gap between what they see during the accelerated testing and what will supposedly happen in the actual operation conditions. Therefore, effective and trustworthy predictive modeling is extremely important (see, for example, [1, 2, 4, 6, 9, 11–15, 23–25, 31, 32, 35, 38, 41, 48, 51, 58, 61, 66]).

The applied stressors (stimuli) include, but might not be limited to, high temperature soaking/storage/baking, aging/dwell, low temperature storage, temperature cycling (most popular today), power cycling, thermal shock, thermal gradients, fatigue tests, mechanical shock, drop shock, sinusoidal or random vibrations, creep/stress relaxation tests, high humidity, electrical current, voltage, and radiation, or their combinations, such as, temperature-humidity bias, fatigue or vibration tests at elevated temperatures, and crack propagation dynamic tests at low temperature conditions. The latter is, in the author's opinion, the most feasible (see Section 10.6). Different types of accelerated tests practiced in today's microelectronics and photonics are shown in Table 10.1.

TABLE 10.1
Accelerated Test Types

Accelerated Test (AT) Type	Product Development Testing (PDT)	Highly Accelerated Life (Stress) Testing HALT	Qualification Testing (QT)	Burn-in Testing (BIT)	Failure-Oriented Accelerated Testing (FOAT)
Objective	Technical feedback to assure acceptability of the taken design approach	Ruggedize the product and tentatively assess its reliability limits	Demonstration that the product is qualified to serve in the expected capacity	Eliminate insufficiently reliable products ("freaks")	Confirm the use of a physically meaningful predictive model
End point	Type, time, level, or the number of observed failures	Predetermined number or percent of failures	Predetermined time and/or the number of cycles until excessive (unexpected) number of failures occurs	The infant mortality portion of the bathtub curve is (hopefully) eliminated	Predetermined number or percent (typically 50%) of failures
Follow-up activity	Failure analysis, design decision	Failure analysis	Pass/fail decision	Shipping the sound devices	Failure- and probabilistic analysis of the test data
Ideal test	Specific definitions	No failures in a long time			Numerous failures in a short time

The objective of product testing development (PDT) is to obtain information on product reliability during design, development, and often also at early manufacturing stages. The tests are intended to pinpoint the weaknesses and limitations of the design, materials, and the manufacturing technology or process. The tests are supposed to evaluate new designs, new processes, the appropriate correction actions, or to compare different products from the standpoint of their reliability. This type of testing is usually limited by time (when almost no failures occur) and is followed by the analysis of the observed failures, or by another in-depth, "independent," investigation. A typical example of PDT is shear-off testing (see Chapter 8). PDTs are, as a rule, destructive tests. Because of that, various and numerous *nondestructive evaluations* (ultrasonic methods, X-raying, Moiré interferometry, etc.) are always useful at the product-development stage. Size and location of a defect, and the loading/stress conditions, should be considered when deciding if a certain defect might result in a reliability problem. Observed (detected) defects should not necessarily be viewed as reliability concerns, but in many cases should rather be considered as quality defects.

The objective of HALT is to obtain the preliminary information about the reliability of the product, and to reveal the principal physics of its failures. HALTs are supposed to determine the weakest links, the bottlenecks of the design, and desirably to obtain the preliminary information about the major modes and mechanisms of failure. HALT may sometime hasten failure mechanisms that are different from those that could be actually observed in service conditions, such as change in materials properties at high or low temperatures, creep or stress relaxation at elevated temperatures, occurrence and movement of dislocations caused by an elevated stress, and so on. It is always necessary to correctly identify the expected modes and mechanisms and to establish the appropriate stress/temperature limits in order to prevent the "shift" in the actual dominant failure mechanisms. HALTs may sometime lead to a bimodal distribution of failures, that is, to a situation when a dual mechanism of failure takes place, such as infant mortality failures might occur concurrently with the anticipated operational failures.

The objective of QTs is to prove that the reliability of the product under test is not lower than a specified level. This level is usually measured as a percentage of failures per lot and/or by the number of failures per unit time (failure rate). QTs are, in general, nondestructive tests. Testing is time limited. The analyst of the test results usually hopes to get as few failures as possible. The pass/fail decision is based on a go/no-go criterion. The typical requirements are no more than a few percent failing parts of the total lot. One should always have in mind that a field failure might occur even if the product passed all the QTs. Although QTs are not supposed to evaluate the actual failure rate, their results can be used to suggest that the actual failure rate is at least not higher than a certain value.

The objective of the burn-in tests (BITs) is to detect and eliminate weak boundaries and delaminations, inclusions and voids, imperfections in geometry and materials, uneven coatings and nonuniform adhesive layers, and current leakage. BITs are supposed to eliminate the infant mortality portion of the bathtub curve. Burn-in is needed to stabilize the performance of the device in use by accelerating the failure mechanism kinetics and cause defective parts to fail, thereby excluding the risk of their failure in the field. For healthy devices, burn-ins are nondestructive tests. Burn-ins can be based on high temperatures, thermal cycling, voltage, current density, high humidity, and so on. Although there is always a possibility that some defects might escape burn-in tests, it is more likely that burn in can introduce some damage to the "healthy" structure, that is, it will consume a certain portion of the useful service life of the product. Some burn-in tests are harmless and will not introduce any damage: high electric fields for dielectric breakdown screening, mechanical stresses below the fatigue limit, and so on.

All these types of accelerated tests, except, perhaps, FOAT, are well established in today's engineering practice. FOAT is, like HALT, an accelerated test, but is also, unlike HALT, the experimental basis of the probabilistic-design-for-reliability (PDfR) concept (see Chapter 11). In this chapter, we address different types of FOAT as possible extension and modification of the 40-years- old and well-established HALT (bad things would not exist for 40 years, would they?) that exists in different modifications, depending on the company, manufacturing technology, product, and application, and is often combined with the BIT.

10.2 FOAT, ITS SIGNIFICANCE, ATTRIBUTES, AND ROLE

FOAT's objectives (Table 10.1) are: understand the underlying physics of failure; confirm usage of a particular reliability physics oriented predictive model, such as multiparametric BAZ model (see Chapter 11) and establish its numerical characteristics (activation energy, sensitivity factors, etc.); predict the (never-zero) probability of failure and the corresponding expected lifetime of the device; and identify parametric degradation and other long-term failure mechanisms, as well as to collect (accumulate) statistical information about the product-under-test through its failures. HALT and QT give no indication on the probability of failure and are not expected to reflect the actual use conditions. FOATs do. The FOAT analyst needs to generate as many failures as feasible and as fast as possible. They can be conducted at the part level, component level, module level, equipment level, or even at a system level. FOAT are terminated when the modes and mechanisms of failure are established, and sufficient failure statistics is collected. The typical acceptable failure ratio is 50%. We always want to understand the physics of failure and its role in creating a reliable product. It goes without saying that if one wants to prevent a premature failure, a physically meaningful highly-focused and highly cost-effective FOAT is a must.

Quite a few FOAT type of accelerated tests were suggested in the past, depending on the particular product and its application. Here are some of them:

- **Power law** is used when the physics of failure is unclear, to describe, for example, degradation in lasers when the injection current or the light output power are used as acceleration parameters, or to address "static fatigue" (delayed fracture) of silica material in optical waveguides;
- **Arrhenius equation** and its numerous extensions and interpretations is used when there is a belief that elevated temperature is the major cause of failure, and that the temperature-induced degradation could be attributed to a combination of physical and chemical processes;
- **Coffin–Manson and related equations (inverse power law)** are used to evaluate the number-of-cycles-to-failure in the low-cycle-fatigue conditions, the MTTF in random vibration tests, or aging in lasers in the cases of current or power acceleration;
- **Paris–Erdogan equation** or other crack growth (Griffith theory-based) equations are used to assess the fracture toughness of brittle materials by establishing the relationship between the fatigue crack growth rate and the variation in the cyclic stress intensity factor; the equation is applicable when the stress intensity factor range is larger than a certain threshold for the given material, below which no crack growth could possibly occur, or below which the crack growth is extremely low;
- **Boltzmann–Arrhenius–Zhurkov (BAZ) equation** considers not only the absolute temperature as an acceleration factor, but the applied "external" stress as well by reducing the activation energy level; this equation underlies the kinetic approach to the evaluation of the strength of materials (see Section 10.3);

- **Eyring equation,** unlike the BAZ constitutive equation, considers the applied "external" stress directly by assuming that the MTTF is inversely proportional to the applied stress (mechanical stress, voltage, humidity, etc.);
- **Peck equation** is, in effect, the Eyring equation modified for modeling the temperature-humidity bias conditions, and is a combination of the inverse power law (for relative humidity) and the original BAZ equation;
- **Fatigue Damage Model (Palmgren-Miner's Rule)** expresses the law of linear accumulation of damages, when the applied stresses and strains are still within the elastic range;
- **Creep rate equations** express the creep rate as a product of the applied stress (at a certain power level), exponential (Arrhenius type) term and, when all the creep stages are important, also the time-dependent term (Graham–Walles equation);
- **Weakest link models** evaluate the TTF in brittle materials with defects and assume that the material (device) failure originates from the weakest point (crack generation and propagation, dielectric breakdown, etc.);
- **Stress–strength interference model** is widely used in various probabilistic problems of structural ("physical") design and considers the interaction of the load (stress) probability distribution function and the bearing capacity (strength) probability distribution function; these two functions could be time-dependent.

FOAT tests are critical as long as there is a need to understand the physics of failure and to quantify the expected lifetime of the product. FOAT tests, their role, and significance has been addressed in numerous publications (see, for example, [67–79]). In this book, a multiparametric version of the BAZ model is employed as part of the PDfR concept in electronic and photonic engineering, including systems that contain solder (see Chapter 11). A highly focused and highly cost-effective FOAT geared to a suitable physically meaningful model, such as multiparametric BAZ equation, is the experimental basis of the PDfR concept. FOAT should be conducted for products whose failure-free performance is critical in addition to, and, in many cases, even instead of, the HALT; especially for new products for which no experience is yet accumulated and no acceptable HALT exists.

FOAT is a "transparent box" and could be viewed as an extension and a modification of HALT, which is a "black box" that does not consider the underlying reliability physics and is unable to determine the never-zero probability of failure. HALT can be used for "rough tuning" of a product's reliability, while FOAT could be employed, when "fine tuning" is needed, such as when there is an intent to quantify, assure and, if possible, even specify the operational reliability of the electronic or photonic package. The FOAT approach could be viewed as a quantified HALT and should be geared to a particular technology, product, and application. The almost 40-years-old HALT tries to "kill many unknown birds with one big stone" and is currently widely employed in different modifications with an intent to:

1. determine a product's reliability weaknesses;
2. tentatively assess, in a qualitative way, the reliability limits;

3. ruggedize the product by applying elevated stresses (not necessarily mechanical and not necessarily limited to the anticipated field stresses) that could cause field failures;
4. provide large (but, actually, unknown) safety margins over expected in-use conditions;
5. precipitate and identify failures of different origins.

HALT does that through a "test-fail-fix" process, in which the applied stresses ("stimuli") are somewhat above the specified operating limits. HALT's end point is defined by the predetermined number or percent of failures, its follow up activity is failure (root cause) analysis, and an ideal HALT is when no failures occur in a long time. FOAT, on the other hand, is aimed at understanding the physics of failure, confirming the use of a particular predictive model, and assessing the probability of failure. An ideal FOAT is the one that is able to generate numerous failures in a short time. A highly focused and highly cost-effective FOAT is the "heart" of the PDfR concept, its experimental foundation, and could/should be conducted in addition to and, in some cases, even instead of HALT. Understanding the underlying reliability physics for the material/device operational performance is critical. If one sets out to understand the physics of failure in an attempt to create a failure-free product, conducting FOAT should be imperative, should it not? FOAT's objectives are to confirm usage of a particular more or less well-established predictive model, such as BAZ, to confirm the anticipated underlying physics of failure and establish the numerical characteristics (activation energy, sensitivity factors, etc.) of the particular governing reliability model. FOAT could be viewed as an extension or a modification of HALT. But while HALT is a "black box", that is, a methodology that can be perceived in terms of its inputs and outputs without a clear knowledge of the underlying physics and the likelihood of failure, and does not measure/quantify reliability, FOAT is a "white/transparent box" and quantifies reliability. Therefore, HALT can be used for "rough tuning" of a product's reliability, while FOAT should be employed when "fine tuning" is needed, that is, when there is a need to quantify, assure, and even specify the operational reliability of a material or a device and could be viewed as a "quantified and reliability physics-oriented HALT."

10.3 MULTIPARAMETRIC BAZ EQUATION

The simplest BAZ equation [80, 81]

$$\tau = \tau_0 \exp\left(\frac{U_0 - \gamma\sigma}{kT}\right) \tag{10.1}$$

was suggested by the Russian physicist Zhurkov in 1957, in application to experimental fracture mechanics as a generalization of the Arrhenius equation [82]

$$\tau = \tau_0 \exp\left(\frac{U_0}{kT}\right) \tag{10.2}$$

introduced by the Swedish chemist Arrhenius in 1889 in the kinetic theory of chemical reactions (1903 Nobel Prize in chemistry). Equations (10.1) and (10.2) consider the role of the ratio $\frac{U_0}{kT}$ of the activation energy U_0 (this term was coined by Arrhenius to characterize material's propensity to get engaged into a chemical reaction) to the thermal energy kT determined as the product of the Boltzmann's constant $k = 8.6173303 \times 10^{-5}$ eV/K and the absolute temperature T. In these equations, τ is interpreted as the MTTF, τ_0 is the time constant, σ is the applied stress per unit volume, and γ is the sensitivity factor.

Equation (10.2) is formally not different of what is known as the Boltzmann or Maxwell–Boltzmann equation [83, 84] in the kinetic theory of gases. This equation postulates that the absolute temperature of an ideal gas, when it is in thermodynamic equilibrium with the environment, is determined by the average probability of the collisions of the gas particles (atoms or molecules). Chemist Arrhenius was member of physicist Boltzmann's team in the University of Graz in Austria in 1887 and suggested that equation (10.2) is used to assess the significance of the energy barrier, the so called activation energy, to be got over in order to commence a chemical reaction. Although the Arrhenius equation has been criticized over the years on several grounds (it has been argued, particularly, that this energy might not be a constant property of a material, but might be time and/or temperature and/or even external loading dependent), it is still widely used, mostly because of its simplicity, in numerous applied science applications, when it is believed that it is the temperature that is primarily responsible for the duration of the useful lifetime of a material or a device of interest. The Arrhenius equation is still sometimes used to assess the MTTF of electronic products.

The effective activation energy U_0 has been determined for many electronic materials and failure mechanisms. For semiconductor device failure mechanisms, the activation energy ranges from 0.3 to 0.6eV. Activation energies for some typical failure mechanisms in semiconductor devices are: for metal migration–1.8eV; for charge injection–1.3eV; for ionic contamination–1.1eV; for Au-Al intermetallic growth, surface charge accumulation and intermetallic diffusion–1.0eV; for humidity induced corrosion and electromigration of Si in Al–0.9eV; for Si junction defects–0.8eV; for charge loss–0.6eV; and for electromigration in Al and metallization defects–0.5eV.

The activation energy

$$U = kT \ln \frac{\tau}{\tau_0} = U_0 - \gamma\sigma \tag{10.3}$$

plays in the BAZ equation (10.1) the same role as the stress-free energy U_0 plays in the Arrhenius equation (10.2). It has been recently shown [6] that these equations can be obtained as steady-state solutions to the Fokker–Planck equation in the theory of Markovian processes (see, for example, [86]), and that these solutions represent the worst case scenarios, so that the predictions based on the steady-state BAZ model (10.1) are reasonably conservative and, hence, advisable in engineering applications.

Zhurkov and his associates used equation (10.1) to determine the fracture toughness of a large number of materials experiencing combined action of elevated temperature and external mechanical loading. While Arrhenius equation (10.2), when used to determine the lifetime of a solid, considers only the effect of the elevated temperature on its lifetime, BAZ equation (10.1) also considers, in addition to that, the role of the applied mechanical stress. While the elevated temperature affects the long-term reliability of the material (its aging/degradation), the mechanical stress might cause its short-term failure.

In Zhurkov's tests, the loading σ was always a constant mechanical tensile stress, and the test specimens were always notched ones. It has been recently suggested [85, 86] that when the performance of an electronic or a photonic material is considered, any other loading of importance (voltage, current, thermal stress, humidity, vibrations, radiation, light output, etc.) can also be used as an appropriate stressor/stimulus, and, since the superposition principle cannot be employed in reliability engineering, that even a combination of relevant stimuli can be considered, so that a multiparametric BAZ equation could be employed.

The use of the BAZ equation has been suggested as a possible physics-of-failure–oriented model in connection with the development of the PDfR (see Chapter 11) concept for electronic or photonic materials, devices, assemblies, packages and systems to quantify, on the probabilistic basis, the lifetime of an electronic or a photonic product in field conditions using the results of an highly-focused and highly cost-effective FOAT [67–79].

The τ value is viewed in the BAZ model (10.1) as the MTTF. This suggests that when the exponential law of reliability

$$P = \exp(-\lambda t) = \exp\left(-\frac{t}{\tau}\right) = \exp\left[-\frac{t}{\tau_0}\exp\left(-\frac{U_0 - \gamma\sigma}{kT}\right)\right] \tag{10.4}$$

that defines the probability of nonfailure is used, the MTTF τ corresponds to the moment of time when the entropy $H(P) = -P\ln P$ of the distribution (10.4) reaches its maximum value. Indeed, from the equation $H(P) = -P\ln P$, it could be found that the function $H(P)$ reaches its maximum $H_{\max} = e^{-1}$ for the probability of nonfailure $P = e^{-1.} = 0.3679$. In such a situation, equation (10.4) yields $t = \tau_0 \exp\left(\frac{U}{kT}\right)$. Comparing this result with equation (10.1), one concludes that the MTTF expressed by this equation corresponds to the moment of time when the entropy $H(P)$ of the process $P = P(t)$ is the largest and is equal to e^{-1} as well.

From (10.4), considering that the probability of failure $Q = 1 - P$, we have

$$\frac{dQ}{dt} = \frac{H(P)}{t} \tag{10.5}$$

This relationship explains the physical meaning of basic equation (10.4): the degree of degradation (aging, damage accumulation) is proportional to the entropy of the process $Q = Q(t)$ and is inversely proportional to time. Note that the aforementioned formulation of entropy is different of both Boltzmann's and Shannon's formulations.

Boltzmann's entropy in thermodynamics [83, 84] is a quantitative measure of disorder, or of the energy in a system to do work. Shannon's entropy [87] in the communication theory is the probability of character number appearing in the stream of characters of the communication message. In our analysis, the entropy that is defined as $H(P) = -P\ln P$ is an integral physical characteristic of a particular distribution, such as, in this case, the exponential law (10.4) of reliability.

When FOAT data about the performance of a particular product in particular known loading conditions is available, the time constant τ_0 in the double-exponential distribution (10.4) could be replaced, for the subsequent reliability evaluations, by a quantity $(\gamma_C Ct)^{-1}$, where t is time, C is a suitable criterion of failure (such as, say, elevated leakage current or high electrical resistance) and γ_C is the corresponding sensitivity factor. Then, the distribution (10.4) can be replaced by the expression

$$P = \exp\left[-\gamma_C Ct \exp\left(-\frac{1}{kT}(U_0 - \gamma\sigma)\right)\right]. \tag{10.6}$$

or, in the case of multiple stressors, as

$$P = \exp\left[-\gamma_C Ct \exp\left(-\frac{1}{kT}\left(U_0 - \sum_{i=1}^{n}\gamma_i\sigma_i\right)\right)\right]. \tag{10.7}$$

It should be emphasized that the sum in this expression does not mean that the superposition principle is used. It is just a convenient way to consider the input of different loading. Equation (10.7) enables to consider the effects of as many stressors. The principle of superposition does not work in reliability engineering, and therefore all the meaningful FOAT stressors should be applied concurrently. This will take care of their possible coupling, nonlinear effects, and so on. The physically meaningful and highly flexible kinetic BAZ approach just helps to bridge the gap between what one "sees" as a result of the appropriate FOAT and what they will supposedly "get" in the actual field conditions.

Let us consider, as a simple example, the action of two stressors: elevated humidity H and elevated voltage V. If the level I_* of the leakage current is accepted as the suitable criterion of the device failure, equation (10.7) can be written as

$$P = \exp\left[-\gamma_I I_* t \exp\left(-\frac{U_0 - \gamma_H H - \gamma_V V}{kT}\right)\right]. \tag{10.8}$$

This equation contains four unknowns: the stress-free activation energy U_0 and three sensitivity factors: the leakage current factor γ_I, the relative humidity factor γ_H and the elevated voltage factor γ_V. These unknowns can be determined by a three-step FOAT.

At the first step, one should conduct the test for two temperatures, T_1 and T_2, while keeping the levels of the relative humidity H and the elevated voltage V unchanged. Assuming a certain level I_* of the monitored/measured leakage current as the physically meaningful criterion of failure, recording the percentages P_1 and P_2 of nonfailed samples, equation (10.8) yields

$$P_{1,2} = \exp\left[-\gamma_I I_* t_{1,2} \exp\left(-\frac{U_0 - \gamma_H H - \gamma_V V}{kT_{1,2}}\right)\right], \tag{10.9}$$

where t_1 and t_2 are the testing times and T_1 and T_2 are the temperatures, at which the failures were observed. Since the numerators in the expression in the parentheses are the same, the equation

$$f(\gamma_I) = \ln\left(\frac{\ln P_1}{I_* t_1 \gamma_I}\right) - \frac{T_2}{T_1}\ln\left(\frac{\ln P_2}{I_* t_2 \gamma_I}\right) = 0 \tag{10.10}$$

must be fulfilled. This equation enables to determine the sensitivity factor γ_I. Let, for example, the following input information has been obtained from the FOAT:

1. After $t_1 = 35h$ of testing at the temperature of $T_1 = 60°\text{C} = 333\text{K}$, the voltage of $\text{V} = 600\text{V}$ and the relative humidity $\text{H} = 0.85$, 10% of the tested samples exceeded the allowable (critical) level of the leakage current of $I_* = 3.5\mu A$ and, hence, failed, so that the probability of nonfailure is $P_1 = 0.9$;
2. After $t_2 = 70h$ of testing at the temperature of $T_2 = 85°\text{C} = 358\text{K}$ at the same voltage and the same relative humidity, 20% of the tested samples reached or exceeded the critical level of the leakage current and, hence, failed, so that the probability of nonfailure is $P_2 = 0.8$ in this case. Then, equation (10.10) results in the following transcendental equation for the leakage current sensitivity factor γ_I:

$$f(\gamma_I) = \ln\left(\frac{0.10536}{\gamma_I}\right) - 1.075075\ln\left(\frac{0.22314}{\gamma_I}\right) = 0.$$

This equation yields $\gamma_I = 4890h^{-1}(\mu A)^{-1}$, so that $\gamma_I I_* = 17115h^{-1}$.

At the second step, tests at two relative humidity levels H_1 and H_2, were conducted for the same temperature and voltage levels. This leads to the relationship

$$\gamma_H = \frac{kT}{H_1 - H_2}\left[\ln\left(-\frac{\ln P_1}{\gamma_I I_* t_1}\right) - \ln\left(-\frac{\ln P_2}{\gamma_I I_* t_2}\right)\right]$$

$$= \frac{kT}{H_1 - H_2}\left[\ln\left(-0.5843\times 10^{-4}\frac{\ln P_1}{t_1}\right) - \ln\left(-0.5843\times 10^{-4}\frac{\ln P_2}{t_2}\right)\right]$$

Let, for example, after $t_1 = 40h$ of testing at the relative humidity of $H_1 = 0.5$ at the given voltage (say, $\text{V} = 600$ V) and temperature (say, $T = 60°\text{C} = 333$ K), 5% of the tested modules failed ($P_1 = 0.95$), and after $t_2 = 55h$ of testing at the same temperature and at the relative humidity of $H_2 = 0.85$, 10% of the tested modules failed

($P_2 = 0.9$). Then, the aforementioned equation for the γ_H value, with the Boltzmann constant $k = 8.61733 \times 10^{-5}$ eV/K, yields

$$\gamma_H = \frac{kT}{H_1 - H_2}\left[\ln\left(-0.5843\times10^{-4}\frac{\ln P_1}{t_1}\right) - \ln\left(-0.5843\times10^{-4}\frac{\ln P_2}{t_2}\right)\right] =$$

$$= \frac{8.6173303\times10^{-5}\times333}{05-0.85}\left[\ln\left(0.5843\times10^{-4}\frac{0.051293}{40}\right)\right.$$

$$\left. - \ln\left(0.5843\times10^{-4}\frac{0.105360}{55}\right)\right] =$$

$$= -0.081988(-16.40676 + 16.00539) = 3.291\times10^{-2}\,\text{eV}$$

At the third step, FOAT at two different voltage levels $V_1 = 600$ V and $V_2 = 1000$ V have been carried out for the same temperature-radiation bias, say, $T = 85°\text{C} = 358$ K and $H = 0.85$, and it has been determined that 10% of the tested devices failed after $t_1 = 40h$ of testing ($P_1 = 0.9$) and 20% of devices failed after $t_2 = 80h$ of testing ($P_2 = 0.8$). The voltage sensitivity factor can be found then as follows:

$$\gamma_V = \frac{kT}{V_1 - V_2}\left[\ln\left(-0.5843\times10^{-4}\frac{\ln P_1}{t_1}\right) - \ln\left(-0.5843\times10^{-4}\frac{\ln P_2}{t_2}\right)\right] =$$

$$= \frac{8.6173303\times10^{-5}\times358}{600-1000}\left[\ln\left(0.5843\times10^{-4}\frac{0.105360}{40}\right)\right.$$

$$\left. - \ln\left(0.5843\times10^{-4}\frac{0.223143}{80}\right)\right] =$$

$$= -0.771251\times10^{-4}(-15.6869 + 15.6297) = 4.4116\times10^{-6}\,\text{eV/V}.$$

After the sensitivity factors of the leakage current, the humidity, and the voltage are found, the stress-free activation energy can be determined for the given temperature and for any combination of loadings (stimuli). From equation (10.8), we find

$$U_0 = -kT\ln\left(\frac{\ln P}{\gamma_I I_* t}\right) + \gamma_H H + \gamma_V V.$$

The input data can be taken from any of the mentioned FOAT steps making sure that these data are consistent.

$$U_0 = -kT_1\ln\left(-\frac{\ln P_1}{\gamma_I I_* t_1}\right) + \gamma_H H_1 + \gamma_V V_1$$

$$= -8.61733\times10^{-5}\times358\ln\left(-\frac{\ln 0.9}{17115\times40}\right) + 0.03291 +$$

$$+ 4.4116\times10^{-6}\times600 = 0.4839 + 0.0329 + 0.00265 = 0.5152\,\text{eV}$$

The first term in the equation for the stress-free activation energy plays the dominant role, so that, in approximate evaluations, only this term could be considered. Calculations indicate that the loading free activation energy in the numerical example (even with the rather tentative, but still realistic, input data) is about $U_0 = 0.4770$ eV. This result is consistent with the existing experimental data. Indeed, for semiconductor device failure mechanisms the activation energy ranges from 0.3 to 0.6 eV, for metallization defects and electro-migration in Al it is about 0.5 eV, for charge loss it is on the order of 0.6 eV, for Si junction defects it is 0.8 eV.

The following expression can be obtained for the probability of nonfailure from equation (10.8) using the calculated data from the numerical example:

$$P = \exp\left[-17115t\exp\left(-\frac{0.5152 - 3.291\times10^{-2}H - 4.4116\times10^{-6}V}{8.6173\times10^{-5}T}\right)\right]$$

If, for example, $H = 0.20, V = 220$, and the operation temperature is $T = 70°\text{C} = 343K$, then this formula yields

$$P = \exp\left[-17115t\exp\left(-\frac{0.5152 - 3.291\times10^{-2}\times0.2 - 4.4116\times10^{-6}\times220}{8.6173\times10^{-5}\times343}\right)\right]$$

$$= \exp(-5.9480\times10^{-4}t)$$

This probability is 0.9858 in 24 hours, 0.9049 in a week, 0.6516 in a month, and only 0.0055 in a year. Clearly, the lifetime of a product, its TTF, is not an independent characteristic of a product's reliability; it depends on the predicted or specified probability of its failure.

10.4 TEMPERATURE CYCLING: PREDICTED TIME-TO-FAILURE

Temperature cycling has become the main accelerated test vehicle in microelectronics (see, for example, [88–98]). While low-cost and short time-to-market are the main driving forces in commercial electronics, operational reliability is of paramount importance in aerospace, military, medical, long-haul communications, and other areas of electronic engineering, when the consequences of failure might be dramatic. Adequate operational reliability is therefore critical in these industries and cannot be assured if the underlying physics of failure is not understood. Since nothing is perfect and the probability of failure is never zero, this probability should be predicted and, if possible, even specified on the probabilistic basis. The recently suggested PDfR concept (see Chapter 11) makes the art of producing reliable electronic products into a reliability-physics and applied-probability–based science. The concept has its experimental foundation in a highly focused and highly cost-effective FOAT designed and conducted for the most vulnerable material(s) and structural element(s) of the device (solder joints, for instance) and geared to a simple and physically meaningful predictive model. The BAZ equation can be employed in this capacity. Let us show how the probability P of nonfailure of a solder joint interconnection experiencing inelastic strains during temperature cycling can be determined using the BAZ equation.

Let us seek this probability, considering the general formula (10.6), as follows:

$$P = \exp\left[-\gamma Rt \exp\left(-\frac{U_0 - nW}{kT}\right)\right]. \tag{10.11}$$

Here, U_0 is the stress-free activation energy viewed as material property and characterizing the material's propensity to fracture, W is the damage caused by a single temperature cycle and measured for the solder material experiencing inelastic deformations, in accordance with Hall's concept [67–69], by the inelastic-strain-energy hysteresis loop area (see Figure 10.1 [99]), T is the absolute temperature (say, the cycle's mean temperature), n is the number of cycles, k is the Boltzmann constant, t is time, R is the measured electrical resistance at the joint locations (this monitored resistance is viewed as a suitable criterion of the material failure), and γ is the sensitivity factor.

The distribution (10.11) makes physical sense. Indeed, the probability P of nonfailure is equal to "one" at the initial moment of time $t = 0$ and/or when the measured electrical resistance R is next-to-zero. This probability decreases with time because of a material's aging and structural degradation (and, not necessarily only because of temperature cycling); is lower for higher electrical resistance; materials with higher activation energy U_0 have a higher probability of nonfailure; the increase in the number n of cycles lowers the effective activation energy $U = U_0 - nW$ and, hence, the probability of nonfailure.

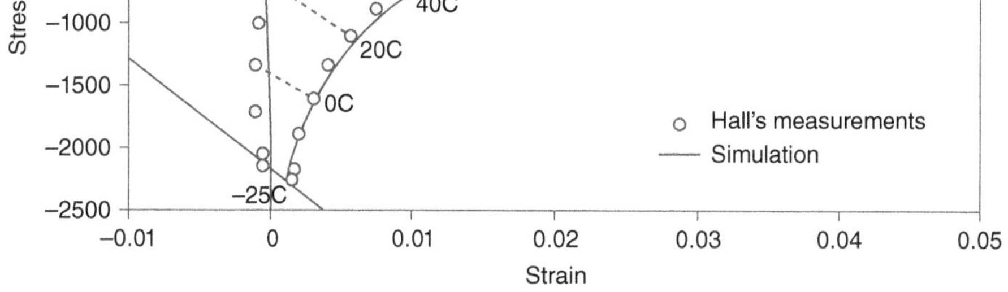

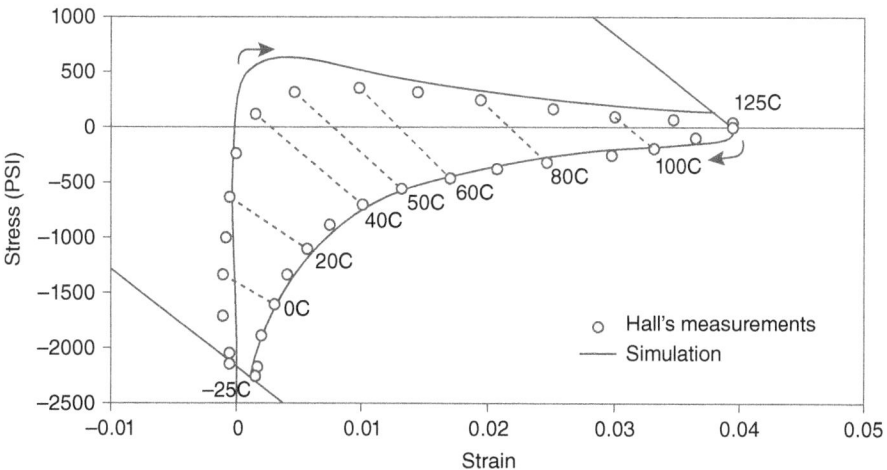

FIGURE 10.1 Hall's hysteresis loop for a single temperature cycle of a solder interconnection. (From C. Hillman, N. Blattau, M. Lacy, "Predicting Fatigue of Solder Joints Subjected to High Number of Power Cycles," IPC APEX EXPO, 2014.)

There is an underlying entropy-related consideration for the distribution (10.11). From (10.11), we have

$$\frac{dP}{dt} = P\exp\left[-\frac{1}{t}\gamma Rt\exp\left(-\frac{U_0 - nW}{kT}\right)\right] = -\frac{H(P)}{t}, \tag{10.12}$$

where $H(P) = -P\ln P$ is the entropy of the distribution (10.11). The result (10.12) indicates that the suggested distribution reflects a physically meaningful assumption that the change in the probability of nonfailure with time is proportional to the entropy (uncertainty) of this distribution and is inversely proportional to the elapsed time. As has been shown earlier, the condition $\frac{dH}{dP} = -\ln P_* - 1 = 0$ yields $P_* = H(P_*) = \frac{1}{e}$. The maximum entropy takes place for the MTTF

$$\tau = \frac{1}{\gamma R}\exp\left(\frac{U_0 - nW}{kT}\right). \tag{10.13}$$

The level of the entropy of the distribution (10.11) is the maximum at this time. Mechanical failure occurs when $n = \frac{U_0}{W}$. In such a situation, the temperature in the denominator in equation (10.12) becomes irrelevant, and this equation results in the following formulas for the probability P_f of nonfailure and the MTTF τ_f of the solder material:

$$P_f = \exp\left(-\frac{t_f}{\tau_f}\right), \quad \tau_f = \frac{1}{\gamma R_f}. \tag{10.14}$$

If, for example, 20 devices have been temperature cycled and the resistance $R_f = 450\Omega$, considered as a suitable indication of failure, was detected in 15 of them, then $P_f = 0.25$. If the number of cycles at failure was, say, $n_f = 2000$ and each cycle lasted, say, for 20 min = 1200 s, then the time at failure is $t_f = 2000 \times 1200 = 24 \times 10^5$ and formulas (10.14) yield

$$\gamma = -\frac{\ln P_f}{R_f t_f} = -\frac{\ln 0.25}{450 \times 24 \times 10^5} = 1.2836 \times 10^{-9}\Omega^{-1}s^{-1},$$

$$\tau_f = \frac{1}{\gamma R_f} = \frac{1}{1.2836 \times 10^{-9} 450} s = 480.9\,\text{h} = 20.0\ \text{days}$$

According to Hall [19], the inelastic strain energy of a single cycle should be evaluated experimentally by conducting a specially designed test.

Let, for example, the measured area of the hysteresis loop in such tests is $W = 4.5 \times 10^{-4}\,\text{eV} = 7.200 \times 10^{-23} J$.

Then the stress-free activation energy is $U_0 = 2000 \times 4.5 \times 10^{-4}\,\text{eV} = 0.9\,\text{eV} = 1.440 \times 10^{-19} J$. When assessing the number of cycles to failure in actual operation conditions, one could assume that the temperature range in these conditions is, say, half the FOAT range, and that the area W of the hysteresis loop is proportional to the temperature range. Then the number of cycles to failure becomes $n_f = \frac{U_0}{W} = \frac{0.9}{4.5 \times 10^{-4}} = 2000$. If the duration of one cycle is one day, then the projected lifetime of the materials (time to failure) is $t_f = 2000$ days $= 5.8$ years. Future work should include, first of all, improved Hall's type measurements, considering today's state of the art in experimental electronic materials science; actual FOAT data; and trustworthy information about the actual operation conditions and lifetimes.

10.5 INCENTIVE FOR MECHANICAL PRESTRESSING OF ACCELERATED TEST SPECIMENS

10.5.1 Background/Incentive

Accelerated testing of soldered assemblies, such as, for example, temperature cycling, is usually conducted in a wider range of temperatures than is expected to take place in actual operation conditions (see, for example, [100–125]). Since the mechanical and electrical properties of electronic materials are temperature sensitive, such testing might generate deviations of the material properties from the properties that they possess in actual operations conditions, and the results of such accelerated testing might be misleading: they might trigger mechanisms and modes of failure that will never occur in the field conditions. In such a situation, mechanical prestressing of the accelerated test specimens [102] might be a promising way to go. FOAT specimens are particularly vulnerable since the temperature range in these tests should be broad enough to eventually lead to failures, and if a shift in the modes and mechanisms of failures takes place during significant temperature excursions, the physics of such failures might be quite different than in actual operation conditions. An appropriate mechanical prestressing can be an effective means for narrowing the range of temperature excursions during accelerated testing and, owing to that, for obtaining consistent and trustworthy test data. If such a prestressing is considered and implemented, the ability to predict/model the thermo-mechanical stresses in the test specimen is certainly a must.

Application of mechanical prestressing could be an effective means for achieving a failure-mode-shift-free "destructive ALT effect" in electronic and photonic devices and micro-electro-mechanical systems (MEMS). A simple, physically meaningful, and easy-to-use analytical ("mathematical") predictive model has been developed to assess the magnitude and the distribution of stresses in a bimaterial assembly subjected to the combined action of thermally induced (considered by the ALT design) and external ("mechanical") prestressing. Such a compressive prestressing is applied to the assembly component that is expected to experience thermal compression. The model is an extension and a modification of the author's

1986 and 1989 "bi-metal thermostat" models suggested as a generalization of the classical 1925 Timoshenko's theory. The objective of the analysis carried out in this section is not so much to add to the existing knowledge in the field of thermal stress modeling, but rather to indicate the feasibility of using mechanically prestressed test specimens, when there is a need to avoid the "shift" in the modes and mechanisms of failure in electronic, photonic or MEMS assemblies subjected to thermal loading during accelerated life testing. When planning and conducting such testing, there is always a temptation to broaden (enhance) as far as possible the temperature range to achieve the maximum "destructive testing effect" in a shortest period of time. There exists, however, one major pitfall: a possible shift in the modes and mechanisms of failure as a result of broadening the temperature range; enhanced accelerated testing conditions may hasten failure mechanisms that are quite different from those that could possibly occur in actual service.

The likely pitfalls include, but might not be limited to, the change in materials properties at high or low temperatures; time-dependent strain due to diffusion; enhanced creep at elevated temperatures; brittle fracture at low temperatures; generation and movement of dislocations caused by an elevated thermal stress; and occurrence of a bimodal distribution of failures. Because of the possibility of such pitfalls, it is always necessary to establish the appropriate narrow enough temperature limits in order to prevent the distortion of the actual (real-life) dominant failure mechanism(s). Therefore, there is an obvious incentive for trying to find ways of increasing the induced stresses without broadening the ALT temperature range. One way to enhance the "destructive ALT effect" without compromising the acceptable temperature limits is to mechanically prestress the test specimens prior to conducting thermal ALT. For instance, a low expansion silicon chip attached to a high-expansion polymeric substrate will experience thermally induced compression, when the bimaterial chip-substrate assembly manufactured at an elevated temperature is subsequently cooled down to a low (say, room) temperature. This compression can be enhanced, and the interfacial stresses will be increased, if the chip is mechanically pre-stressed in compression. The objective of our analysis is to develop a simple, easy-to-use, and physically meaningful predictive model for the evaluation of the thermo-mechanical stresses in a mechanically prestressed bimaterial specimen. Although our model is an approximate engineering model that is not aimed at the most accurate prediction of the induced stresses, it seems nonetheless acceptable for engineering evaluations and applications. Some additional details of the suggested methodology are given in Appendices A and B.

10.5.2 Basic Equations

Let an elongated bimaterial adhesively bonded or soldered assembly be manufactured at an elevated temperature then cooled down to a low (say, room) temperature and then, prior to ALT testing, subjected to mechanical compression applied to the assembly component with the lower coefficient of thermal expansion (contraction), as schematically shown in Figure 10.2. It is this component that will experience compressive thermal stress.

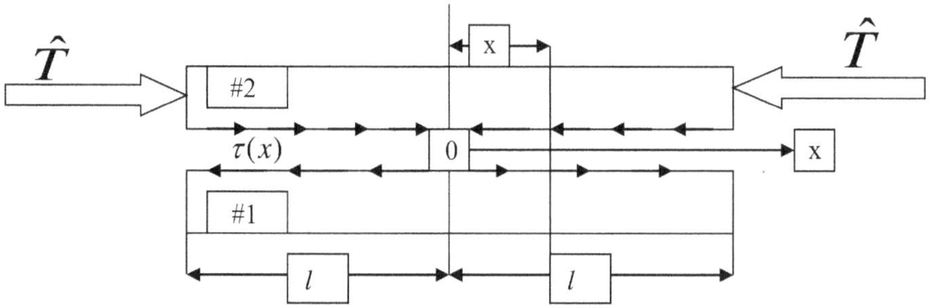

FIGURE 10.2 Bimaterial assembly subjected to the combined action of thermally induced and compressive external forces applied to the assembly component that will experience compressive thermal stresses during accelerated testing.

The longitudinal interfacial displacements $u_1(x)$ and $u_2(x)$ in the assembly components #1 and #2 can be evaluated, in an approximate analysis, by the following formulas based on the concept of interfacial compliance:

$$\left.\begin{aligned} u_1(x) &= -\alpha_1 \Delta t x + \lambda_1 \int_0^x [\hat{T} - T(\xi)]d\xi + \kappa_1 \tau(x) - \frac{h_1}{2} w_1'(x) \\ u_2(x) &= -\alpha_2 \Delta t x + \lambda_2 \int_0^x T(\xi)d\xi - \kappa_2 \tau(x) + \frac{h_2}{2} w_2'(x) \end{aligned}\right\}, \tag{10.15}$$

where α_1 and α_2 are the coefficients of thermal expansion (CTE) of the component materials, Δt is the change in temperature,

$$\lambda_1 = \frac{1-\nu_1}{E_1 h_1}, \quad \lambda_2 = \frac{1-\nu_2}{E_2 h_2}, \tag{10.16}$$

are the axial compliances of the components, E_1 and E_2 are Young's moduli of the materials, ν_1 and ν_2 are their Poisson's ratios, h_1 and h_2 are the thicknesses of the components, $\hat{T}$ are the external "mechanical" compressive forces (per unite assembly width) acting on the low expansion component #1,

$$T(x) = \int_{-l}^{x} \tau(\xi)d\xi \tag{10.17}$$

is the force acting in the cross section x of component #2, l is half the assembly length, $\tau(x)$ is the interfacial shearing stress,

$$\kappa_1 = \frac{h_1}{3G_1}, \quad \kappa_2 = \frac{h_2}{3G_2} \tag{10.18}$$

are the longitudinal interfacial compliances of the assembly components in the case of a long enough and/or stiff enough assembly,

$$G_1 = \frac{E_1}{2(1+\nu_1)}, \quad G_2 = \frac{E_2}{2(1+\nu_2)}, \tag{10.19}$$

are the shear moduli of the materials, and $w_1(x)$ and $w_2(x)$ are the component deflections. The origin of coordinate x is in the mid-cross section of the assembly at the interface. The first terms in (10.15) are stress-free thermal contractions. The second terms are evaluated based on Hooke's law assuming that the longitudinal displacements are the same for all the points of the given cross section. The third terms are "corrections" to this assumption and account for the fact that the interfacial longitudinal displacements are somewhat larger than the displacements of the inner points of the cross section. The fourth terms are due to bending.

The condition of the compatibility of the displacements (10.15) can be written as

$$u_1(x) = u_2(x) - \kappa_0 \tau(x), \tag{10.20}$$

where

$$\kappa_0 = \frac{h_0}{G_0} \tag{10.21}$$

is the longitudinal interfacial compliance of the bonding layer, h_0 is its thickness,

$$G_0 = \frac{E_0}{2(1+\nu_0)} \tag{10.22}$$

is the shear modulus of the bonding material, and E_0 and ν_0 are its elastic constants. Introducing formulas (10.15) into condition (10.20), we obtain the following integral equation for the shearing stress function, $\tau(x)$:

$$\kappa\tau(x) - (\lambda_1 + \lambda_2)\int_0^x T(\xi)d\xi - \frac{h_1}{2}w_1'(x) - \frac{h_2}{2}w_2'(x) = -(\Delta\alpha\Delta t + \lambda_1 \hat{T})x, \tag{10.23}$$

where $\Delta\alpha = \alpha_2 - \alpha_1$ is the difference in the CTE of the component materials, and

$$\kappa = \kappa_0 + \kappa_1 + \kappa_2 \tag{10.24}$$

is the total longitudinal interfacial compliance of the assembly. As evident from equation (10.23), the "external" thermal strain $\Delta\alpha\Delta t$ can be enhanced, without broadening the temperature range, by mechanical prestressing one of the assembly components. Equation (10.23) indicates also that such an enhancement increases with an increase in the axial compliance of the compressed component. From (10.23), we find by differentiation:

$$\kappa\tau'(x)-(\lambda_1+\lambda_2)T(x)-\frac{h_1}{2}w_1''(x)-\frac{h_2}{2}w_2''(x)=-(\Delta\alpha\Delta t+\lambda_1\hat{T}), \tag{10.25}$$

$$\kappa\tau''(x)-(\lambda_1+\lambda_2)\tau(x)-\frac{h_1}{2}w_1'''(x)-\frac{h_2}{2}w_2'''(x)=0, \tag{10.26}$$

$$\kappa\tau'''(x)-(\lambda_1+\lambda_2)\tau'(x)-\frac{h_1}{2}w_1^{IV}(x)-\frac{h_2}{2}w_2^{IV}(x)=0. \tag{10.27}$$

Treating the assembly components as elongated rectangular plates, we proceed from the following equations of bending (equilibrium):

$$\int_{-l}^{x}\int_{-l}^{x}p(\xi)d\xi d\xi_1=-D_1w_1''(x)+\frac{h_1}{2}T(x)=D_2w_2''(x)-\frac{h_2}{2}T(x), \tag{10.28}$$

where $p(x)$ is the interfacial peeling stress (i.e., normal interfacial stress acting in the through-thickness direction of the assembly), and

$$D_1=\frac{E_1h_1^3}{12(1-\nu_1^2)},\quad D_1=\frac{E_1h_2^3}{12(1-\nu_2^2)} \tag{10.29}$$

are the flexural rigidities of the components. From (10.28), we find by differentiation

$$\int_{-l}^{x}p(\xi)d\xi=-D_1w_1'''(x)+\frac{h_1}{2}\tau(x)=D_2w_2'''(x)-\frac{h_2}{2}\tau(x), \tag{10.30}$$

$$p(x)=-D_1w_1^{IV}(x)+\frac{h_1}{2}\tau'(x)=D_2w_2^{IV}(x)-\frac{h_2}{2}\tau'(x). \tag{10.31}$$

Solving equations (10.31) for the fourth derivatives of the deflection functions and substituting the obtained expressions into equation (10.27), we obtain the following equation:

$$\kappa\tau'''(x)-\lambda\tau'(x)=-\mu p(x), \tag{10.32}$$

that couples the interfacial shearing stress $\tau(x)$ and the interfacial peeling stress $p(x)$. In the obtained equation,

$$\lambda=\lambda_1+\lambda_2+\frac{h_1^2}{4D_1}+\frac{h_2^2}{4D_2} \tag{10.33}$$

is the total axial compliance of the assembly (with consideration of the effect of bending), and

$$\mu = \frac{h_1}{2D_1} - \frac{h_2}{2D_2} \tag{10.34}$$

is the factor of the peeling stress. As evident from formula (10.33), the total axial compliance of the assembly increases with a decrease in the flexural rigidities of its components. As to the factor (10.34), it is the lowest for adherends with close flexural rigidities and is the highest in the case of considerably different rigidities.

We assume that the peeling stress is related to the deflections $w_1(x)$ and $w_2(x)$ as

$$p(x) = K[w_1(x) - w_2(x)], \tag{10.35}$$

where K is the through-thickness stiffness of the assembly. Formula (10.35) reflects an assumption that no peeling stress could possibly occur in the given cross section, if the deflections $w_1(x)$ and $w_2(x)$ are the same in this cross section. In an approximate analysis, by analogy with the longitudinal interfacial stiffness (compliance), one could assume

$$K = \frac{1}{\frac{(1-\nu_1)h_1}{3E_1} + \frac{(1-\nu_2)h_2}{3E_2} + \frac{(1-\nu_0)h_0}{E_0}}. \tag{10.36}$$

This formula indicates that while the entire bonding layer experiences stresses acting in the thorough-thickness direction of the assembly, only the inner portions of the assembly components, that is, the regions adjacent to the interface, are in the state of appreciable stress. From (10.35), we have

$$p^{IV}(x) = K[w_1^{IV}(x) - w_2^{IV}(x)]. \tag{10.37}$$

Solving equations (10.31) for the fourth derivatives of the deflection functions and substituting the obtained expressions into equation (10.37), we come up with another equation that couples the interfacial shearing, $\tau(x)$, and the interfacial peeling, $p(x)$, stresses

$$p^{IV}(x) + 4\beta^4 p(x) = \mu K \tau'(x), \tag{10.38}$$

where

$$\beta = \sqrt[4]{K\frac{D_1 + D_2}{4D_1D_2}} \tag{10.39}$$

is the parameter of the interfacial peeling stress. Equation (10.38) indicates that the longitudinal gradient of the interfacial shearing stress plays the role of the external loading for the peeling stress: the greater this gradient, the higher is the peeling stress. It is noteworthy that equation (10.38) has the form of the equation of bending of a beam lying on a continuous elastic foundation. In the engineering theory of such beams, this equation is being written, however, for the deflection function, and not for the peeling stress.

Equations (10.32) and (10.38) are the two basic equations in the problem in question. These equations indicate that the two types of the interfacial stresses are coupled. Separating the functions $\tau(x)$ and $p(x)$ in equations (10.32) and (10.38), we find that these two functions could be determined, in effect, from the same equation:

$$\tau^{VI}(x)-k^2\tau^{IV}(x)+4\beta^4\left[\tau''(x)-k^2\tau(x)+\frac{\mu^2}{\kappa}\frac{D_1D_2}{D_1+D_2}\tau(x)\right]=0, \tag{10.40}$$

or

$$p^{VI}(x)-k^2p^{IV}(x)+4\beta^4\left[p''(x)-k^2p(x)+\frac{\mu^2}{\kappa}\frac{D_1D_2}{D_1+D_2}p(x)\right]=0. \tag{10.41}$$

Here,

$$k=\sqrt{\frac{\lambda}{\kappa}} \tag{10.42}$$

is the parameter of the interfacial shearing stress. The solution to equation (10.40) should be sought, however, in an antisymmetric form and should contain only odd functions, while the solution to equation (10.41) should be symmetric with respect to the mid-cross section of the assembly and should contain therefore only even functions. This circumstance is reflected by the appropriate boundary conditions.

10.5.3 Boundary Conditions

Since there are no external forces acting at the ends of component #2, the force $T(x)$ must be zero at the end $x=l$ of this component:

$$T(l)=0. \tag{10.43}$$

Since no concentrated bending moments act at the assembly ends, the curvatures $w_1''(x)$ and $w_2''(x)$ must be zero at these ends:

$$w_1''(l)=w_2''(l)=0. \tag{10.44}$$

As to the lateral forces, the following boundary conditions have to be fulfilled:

$$D_1w_1'''(l)+\hat{T}w'(l)=0, \quad w_2'''(l)=0. \tag{10.45}$$

The first condition in (10.45) indicates that the lateral projection of the external force $\hat{T}$ should be equilibrated by the elastic force. Considering, however, that a typical electronic or photonic assembly is stiff enough, so that the angle of rotation $w'(l)$ is small, and, in addition, that the external force $\hat{T}$ should be sufficiently low (actually, well below its critical value), one could assume, in an approximate analysis, that the third derivative $w_1'''(l)$ of the deflection function for the component #1 can also be put equal to zero. Then, conditions (10.45) can be substituted, in an approximate analysis, with the following conditions:

$$w_1'''(l) \approx w_2'''(l) = 0. \tag{10.46}$$

The peeling stress $p(x)$ must be self-equilibrated. This means that the following equilibrium conditions are to be fulfilled:

$$\int_{-l}^{l} p(\xi)d\xi = 0, \quad \int_{-l}^{l}\int_{-l}^{x} p(\xi)d\xi d\xi_1 = 0. \tag{10.47}$$

Equation (10.25), considering conditions (10.43) and (10.44), results in the following boundary condition for the shearing stress function:

$$\tau'(l) = -\frac{\Delta\alpha\Delta t + \lambda_1\hat{T}}{\kappa}. \tag{10.48}$$

Equation (10.26), considering (10.46), yields

$$\kappa\tau''(l) - (\lambda_1 + \lambda_2)\tau(l) = 0. \tag{10.49}$$

Note that equation (10.28), considering the boundary conditions (10.44), (10.46), and the second condition in (10.47), is always fulfilled at the assembly ends.

Equation (10.30), taking into account conditions (10.46) and the first condition in (10.47), yields

$$\tau(l) = 0. \tag{10.50}$$

Then, formula (10.49) also results in a zero-boundary condition for the second derivative of the shearing stress function:

$$\tau''(l) = 0. \tag{10.51}$$

As to the peeling stress, we obtain from (10.35), considering (10.44) and (10.46), the following conditions:

$$p''(l) = 0, \quad p'''(l) = 0. \tag{10.52}$$

10.5.4 Solutions to the Basic Equations

We seek the interfacial shearing stress function $\tau(x)$ in the form

$$\tau(x) = C_1 \sinh\gamma_1 x \cos\gamma_2 x + C_3 \cosh\gamma_1 x \sin\gamma_2 x + C_5 \sinh\gamma x \tag{10.53}$$

By differentiation, we find

$$\tau'(x) = (\gamma_1 C_1 + \gamma_2 C_3)\cosh\gamma_1 x\cos\gamma_2 x + (-\gamma_2 C_1 + \gamma_1 C_3)\sinh\gamma_1 x\sin\gamma_2 x + \gamma C_5\cosh\gamma x, \tag{10.54}$$

$$\tau''(x) = [(\gamma_1^2 - \gamma_2^2)C_1 + 2\gamma_1\gamma_2 C_3]\sinh\gamma_1 x\cos\gamma_2 x + \\ +[(\gamma_1^2 - \gamma_2^2)C_3 - 2\gamma_1\gamma_2 C_1]\cosh\gamma_1 x\sin\gamma_2 x + \gamma^2 C_5\sinh\gamma x, \tag{10.55}$$

$$\tau'''(x) = [\gamma_1(\gamma_1^2 - 3\gamma_2^2)C_1 - \gamma_2(\gamma_2^2 - 3\gamma_1^2)C_3]\cosh\gamma_1 x\cos\gamma_2 x + \\ +[\gamma_2(\gamma_2^2 - 3\gamma_1^2)C_1 + \gamma_1(\gamma_1^2 - 3\gamma_2^2)C_3]\sinh\gamma_1 x\sin\gamma_2 x + \gamma^3 C_5\cosh\gamma x, \tag{10.56}$$

$$\tau^{IV}(x) = [(\gamma_1^4 - 6\gamma_1^2\gamma_2^2 + \gamma_2^4)C_1 + 4\gamma_1\gamma_2(\gamma_1^2 - \gamma_2^2)C_3]\sinh\gamma_1 x\cos\gamma_2 x + \\ +[(\gamma_1^4 - 6\gamma_1^2\gamma_2^2 + \gamma_2^4)C_3 - 4\gamma_1\gamma_2(\gamma_1^2 - \gamma_2^2)C_1]\cosh\gamma_1 x\sin\gamma_2 x + \gamma^4 C_5\sinh\gamma x, \tag{10.57}$$

$$\tau^{V}(x) = [\gamma_1(\gamma_1^4 - 10\gamma_1^2\gamma_2^2 + 5\gamma_2^4)C_1 + \gamma_2(\gamma_2^4 - 10\gamma_1^2\gamma_2^2 + 5\gamma_1^4)C_3]\cosh\gamma_1 x\cos\gamma_2 x + \\ +[-\gamma_2(\gamma_2^4 - 10\gamma_1^2\gamma_2^2 + 5\gamma_1^4)C_1 + \gamma_1(\gamma_1^4 - 10\gamma_1^2\gamma_2^2 + 5\gamma_2^4)C_3]\sinh\gamma_1 x\sin\gamma_2 x \\ + \gamma^5 C_5\cosh\gamma x, \tag{10.58}$$

$$\tau^{VI}(x) = [(\gamma_1^6 - \gamma_2^6 + 15\gamma_1^2\gamma_2^4 - 15\gamma_1^4\gamma_2^2)C_1 \\ + 2\gamma_1\gamma_2(3\gamma_1^4 + 3\gamma_2^4 - 10\gamma_1^2\gamma_2^2)C_3]\sinh\gamma_1 x\cos\gamma_2 x + \\ +[2\gamma_1\gamma_2(-3\gamma_1^4 - 3\gamma_2^4 + 10\gamma_1^2\gamma_2^2)C_1 + \\ (\gamma_1^6 - \gamma_2^6 + 15\gamma_1^2\gamma_2^4 - 15\gamma_1^4\gamma_2^2)C_3]\cosh\gamma_1 x\sin\gamma_2 x + \\ + \gamma^6 C_5\sinh\gamma x \tag{10.59}$$

Then, equation (10.40) results in the following three equations for the factors γ_1, γ_2 and γ:

$$\gamma_1^6 - \gamma_2^6 + 15\gamma_1^2\gamma_2^4 - 15\gamma_1^4\gamma_2^2 - k^2(\gamma_1^4 - 6\gamma_1^2\gamma_2^2 + \gamma_2^4) + 4\beta^4[\gamma_1^2 - \gamma_2^2 - k^2(1-\delta)] = 0, \tag{10.60}$$

$$3\gamma_1^4 + 3\gamma_2^4 - 10\gamma_1^2\gamma_2^2 - 2k^2(\gamma_1^2 - \gamma_2^2) + 4\beta^4 = 0, \tag{10.61}$$

$$\gamma^6 - k^2\gamma^4 - 4\beta^4\gamma^2 + 4\beta^4 k^2(1-\delta) = 0, \tag{10.62}$$

where

$$\delta = \frac{\mu^2}{\lambda}\frac{D_1 D_2}{D_1 + D_2} \tag{10.63}$$

is the parameter of coupling of the interfacial stresses. This parameter, as follows from formula (10.34), is very small if the assembly components have close flexural rigidities.

Introducing new unknowns, ξ and η, as

$$\xi = \gamma_1^2 - \gamma_2^2, \quad \eta = 2\gamma_1\gamma_2, \tag{10.64}$$

we obtain equations (10.60) and (10.61) in the form

$$\xi^3 - k^2\xi^2 - 3\eta^2\xi + 4\beta^4\xi + k^2\eta^2 - 4\beta^4k^2(1-\delta) = 0, \tag{10.64}$$

$$\eta^2 = 3\xi^2 - 2k^2\xi + 4\beta^4. \tag{10.66}$$

Introducing the η^2 value from equation (10.66) into equation (10.65), we obtain the following cubic equation for the unknown ξ:

$$\xi^3 - k^2\xi^2 + \frac{1}{4}(k^4 + 4\beta^4)\xi - \frac{1}{2}k^2\beta^4\delta = 0. \tag{10.67}$$

After the ξ value is found, the η value can be determined from equation (10.66), and then the γ_1 and γ_2 values could be evaluated as

$$\gamma_{1,2} = \sqrt{\frac{\xi}{2}\left(\pm 1 + \sqrt{1 + \frac{\eta^2}{\xi^2}}\right)}. \tag{10.68}$$

These values are close to each other, if the ξ value is small, and the η value is large. In such a case, as one could see from equation (10.66), $\eta = 2\beta^2$. Note that the result (10.68) could be obtained, if equation (10.41) and the particular solution

$$p(x) = C_0 \cosh\gamma_1 x \cos\gamma_2 x + C_2 \sinh\gamma_1 x \sin\gamma_2 x + C_4 \cosh\gamma x \tag{10.69}$$

for the peeling stress $p(x)$ were considered.

10.5.5 Constants of Integration

The constants C_1, C_3, and C_5 of integration in the expression (10.53) for the interfacial shearing stress can be found, based on boundary conditions (10.58), (10.50), and (10.51) from the following system of equations:

$$\left.\begin{aligned}
&(\sinh u_1 \cos u_2)C_1 + (\cosh u_1 \sin u_2)C_3 + (\sinh u)C_5 = 0\\
&(u_1 \cosh u_1 \cos u_2 - u_2 \sinh u_1 \sin u_2)C_1 + (u_2 \cosh u_1 \cos u_2 + u_1 \sinh u_1 \sin u_2)C_3 +\\
&\quad + (u \cosh u)C_5 = -l\frac{\Delta\alpha\Delta t + \lambda_1\hat{T}}{\kappa}\\
&[(u_1^2 - u_2^2)\sinh u_1 \cos u_2 - 2u_1u_2 \sinh u_1 \sin u_2]C_1 +\\
&+ [(u_1^2 - u_2^2)\cosh u_1 \sin u_2 + 2u_1u_2 \sinh u_1 \cos u_2]C_3 + (u^2 \sinh u)C_5 = 0
\end{aligned}\right\} \tag{10.70}$$

Here, the following notation is used:

$$u_1 = \gamma_1 l, \quad u_2 = \gamma_2 l, \quad u = \gamma l. \tag{10.71}$$

As to the constants C_0, C_2, and C_4 in expression (10.69) for the interfacial peeling stress, $p(x)$, they can be determined after substituting solutions (10.53) and (10.69) into equation (10.32) or into equation (10.38) and comparing the expressions at the left and the right parts of the obtained relationships. This leads to the following formulas for the constants C_0, C_2, and C_4 of integration:

$$\left.\begin{aligned} C_0 &= -\frac{\kappa}{\mu}\left[\gamma_1(\gamma_1^2 - 3\gamma_2^2 - k^2)C_1 - \gamma_2(\gamma_2^2 - 3\gamma_1^2 + k^2)C_3\right] \\ C_2 &= -\frac{\kappa}{\mu}\left[\gamma_1(\gamma_1^2 - 3\gamma_2^2 - k^2)C_3 + \gamma_2(\gamma_2^2 - 3\gamma_1^2 + k^2)C_1\right] \\ C_4 &= -\frac{\kappa}{\mu}\gamma(\gamma^2 - k^2)C_5 \end{aligned}\right\} \tag{10.72}$$

10.5.6 Numerical Example

Input data

Component #1: Young's modulus: $E_1 = 12300$ kg/mm^2; Poisson's ratio: $\nu_1 = 0.24$; CTE: $\alpha_1 = 2.2\times10^{-6}$ 1/°C; Thickness: $h_1 = 0.5$ mm; External force: $\hat{T} = 8.0$ kg/mm;

Component #2: Young's modulus: $E_2 = 2000$ kg/mm^2; Poisson's ratio: $\nu_2 = 0.30$;

CTE: $\alpha_2 = 13.2\times10^{-6}$ 1/°C; Thickness: $h_2 = 1.5$ mm;

Bonding layer: Young's modulus: $E_0 = 200$ kg/mm^2; Poisson's ratio: $\nu_0 = 0.40$; Thickness: $h_0 = 0.05$ mm;

Change in temperature: $\Delta t = 100$°C; Assembly length: $2l = 20$ mm

Computed data

Thermal strain: $\Delta\alpha\Delta t = (\alpha_2 - \alpha_1)\Delta t = 11\times10^{-6}\times100 = 0.0011$.

Axial compliances, as predicted by formulas (10.16):

$$\lambda_1 = \frac{1-\nu_1}{E_1 h_1} = \frac{1-0.24}{12300\times0.5} = 1.2358\times10^{-4}\,\text{mm/kg}\ \lambda_2 = \frac{1-\nu_2}{E_2 h_2} = \frac{1-0.30}{2000\times1.5}$$

$$= 2.3333\times10^{-4}\,\text{mm/kg}$$

Shear moduli, as predicted by formulas (10.22) and (10.19):

$$G_0 = \frac{E_0}{2(1+\nu_0)} = \frac{200}{2\times1.4} = 71.4\ \text{kg/mm}^2;$$

$$G_1 = \frac{E_1}{2(1+\nu_1)} = \frac{12300}{2\times1.24} = 4960\ \text{kg/mm}^2;$$

$$G_2 = \frac{E_2}{2(1+\nu_2)} = \frac{2000}{2\times1.30} = 769\ \text{kg/mm}^2$$

Interfacial shearing compliances, as predicted by formulas (10.21), (10.18), and (10.24):

$$\kappa_0 = \frac{h_0}{G_0} = \frac{0.05}{71.4} = 7.00 \times 10^{-4}\,\text{mm}^3/\text{kg};$$

$$\kappa_1 = \frac{h_1}{3G_1} = \frac{0.5}{3 \times 4960} = 0.3360 \times 10^{-4}\,\text{mm}^3/\text{kg};$$

$$\kappa_2 = \frac{h_2}{3G_2} = \frac{1.5}{3 \times 769} = 6.5020 \times 10^{-4}\,\text{mm}^3/\text{kg};$$

$$\kappa = \kappa_0 + \kappa_1 + \kappa_2 = 7.00 \times 10^{-4} + 0.3360 \times 10^{-4} + 6.5020 \times 10^{-4}$$
$$= 13.838 \times 10^{-4}\,\text{mm}^3/\text{kg}$$

Boundary condition for the interfacial shearing stress, as given by formula (10.48):

$$\tau'(l) = -\frac{\Delta\alpha\Delta t + \lambda_1 \hat{T}}{\kappa} = -\frac{0.0011 + 0.0009886}{13.838 \times 10^{-4}} = -1.5093\ \text{kg/mm}^3$$

Note that because of the prestressing of the low expansion component of the assembly, the interfacial stresses in the assembly increase by a factor of 1.9. Flexural rigidities of the assembly components (treated as elongated rectangular plates) are evaluated on the basis of formulas (10.29):

$$D_1 = \frac{E_1 h_1^3}{12(1-\nu_1^2)} = \frac{12300 \times 0.5^3}{12 \times 0.9424} = 136.0\ \text{kg/mm}$$

$$D_2 = \frac{E_2 h_2^3}{12(1-\nu_2^2)} = \frac{2000 \times 1.5^3}{12 \times 0.9100} = 618.1\ \text{kg/mm}$$

Total axial compliance of the assembly, as predicted by formula (10.33):

$$\lambda = \lambda_1 + \lambda_2 + \frac{h_1^2}{4D_1} + \frac{h_2^2}{4D_2} = 1.2358 \times 10^{-4} + 2.3333 \times 10^{-4} + 4.5956 \times 10^{-4}$$
$$+ 9.1005 \times 10^{-4}$$
$$= 17.2652 \times 10^{-4}\,\text{mm/kg}$$

Parameter of the interfacial shearing stress, as given by formula (10.42):

$$k = \sqrt{\frac{\lambda}{\kappa}} = \sqrt{\frac{17.2652 \times 10^{-4}}{13.838 \times 10^{-4}}} = 1.117\ \text{mm}^{-1}$$

Factor of the peeling stress, as given by formula (10.34):

$$\mu = \frac{h_1}{2D_1} - \frac{h_2}{2D_2} = \frac{0.5}{272} - \frac{1.5}{618.1} = 18.382 \times 10^{-4} - 24.268 \times 10^{-4} = -5.886 \times 10^{-4}\,\text{kg}^{-1}$$

Through-thickness stiffness, as predicted by formula (10.36):

$$K = \frac{1}{\frac{(1-\nu_1)h_1}{3E_1} + \frac{(1-\nu_2)h_2}{3E_2} + \frac{(1-\nu_0)h_0}{E_0}} = \frac{1}{0.103\times10^{-4} + 1.750\times10^{-4} + 1.500\times10^{-4}}$$
$$= 2982 \text{ kg/mm}^3$$

Parameter of the peeling stress, as predicted by formula (10.39):

$$\beta = \sqrt[4]{K\frac{D_1 + D_2}{4D_1D_2}} = \sqrt[4]{2982\frac{754.1}{336246.4}} = 0.7441 \text{ mm}^{-1}$$

Parameter of coupling of the interfacial stresses, as predicted by formula (10.63):

$$\delta = \frac{\mu^2}{\lambda}\frac{D_1D_2}{D_1 + D_2} = \frac{34.6450\times10^{-8}}{17.2652\times10^{-4}}\frac{84061.6}{754.1} = 0.02237$$

Equation (10.67) for the unknown ξ value yields

$$\xi^3 - 1.2477\xi^2 + 0.6957\xi - 0.004278 = 0,$$

and has the following root: $\xi = 0.00615$. Then, equation (10.66) yields $\eta = 1.10047$. From formula (10.68) we find

$$\gamma_1 = 0.74385 \text{ mm}^{-1}; \quad \gamma_2 = 0.73970 \text{ mm}^{-1}.$$

Note that the obtained two values are very close to each other and to the β value. This is because the η value turned out to be substantially larger than the ξ value, which is the case for adherends with not-very-much-different flexural rigidities. Equation (10.62) for the γ value yields

$$\gamma^6 - 1.2477\gamma^4 - 1.22627\gamma^2 + 1.495788 = 0$$

Its root is $\gamma = 1.0165 \text{ mm}^{-1}$ and is not very much different from the k value.

The parameters $u_1 = \gamma_1 l$, $u_2 = \gamma_2 l$, and $u = \gamma l$, expressed by formulas (10.71), are as follows:

$$u_1 = \gamma_1 l = 0.74385\times10 = 7.4385; \quad u_2 = \gamma_2 l = 0.73970\times10 = 7.3970;$$
$$u = \gamma l = 1.0165\times10 = 10.165.$$

These parameters are large enough, so that the assembly can be treated as an elongated one. Equations (10.70) for the constants C_1, C_3, and C_5 of integration yield

$$0.028876C_1 + 0.058733C_3 + C_5 = 0,$$
$$0.021609C_1 - 0.063992C_3 - C_5 = 0.00011431,$$
$$0.062380C_1 - 0.03110C_3 - C_5 = 0.$$

TABLE 10.2
Calculated Interfacial Shearing Stresses

x, mm	8.0	8.4	8.5	8.75	9.0	9.5	9.7	9.9	10
$\tau(x)$, kg/mm^2	**0.0813**	**0.2164**	**0.2584**	**0.2518**	**0.2313**	**0.1422**	**0.0894**	**0.0297**	**0**
$\tau_1(x)$, kg/mm^2	**0.1441**	**0.2262**	**0.2530**	**0.3345**	**0.4422**	**0.7730**	**0.9665**	**1.2085**	**1.351**

These equations have the following solutions:

$$C_1 = 0.0003488 \text{ kg/mm}^2; \quad C_3 = -0.0011520 \text{ kg/mm}^2; \quad C_5 = 0.00005759 \text{ mm}^2.$$

The shearing stress in the region close to the assembly ends can be computed by formula:

$$\tau(x) = e^{0.7438x}(0.0001744\cos 0.7397x - 0.0005770\sin 0.7397x) + 0.000028795e^{1.0165x},$$

which can be obtained from solution (10.53). The calculated stresses are shown in Table 10.2:

At the bottom line, the stresses calculated using the simplified formula

$$\tau_1(x) = k\frac{\Delta\alpha\Delta t + \lambda_1\hat{T}}{\lambda} \tag{10.73}$$

are indicated. The calculated data indicate that this formula can be used for conservative engineering assessments. For the interfacial peeling stress, we obtain on the basis of solution (10.69),

$$p(x) = e^{-0.7438x}[-0.001150\cos(0.7397x) + 0.002221\sin(0.7397x)] \\ - 0.000014750e^{-1.0165x}$$

The computed peeling stress is shown in Table 10.3. These data indicate that this stress is indeed self-equilibrated.

We conclude that the application of mechanical prestressing provides an effective means for achieving a failure-mode-shift-free "destructive ALT effect" in electronic and photonic devices. A simple and easy-to-use analytical predictive model has been

TABLE 10.3
Calculated Peeling Stresses

x, mm	4.0	5.0	6.0	6.5	7.0	7.5
$p(x)$, kg/mm^2	**0.0285**	**–0.0910**	**–0.1650**	**–0.3032**	**–0.4744**	**–0.6502**
x, mm	**8.0**	**8.25**	**8.5**	**9.0**	**9.5**	**10.0**
$p(x)$, kg/mm^2	**–0.7670**	**–0.7719**	**–0.7179**	**–0.3467**	**0.5419**	**2.1439**

developed to assess the magnitude and the distribution of the interfacial shearing and peeling stresses in a bimaterial assembly subjected to the combined action of thermally induced and external ("mechanical") loading during ALTs. A simplified predictive model developed earlier [11] can be used for conservative engineering assessment of the level of the induced interfacial shearing stress. The peeling stress should be evaluated, however, based on a more accurate model of the type suggested in [11] and modified in this paper. We would like to point out that our objective was not to come up with the most accurate predictions possible, but rather to obtain approximate, although still physically meaningful solutions, suitable for the application in engineering practice.

Note that our model has applications in prognostics and systems health management (PHM). This is a process of predicting the future reliability of the system by assessing the extent of deviation or degradation of a product from its expected normal operating conditions in a preemptive and opportunistic manner to the anticipation of failures. This can enable continuous, autonomous, real-time monitoring of the health conditions of a system by means of embedded or attached sensors with minimum manual intervention to evaluate its actual life-cycle conditions, to determine the advent of failure, and to mitigate system risks. The benefits of PHM include: (1) providing advance warning of failures; (2) minimizing unscheduled maintenance, extending maintenance cycles, and maintaining effectiveness through timely repair actions; (3) reducing the life-cycle cost of equipment by decreasing inspection costs, downtime, and inventory; and (4) improving qualification and assisting in the design and logistical support of fielded and future systems.

10.6 ACCELERATED TESTING OF SOLDER JOINT INTERCONNECTIONS: INCENTIVE FOR USING A LOW-TEMPERATURE/RANDOM-VIBRATIONS BIAS

10.6.1 Background/Incentive

Although promising ways exist to avoid inelastic strains in solder joints of the second-level interconnections in IC package designs, it still appears more typical than not that the peripheral joints of a package/PCB assembly experience inelastic strains. This takes place at low-temperature conditions, when the deviation from the high fabrication temperature is the largest and the induced thermal stresses are the highest. On the other hand, it is well known that it is the combination of low temperatures and repetitive dynamic loading that accelerate dramatically the propagation of fatigue cracks, whether elastic or inelastic. Accordingly, a modification of the recently suggested multiparametric BAZ model is developed for the evaluation of the remaining useful lifetime (RUL) of the second-level solder joint interconnections whose peripheral joints experience inelastic strains. The experimental basis of the approach is the highly focused and highly cost-effective FOAT. The FOAT specimens have been subjected in our methodology (which is "reduced to practice") to the combined action of low temperatures (not to elevated temperatures, as in the classical Arrhenius model) and random vibrations with the given input energy spectrum.

10.6.2 Methodology

The suggested methodology is viewed as a possible, promising, effective, and attractive alternative to temperature cycling tests. As long as inelastic deformations are inevitable, it is assumed that it is these deformations that determine the fatigue lifetime of the solder material, and the state of stress in the elastic midportion of the assembly does not have to be accounted for. The roles of the size and stiffness of this midportion have to be considered, however, when determining the very existence the inelastic zones at the peripheral portions of the design and establishing their size. The general concept is illustrated by a numerical example. Although this example is carried out for a ball-grid array (BGA) design, it is applicable to highly popular column-grid array (CGA) and quad-flat no-lead (QFN) designs as well. It is noteworthy that it is much easier to avoid inelastic strains in CGA and QFN structures than in the addressed DGA design. The random vibrations are considered in the developed methodology as a white noise of the given ratio of the acceleration amplitudes squared to the vibration frequency.

10.6.3 Reduction to Practice

The suggested model is confirmed by accelerated testing. Testing was carried out for two PCBs, with surface-mounted packages on them, at the same level (with the mean value of 50 g) of 3D random vibrations. One board was subjected, concurrently with random vibrations, to the low temperature of –20°C = 253K and another one – to –100°C = 173K. It has been predicted, by preliminary calculations using the developed model that the solder joints at the –20°C will still perform within the elastic range, while the solder joints at –100°C will experienced static inelastic strains. No wonder that no failures were detected in the solder joints of the board tested at –20°C while the joints of the board tested at –100°C failed after several hours of testing. Some results of such an accelerated testing are addressed, described, and commented on. Here is how FOAT could be implemented in the problem in question.

10.6.4 Calculation Procedure

Let us assume that the failure rate of the solder material, which characterizes the rate of propensity of the material or the device to failure, could be monitored determined by the level of the measured electrical resistance: $\gamma = \gamma_R R$. Using the BAZ model (10.7) and considering the combined action of low temperature T (that supposedly leads to elevated thermal stresses in the solder material) and external random vibrations characterized by their spectrum S, one can seek the probability of the material nonfailure after FOAT for the time t in the form

$$P = \exp\left[-\gamma_R Rt \exp\left(-\frac{U_0 - \gamma_S S}{kT}\right)\right], \tag{10.74}$$

where the γ values reflect the sensitivities of the material to the corresponding stimuli (stressors), and R is the continuously measured/monitored electrical resistance for the peripheral joints. Although only two stimuli (stressors) were selected in this model—low temperature and random vibrations—the model can be easily made multiparametric, that is, generalized for as many stimuli as necessary. The units for the sensitivity parameter γ_R are obviously $\Omega^{-1}h^{-1}$ if the measured electrical resistance of the peripheral solder joints is measured in ohms, and the elapsed time t is measured in hours. The unites of the sensitivity parameter γ_S are $eVm^{-2}s^{-3}$ if the stress-free activation energy U_0 is measured in eV and the power spectral density (PSD) amplitudes are measured in $(\text{m/s}^2)^2/\text{Hz} = m^2s^3$. The physical meaning of this distribution could be seen from the formulas

$$\frac{\partial P}{\partial R} = -\frac{H(P)}{R}, \quad \frac{\partial P}{\partial t} = -\frac{H(P)}{t}, \quad \frac{\partial P}{\partial S} = -\frac{H(P)}{kT}\gamma_S, \tag{10.75}$$

where $H(P) = -P\ln P$ is the entropy of the probability P of nonfailure. Thus, the change in the probability of nonfailure always increases with an increase in the entropy (uncertainty) of the distribution and decreases with an increase in the monitored (measured) electrical resistance and the elapsed time. As to the sensitivity factor γ_S, it can be found as the ratio

$$\gamma_S = -\frac{\dfrac{\partial P}{\partial S}}{\dfrac{H(P)}{kT}} \tag{10.76}$$

of the (negative) derivative $\dfrac{\partial P}{\partial S}$ of the probability of nonfailure with respect to the level of the vibration excitation (power spectrum) to the ratio of the entropy of the probability of nonfailure to the level of the thermal energy kT. It should be emphasized that the temperature T in the aforementioned formulas is, unlike in Boltzmann's statistics or in the Arrhenius formula, a parameter, not an argument. It is the threshold of the low temperature, below which the inelastic strains in the peripheral solder joints occur. This temperature/threshold should be determined and established based on the procedures addressed in the "Inelastic strains in solder material" section in chapter 4. The expression for the probability of nonfailure contains three empirical parameters: the stress-free activation energy U_0 and two sensitivity factors, γ_R and γ_S. Here is how these parameters can be obtained from the conducted highly focused and highly cost effective FOAT data.

At the first step, one should run the FOAT for two different temperatures T_1 and T_2, keeping their levels unchanged during the experiment. Unlike in the original Arrhenius or Zhurkov's experiments, these levels should be established and kept low enough so that inelastic strains in the peripheral solder joints of the package/PCB assembly could occur. These temperatures could/should be obtained from the preliminary thermal stress analysis already described. Recording the percentages (values) P_1 and P_2 of nonfailed samples (or values $Q_1 = 1 - P_1$ and $Q_2 = 1 - P_2$ of the failed samples) and assuming a certain criterion of failure (say, when the level of the measured electrical resistance, because of the "opens" in the failed joints, exceeds a certain level R_*) one could obtain the following two relationships:

$$P_1 = \exp\left[-\gamma_R R_* t_1 \exp\left(-\frac{U_0 - \gamma_S S}{kT_1}\right)\right], \quad P_2 = \exp\left[-\gamma_R R_* t_2 \exp\left(-\frac{U_0 - \gamma_S S}{kT_2}\right)\right]. \tag{10.77}$$

Since the numerators $U_0 - \gamma_S S$ (effective activation energies) in these relationships are kept the same during the FOAT, the following equation must be fulfilled for the sought sensitivity factor γ_R of the electrical resistance:

$$\ln\left(-\frac{\ln P_1}{R_* t_1 \gamma_R}\right) - \frac{T_2}{T_1}\ln\left(-\frac{\ln P_2}{R_* t_2 \gamma_R}\right) = 0. \tag{10.78}$$

Here, t_1 and t_2 are times at which the failures defined as the moments of time when the level R_* of the continuously measured electrical resistance were observed. Equation (10.78) has the following solution:

$$\gamma_R = \exp\left[\frac{\frac{T_2}{T_1}\ln\left(-\frac{\ln P_2}{R_* t_2}\right) - \ln\left(-\frac{\ln P_1}{R_* t_1}\right)}{\frac{T_2}{T_1} - 1}\right] \tag{10.79}$$

It is expected that more than two series of FOAT tests and at more than two temperature levels should be conducted, so that the sensitivity parameter γ_R could be established with a high enough degree of accuracy and certainty.

At the second step, FOAT tests at two spectra levels S_1 and S_2 should be conducted for the same temperature T. This leads to the relationship

$$\gamma_S = \frac{kT}{S_1 - S_2}\left[\ln\left(-\frac{\ln P_1}{R_* t_1 \gamma_R}\right) - \ln\left(-\frac{\ln P_2}{R_* t_2 \gamma_R}\right)\right] = \frac{kT}{S_1 - S_2}\ln\left(\frac{t_2 \ln P_1}{t_1 \ln P_2}\right) \tag{10.80}$$

Note, that the γ_S value is independent, in this approach, of the resistance R_* threshold and the sensitivity factor γ_R. Finally, the stress-free activation energy can be computed, for the determined factors γ_R and γ_S as

$$U_0 = \gamma_S S - kT\ln\left(-\frac{\ln P}{R_* t \gamma_R}\right) \tag{10.81}$$

for any consistent vibration spectrum level, temperature threshold and time values. After the sensitivity factors and the loading (stressor) free activation energy are determined for the tested combinations of the input data, the aforementioned formula could be used, but should be checked (validated), of course, for other physically meaningful combinations of the FOAT parameters. The fatigue lifetime can be found for the induced temperature below the temperature, at which the inelastic strains occur from the basic formula for the probability of nonfailure as follows:

$$t = RUL = \frac{\ln P}{\gamma_R R_*} \exp\left(\frac{U_0 - \gamma_S S}{kT}\right) \tag{10.82}$$

This formula makes physical sense. Indeed, the RUL increases with an increase in the probability P of nonfailure, and with an increase in the level of the effective activation energy $U = U_0 - \gamma_s S$. The RUL decreases with an increase in the acceptable level R_* of the electrical resistance of the damaged joints, with an increase in the sensitivity factor γ_R and the level kT of the thermal energy. This level is higher for lower thermal energies.

10.6.5 Numerical Example

Input data:

Structural Element	Package	PCB	Solder (96.5%Ag3.5%Sn)
Element's Number	1	2	0
Effective Young's Modulus, E, kg/mm^2	8775.5	2321.4	1939.0
Poisson's Ratio, ν	0.30	0.30	0.30
Shear Modulus, G, kg/mm^2	3367.3	892.7	1958.8
CTE, 1/°C	6.5×10^{-6}	15.0×10^{-6}	xxxx
Thickness, mm	2.0	1.5	0.2

Estimated yield stress of the solder material in shear: $\tau_Y = 1.825$ kg/mm^2
Soldering Temperature: 158°C = 431°K
Testing Temperatures: $T_1 = -20°C = 253°K$, $T_2 = -100°C = 173°K$
Changes in Temperature: $\Delta t_1 = 178°C = 178°K$, $\Delta t_2 = 258°C = 258°K$
The "external" thermal strains: $\varepsilon_1 = \Delta\alpha\Delta t_1 = 151.3\times10^{-5}$, $\varepsilon_2 = \Delta\alpha\Delta t_2 = 219.3\times10^{-5}$
Half Package Length $l = 15$ mm;
Electrical Resistance Threshold at Solder Failure [52]: $R_* = 450\Omega$

Computed data:

Axial compliances of the assembly components:

$$\lambda_1 = \frac{1-\nu_1}{E_1 h_1} = \frac{1-0.3}{8775.5\times2.0} = 3.9884\times10^{-5}\,\text{mm/kg};$$

$$\lambda_2 = \frac{1-\nu_2}{E_2 h_2} = \frac{1-0.3}{2321.4\times1.5} = 20.1028\times10^{-5}\,\text{mm/kg};$$

Flexural rigidities of the assembly components:

$$D_1 = \frac{E_1 - h_1^3}{12(1-v_1^2)} = \frac{8775.5 \times 2^3}{12(1-0.3^2)} = 6428.9377 \text{ kgmm}$$

$$D_2 = \frac{E_2 - h_2^3}{12(1-v_2^2)} = \frac{2321.4 \times 1.5^3}{12(1-0.3^2)} = 717.4657 \text{ kgmm;}$$

Total axial compliance of the assembly:

$$\lambda = \lambda_1 + \lambda_2 + \frac{h_1}{4D_1} + \frac{h_2}{4D_2} = (3.9884 + 20.1028 + 7.7773 + 52.2673)10^{-5}$$
$$= 84.1358 \times 10^{-5} \text{mm/kg}$$

Interfacial compliances:

$$\kappa_1 = \frac{h_1}{3G_1} = \frac{2.0}{3 \times 3367.3} = 19.7983 \times 10^{-5} \text{mm}^3/\text{kg},$$

$$\kappa_2 = \frac{h_2}{3G_{12}} = \frac{1.5}{3 \times 892.7} = 56.0100 \times 10^{-5} \text{mm}^3/\text{kg}$$

$$\kappa_0 = \frac{h_0}{G_0} = \frac{0.2}{1958.8} = 10.2103 \times 10^{-5} \text{mm}^3/\text{kg}$$

$$\kappa = \kappa_0 + \kappa_1 + \kappa_2 = 86.0186 \times 10^{-5} \text{mm}^3/\text{kg}$$

Parameter of the interfacial shearing stress

$$k = \sqrt{\frac{\lambda}{\kappa}} = \sqrt{\frac{84.1358 \times 10^{-5}}{86.0186 \times 10^{-5}}} = 0.9890 \text{ mm}^{-1}$$

The product $kl = 0.9890 \times 15.0 = 14.8350$ is significant, and therefore the maximum interfacial shearing stress can be evaluated assuming infinitely large assembly.

For the board tested at –20°C this stress is

$$\tau_{max}^{\infty} = k\frac{\Delta\alpha\Delta t}{\lambda} = 0.9890\frac{151.3 \times 10^{-5}}{84.1358 \times 10^{-5}} = 1.7785 \text{ kg/mm}^2$$

and is somewhat below the yield stress of the solder material, so that no inelastic strains are likely to occur. For the board tested at –100°C this stress is

$$\tau_{max}^{\infty} = k\frac{\Delta\alpha\Delta t}{\lambda} = 0.9890\frac{219.3 \times 10^{-5}}{84.1358 \times 10^{-5}} = 2.5778 \text{ kg/mm}^2,$$

and the lengths of the inelastic zones in this case are

$$l_Y = l - l_e = \frac{1}{k}\left(\frac{\tau_{max}^{\infty}}{\tau_Y} - 1\right) = \frac{1}{0.9890}\left(\frac{2.5778}{1.825} - 1\right) = 0.4171 \text{ mm.}$$

The temperature boundary between the elastic and inelastic states of stress is characterized by the temperature change of

$$\Delta t = \frac{\lambda \tau_Y}{k\Delta\alpha} = \frac{84.1358\times10^{-5}\times1.825}{0.9890\times8.5\times10^{-6}} = 182.6°\text{C},$$

and, with the soldering temperature of −158°C, is −24.6°C.

Here is a hypothetical example of how the parameters of the BAZ equation can be determined when testing is conducted until failure. Note that has not been the case for the two PCBs whose testing is described in Section 10.6.6, since only the solder joints in the PCB tested at −100°C have failed, while the joints in the board tested at −20°C have not exhibit any failure after many hours of testing. Let, for example, the FOAT is carried out until the resistance threshold is reached. Half of the specimen population failed at the first stage of testing at the temperature of $T_1 = -30°\text{C} = 243°\text{K}$ after t_1 = 200 hrs of testing. When testing was conducted at the temperature of $T_2 = -10°\text{C} = 263°\text{K}$, half of the specimen population failed after $t_2 = 300$ hrs of testing. The level of the vibration power spectrum density S was kept the same in both sets of the tests and therefore did not affect the factor γ_R. Then, the equation for the sensitivity parameter γ_R yields

$$\gamma_R = \exp\left[\frac{\frac{T_2}{T_1}\ln\left(\frac{\ln P_2}{R_* t_2}\right) - \ln\left(\frac{\ln P_1}{R_* t_1}\right)}{\frac{T_2}{T_1} - 1}\right] = \exp\left[\frac{\frac{263}{243}\ln\left(\frac{\ln 0.5}{450\times300}\right) - \ln\left(-\frac{\ln 0.5}{450\times200}\right)}{\frac{T_2}{T_1} - 1}\right]$$

$$= 1.9692\times10^{-6}\,\Omega^{-1}\text{hr}^{-1}$$

The thermal energy is $kT_1 = 8.6176\times10^{-5}\times243° = 2.0941\times10^{-2}\,\text{eV}$, when testing is carried out at the temperature $T_1 = -30°\text{C} = 243°\text{K}$, and is $kT_2 = 8.6176\times10^{-5}\times263° = 2.2664\times10^{-2}\,\text{eV}$, when testing is carried out at the temperature $T_2 = -10°\text{C} = 263°\text{K}$. Let the testing at the second stage of testing be carried out until 99% of the tested specimens failed, so that $P = 0.01$, and that this took place after t_1 = 500 hrs of testing at the temperature of $T_1 = -30°\text{C} = 243°\text{K}$ and at the vibration level of $S_1 = 2\times10^{-6}\,\text{mm}^2\,\text{sec}^{-3}$ and after t_1 = 650 hrs of testing at the temperature of $T_2 = -10°\text{C} = 263°\text{K}$ and at the vibration level of $S_2 = 10^{-6}\,\text{mm}^2\,\text{sec}^{-3}$. The effective activation energy is

$$U_1 = U_0 - \gamma_S S_2 = -kT_1\ln\left(-\frac{\ln P}{R_* t_1 \gamma_R}\right)$$

$$= -2.0941\times10^{-2}\ln\left(-\frac{\ln 0.01}{450\times500\times1.9692\times10^{-6}}\right) = 0.049027\,\text{eV},$$

when testing was carried out at the temperature $T_1 = -30°C = 243°K$, and is

$$U_2 = U_0 - \gamma_S S_2 = -kT_2 \ln\left(-\frac{\ln P}{R_* t_2 \gamma_R}\right)$$

$$= -2.2664 \times 10^{-2} \ln\left(-\frac{\ln 0.01}{450 \times 650 \times 1.9692 \times 10^{-6}}\right) = 0.0918687 \text{ eV}$$

when testing was performed at the temperature $T_2 = -10°C = 263°K$. Clearly, since the thermal strain and/or the region occupied by the inelastic stresses in the solder material are higher at the lower temperature condition, the remaining effective activation energy is lower at this temperature.

From the last two equations, considering that the zero-stress activation energy should be loading independent, we have the following formula for the vibration sensitivity factor:

$$\gamma_S = \frac{U_2 - U_1}{S_1 - S_2} = \frac{0.0918687 - 0.049027}{2 \times 10^6 - 10^6} = 4.2842 \times 10^{-8} \text{eV mm}^{-2} \text{ sec}^3$$

Then, the stress-free activation energy can be computed as

$$U_0 = U_1 + \gamma_S S_1 = 0.049027 + 4.2842 \times 10^{-8} \times 2 \times 10^6 = 0.1347 \text{ eV}$$

or as

$$U_0 = U_2 + \gamma_S S_2 = 0.0918687 + 4.2842 \times 10^{-8} \times 10^6 = 0.1347 \text{ eV}$$

The RUL can be computed for any probability of nonfailure, low temperature, and vibration spectral density as

$$t = RUL = -\frac{\ln P}{\gamma_R R_*} \exp\left(\frac{U_0 - \gamma_S S}{kT}\right)$$

$$= -\frac{\ln P}{1.9692 \times 10^{-6} \times 450} \exp\left(\frac{0.1347 - 4.2842 \times 10^{-8} S}{8.6176 \times 10^{-5} T}\right)$$

If, for example, $P = 0.9$, $T = -20°C = 253°K$, and $S = 10^3 \text{mm}^2 \text{ sec}^{-3}$, then the predicted RUL of the solder material is

$$RUL = \frac{0.1054}{1.9692 \times 10^{-6} \times 450} \exp\left(\frac{0.1347 - 4.2842 \times 10^{-8} \times 10^3}{8.6176 \times 10^{-5} \times 253}\right) = 57231.5502 \text{ hrs}$$

$$= 6.5333 \text{ years}$$

10.6.6 Testing Facility and Procedure

The actual testing has been carried out at the Reliant Labs, Inc., 925 Thompson Place, Sunnyvale, California. Two PCB boards, serial numbers QFN-P-07 and QFN-P-08, provided by the customer, were tested. Qualmark OVS 2.5LF HALT/HASS Chamber (model # 2.5LF) was used to accommodate the test specimens (one at a time). Omega thermocouples were used to measure temperature, and a Dytran accelerometer control was used to measure the applied accelerations. All test equipment that requires periodic calibration was in current calibration at time of test.

The test results could be summarized as follows. Board #1 was tested at the temperature of −20°C and the (identical) board #2 at the temperature of −100°C. In both cases the established level of the random 3D vibrations was 50 g.

The reason why these temperatures were chosen, is that, according to these calculations, the −20°C temperatures were not expected to lead to inelastic static strains, while the −100°C temperature was supposed to result in appreciable plastic deformations and, hence, in a considerably shorter fatigue life of the solder material. Electrical resistance was continuously measured in four corner packages of each board. Prior to testing, all the joints showed resistance of about 0.15μΩ For board #1 (tested at −20°C), this level of resistance has not changed after 5 hours of testing. For board #2 (tested at −100°C), opens in two packages were detected after about 1.5 hours of testing, and an increase in the resistance to about 0.9Ω was detected for the remaining two corner packages after about 3.5 hours of testing. The total time of testing of board #2 was about 4 hours. Hence, the test results have confirmed the general concept that low temperatures in combination with random vibrations might be an attractive accelerated test vehicle for electronic materials and packages, and that there is a significant difference in the fatigue lifetime (RUL) for the solder material that remains within the elastic region (when subjected to moderately low temperatures) and the material that is stressed above this region at significant low temperatures. The tests were not continued beyond the above times, since no substantial new information was expected if they would be. It should be emphasized, however, that the FOAT should be always conducted if there is an intent to quantify the RUL. For materials that failed within the elastic region, the probabilistic Palmgren-Miner "rule of the linear accumulation of damages" can be used to predict the RUL.

10.6.7 Conclusion

The carried out analyses explain how the predictive modeling approach can be used in predicting and prevention of failures of solder joints in electronic products in which high reliability is imperative.

10.7 POSSIBLE NEXT-GENERATION QT

The next-generation QT could be viewed as a "quasi-FOAT," "mini-FOAT," a sort-of an "initial stage of FOAT" that more or less adequately replicates the initial nondestructive, yet full-scale, stage of FOAT. The duration and conditions of such a "mini-FOAT" QT could and should be established based on the observed and recorded results of the actual FOAT, and should be limited to the stage when no failures, or a predetermined and acceptable small number of failures in the actual full-scale FOAT, were observed. PHM technologies ("canaries") could be concurrently tested to make sure that the safe limit is established correctly and is not exceeded. Such an approach to qualify electronic devices into products will enable the industry to specify, and the manufacturers to assure, a predicted and adequate probability of failure for an electronic product that passed the QT and is expected to be operated in the field under the given conditions for the given time. FOAT should be thoroughly designed, implemented, and analyzed so that the QT is based on the trustworthy FOAT data.

APPENDIX A ELASTIC STABILITY OF THE SPECIMEN AS A WHOLE

A long enough test specimen as whole can be treated as a cantilever elongated plate of the total flexural rigidity $D = D_1 + D_2$ length l subjected to the compressive force $\hat{T}$. The critical force for such a specimen is

$$T_e = \frac{\pi^2 D}{4l^2}. \tag{A.1}$$

Using the input data from the numerical example, we obtain

$$T_e = \frac{\pi^2 D}{4l^2} = \left(\frac{\pi}{2l}\right)^2 D = \left(\frac{\pi}{20}\right)^2 754.1 = 18.6067 \text{ kg/mm} \tag{A.2}$$

This value is significantly higher than the $\hat{T} = 8$ kg/mm accepted in this example.

APPENDIX B APPROXIMATE FORMULA FOR THE INTERFACIAL PEELING STRESS AND ELASTIC STABILITY OF THE COMPRESSED COMPONENT #1

When evaluating the elastic stability of compressed component #1 of the assembly, we proceed from the simplifying assumption that the interfacial shearing stress can be evaluated without considering its coupling with the interfacial peeling stress. With this assumption, we replace equation (10.1) with the following simplified equations

$$\left.\begin{aligned} u_1(x) &= -\alpha_1 \Delta t x + \lambda_1 \int_0^x [\hat{T} - T(\xi)] d\xi + \kappa_1 \tau(x) \\ u_2(x) &= -\alpha_2 \Delta t x + \lambda_2 \int_0^x T(\xi) d\xi - \kappa_2 \tau(x) \end{aligned}\right\}, \tag{B.1}$$

in which the effect of bending is not accounted for. Then, equation (10.9) for the shearing stress function $\tau(x)$ results in the following simplified equation:

$$\kappa\tau(x) - (\lambda_1 + \lambda_2) \int_0^x T(\xi) d\xi = -(\Delta\alpha\Delta t + \lambda_1 \hat{T})x, \tag{B.2}$$

where the force $T(x)$ is expressed by formula (10.3). By differentiation, we find

$$\kappa\tau'(x)-(\lambda_1+\lambda_2)T(x)=-(\Delta\alpha\Delta t+\lambda_1\hat{T}). \tag{B.3}$$

We seek the solution to equation (B.3) in the form

$$\tau(x)=C\frac{\sinh kx}{\cosh kl} \tag{B.4}$$

Introducing this solution into equation (B.3) and considering formula (B.2), we conclude that equation (B.3) is fulfilled if the following relationships take place:

$$k=\sqrt{\frac{\lambda_1+\lambda_2}{\kappa}},\quad C=-k\frac{\Delta\alpha\Delta t+\lambda_1\hat{T}}{\lambda_1+\lambda_2} \tag{B.5}$$

Then we have

$$\tau(x)=-k\frac{\Delta\alpha\Delta t+\lambda_1\hat{T}}{\lambda_1+\lambda_2}\frac{\sinh kx}{\cosh kl},\quad T(x)=\frac{\Delta\alpha\Delta t+\lambda_1\hat{T}}{\lambda_1+\lambda_2}\left(1-\frac{\cosh kx}{\cosh kl}\right) \tag{B.6}$$

The first formula in (B.6) satisfies the boundary condition (10.35) in the main text, and the second formula satisfies condition (10.30).

Introducing the first formula in (B.6) into equation (10.25), we obtain the following equation for the interfacial peeling stress $p(x)$:

$$p^{IV}(x)+4\beta^4p(x)=-4\beta^4\frac{D_1D_2}{D}\mu\frac{\Delta\alpha\Delta t+\lambda_1\hat{T}}{\kappa}\frac{\cosh kx}{\cosh kl} \tag{B.7}$$

We seek the solution to this equation in the form

$$p(x)=A_0V_0(\beta x)+A_2V_2(\beta x)+p_0\frac{\cosh kx}{\cosh kl}, \tag{B.8}$$

where the functions $V_i(\beta x)$, $i=0,1,2,3$, and their derivatives are expressed as follows:

$$\begin{gathered}V_0(\beta x)=\cosh\beta x\cos\beta x,\\ V_2(\beta x)=\sinh\beta x\sin\beta x,\\ V_{1,3}(\beta x)=\frac{1}{\sqrt{2}}(\cosh\beta x\sin\beta x\pm\sinh\beta x\cos\beta x),\\ V_0'(\beta x)=-\beta\sqrt{2}V_3(\beta x),\quad V_1'(\beta x)=\beta\sqrt{2}V_0(\beta x),\\ V_2'(\beta x)=\beta\sqrt{2}V_1(\beta x),\quad V_3'(\beta x)=\beta\sqrt{2}V_2(\beta x).\end{gathered} \tag{B.9}$$

Introducing (B.8) into equation (B.7), we obtain

$$p_0 = -\mu \frac{\frac{D_1 D_2}{D}}{1+\left(\frac{k}{\beta\sqrt{2}}\right)^4} \frac{\Delta\alpha\Delta t + \lambda_1 \hat{T}}{\kappa} \tag{B.10}$$

Boundary conditions (10.52) result in the following equations for the constants A_0 and A_2 of integration:

$$\left.\begin{aligned} V_2(u)A_0 - V_0(u)A_2 &= \left(\frac{k}{\beta\sqrt{2}}\right)^2 p_0, \\ V_1(u)A_0 + V_3(u)A_2 &= \left(\frac{k}{\beta\sqrt{2}}\right)^3 p_0 \tanh kl \end{aligned}\right\} \tag{B.11}$$

where $u = \beta l$. These equations result in the following solutions:

$$A_0 = \left(\frac{k}{\beta}\right)^2 p_0 \frac{\frac{k}{\beta}\cosh u \cos u + \cosh u \sin u - \sinh u \cos u}{\sinh 2u + \sin 2u}, \tag{B.12}$$

$$A_2 = \left(\frac{k}{\beta}\right)^2 p_0 \frac{\frac{k}{\beta}\sinh u \sin u - \cosh u \sin u - \sinh u \cos u}{\sinh 2u + \sin 2u},$$

and solution (B.8) yields

$$p(x) = p_0 \left\{ \left(\frac{k}{\beta}\right)^2 \left[\frac{\frac{k}{\beta}\cosh u \cos u + \cosh u \sin u - \sinh u \cos u}{\sinh 2u + \sin 2u} \cosh \beta x \cos \beta x + \right.\right.$$

$$\left.\left. + \frac{\frac{k}{\beta}\sinh u \sin u - \cosh u \sin u - \sinh u \cos u}{\sinh 2u + \sin 2u} \sinh \beta x \sin \beta x \right] + \frac{\cosh kx}{\cosh kl} \right\} \tag{B.13}$$

For long enough assembly, this formula yields

$$p(x) = p_0 \left[\left(\frac{k}{\beta}\right)^2 e^{-\beta(l-x)} \left(\left(\frac{k}{\beta} - 1\right) \cos[\beta(l-x)] + \sin[\beta(l-x)] \right) + e^{-k(l-x)} \right] \tag{B.14}$$

The peeling stress at the specimen's ends is

$$p(l) = p_0\left[\left(\frac{k}{\beta}\right)^2\left(\frac{k}{\beta}-1\right)+1\right] \tag{B.15}$$

If the origin is placed at the specimen's end and directed inward, the specimen, then equation (B.14), could be written in the following simple way:

$$p(x) = p_0\left[\left(\frac{k}{\beta}\right)^2 e^{-\beta x}\left(\left(\frac{k}{\beta}-1\right)\cos\beta x+\sin\beta x\right)+e^{-kx}\right] \tag{B.16}$$

If this stress is low enough, no delamination buckling could possibly occur. This circumstance should be verified experimentally, prior to conducting accelerated testing.

REFERENCES

1. S.P. Timoshenko, "Analysis of Bi-Metal Thermostats," *Journal of the Optical Society of America*, No. 11, 1925.
2. B.J. Aleck, "Thermal Stresses in a Rectangular Plate Clamped Along an Edge," *ASME Journal of Applied Mechanics*, vol. 16, 1949.
3. R. Zeyfang, "Stresses and Strains in a Plate Bonded to a Substrate: Semiconductor Devices," *Solid State Electronics*, vol. 14, 1971.
4. B.A. Boley, J.H. Weiner, *Theory of Thermal Stresses*, Quantum Publishers, 1974.
5. K. Roll, "Analysis of Stress and Strain Distribution in Thin Films and Substrates," *Journal of Applied Physics*, vol. 47, No. 7, 1976.
6. G.H. Olsen, M. Ettenberg, "Calculated Stresses in Multilayered Heteroepitaxial Structures," *Journal of Applied Physics*, vol. 48, No. 6, 1977.
7. S.M. Hu, "Film-Edge Induced Stresses in Substrates," *ASME Journal of Applied Mechanics*, vol. 50, No. 7, 1979.
8. R.A. Riddle, "The Application of the J Integral to Fracture Under Mixed Mode Loading," Lawrence Livermore National Laboratory, UCRL-53182, 1981.
9. J. Vilms, D. Kerps, "Simple Stress Formula for Multilayered Thin Films on a Thick Substrate," *Journal of Applied Physics*, vol. 53, No. 3, 1982.
10. F.-V. Chang, "Thermal Contact Stresses of Bi-Metal Strip Thermostat," *Applied Mathematics and Mechanics*, vol. 4, No. 3, Tsing Hua Univ., Beijing, China, 1983.
11. E. Suhir, "Stresses in Bi-Metal Thermostats," *ASME Journal of Applied Mechanics*, vol. 53, No. 3, Sept. 1986.
12. E. Suhir, "An Approximate Analysis of Stresses in Multilayer Elastic Thin Films," *ASME Journal of Applied Mechanics*, vol. 55, No. 3, 1988.
13. A.Y. Kuo, "Thermal Stresses at the Edge of a Bimetallic Thermostat," *ASME Journal of Applied Mechanics*, vol. 56, 1989.
14. E. Suhir, "Interfacial Stresses in Bi-Metal Thermostats," *ASME Journal of Applied Mechanics*, vol. 56, No. 3, Sept. 1989.
15. J.H. Lau, "A Note on the Calculation of Thermal Stresses in Electronic Packaging by Finite-Element Method," *ASME Journal of Electronic Packaging*, vol. 111, No. 12, 1989.

16. E. Suhir, "Twist-off Testing of Solder Joint Interconnections," *ASME Journal of Electronic Packaging*, vol. 111, No. 3, Sept. 1989.
17. E. Suhir, "Can Power Cycling Life of Solder Joint Interconnections Be Assessed on the Basis of Temperature Cycling Tests?" *ASME Journal of Electronic Packaging*, vol. 111, No. 4, Dec. 1989.
18. D. Post, J.D. Wood, "Determination of Thermal Strains by Moire Interferometry," *Experimental Mechanics*, vol. 29, No. 3, 1989.
19. T.-Y. Pan, Y.-H. Pao, "Deformation of Multilayer Stacked Assemblies," *ASME Journal of Electronic Packaging*, vol. 112, No. 1, 1990.
20. J.C. Glaser, "Thermal Stresses in Compliantly Joined Materials," *ASME Journal of Electronic Packaging*, vol. 112, No. 1, 1990.
21. P.M. Hall, et al., "Strains in Aluminum-Adhesive-Ceramic Trilayers," *ASME Journal of Electronic Packaging*, vol. 112, No. 4, 1990.
22. A.Y. Kuo, "Thermal Stress at the Edge of a Bi-Metallic Thermostat," *ASME Journal of Applied Mechanics*, vol. 57, 1990.
23. J.W. Eischen, C. Chung, J.H. Kim, "Realistic Modeling of the Edge Effect Stresses in Bimaterial Elements," *ASME Journal of Electronic Packaging*, vol. 112, No. 1, 1990.
24. J.T. Gillanders, R.A. Riddle, R.D. Streit, I. Finnie, "Methods for Determining the Mode I and Mode II Fracture Toughness of Glass Using Thermal Stresses," *ASME Journal of Engineering Materials and Technology*, vol. 112, 1990.
25. W. Nelson, *Accelerated Testing: Statistical Models, Test Plans and Data Analyses*, John Wiley & Sons, Inc., 1990.
26. P.M. Hall, F.L. Howland, Y.S. Kim, L.H. Herring, "Strains in Aluminum-Adhesive-Ceramic Trylayers," *Journal of Electronic Packaging*, vol. 112, No. 4, Dec. 1990.
27. A.O. Cifuentes, "Elastoplastic Analysis of Bimaterial Beams Subjected to Thermal Loads," *ASME Journal of Electronic Packaging*, vol. 113, No. 4, 1991.
28. H.S. Morgan, "Thermal Stresses in Layered Electrical Assemblies Bonded with Solder," *ASME Journal of Electronic Packaging*, vol. 113, No. 4, 1991.
29. T. Hatsuda, H. Doi, T. Hayasida, "Thermal Strains in Flip-Chip Joints of Die-Bonded Chip Packages," Procedure of the EPS Conference, San Diego, California, 1991.
30. W.A. Strifler, C.W. Bates, Jr., "Stress in Evaporated Films Used in GaAs Processing," *Journal of Materials Research*, vol. 6, No. 3, 1991.
31. E. Suhir, "Stress Relief in Solder Joints Due to the Application of a Flex Circuit," *ASME Journal of Electronic Packaging*, vol. 113, No. 3, 1991.
32. E. Suhir, "Approximate Evaluation of the Elastic Interfacial Stresses in Thin Films with Application to High-Tc Superconducting Ceramics," *International Journal of Solids and Structures*, vol. 27, No. 8, 1991.
33. E. Suhir, "Mechanical Behavior and Reliability of Solder Joint Interconnections in Thermally Matched Assemblies," 42nd ECTC, San Diego, California, May 1992.
34. R. John, G.A. Hartman, J.P. Gallagher, "Crack Growth Induced by Thermal-Mechanical Loading," *Experimental Mechanics*, vol. 32, No. 2, 1992.
35. R. Darveaux, K. Banerji, "Constitutive Relations for Tin-Based Solder Joints", 50th ECTC, 1992.
36. J.H. Lau (ed.), *Thermal Stress and Strain in Microelectronics Packaging*, Van-Nostrand Reinhold, 1993.
37. R.A. Riddle, "Thermal Stresses in the Microchannel Heatsink Cooled by Liquid Nitrogen": SPIE, High Heat Flux Engineering II, 1993.
38. E. Suhir, "Thermally Induced Stresses in an Optical Glass Fiber Soldered into a Ferrule," *IEEE/OSA Journal of Lightwave Technology*, vol. 12, No. 10, 1994.

39. E. Suhir, "Approximate Evaluation of the Elastic Thermal Stresses in a Thin Film Fabricated on a Very Thick Circular Substrate," *ASME Journal of Electronic Packaging*, vol. 116, No. 3, 1994.
40. M. Pecht, X. Wu, K.W. Paik, S.N. Bhandarkar, "To Cut or Not to Cut: A Thermomechanical Stress Analysis of Polyimide Thin-Film on Ceramics Structures," *IEEE CPMT Transactions Part B*, vol. 18, No. 1, 1995.
41. E. Suhir, "Predicted Thermally Induced Stresses in, and the Bow of, a Circular Substrate/Thin-Film Structure," *Journal of Applied Physics*, vol. 88, No. 5, 2000.
42. K. Noor, M. Malik, "An Assessment of Five Modelling Approaches for Thermo-Mechanical Stress Analysis of Laminated Composite Panels," *Computational Mechanics*, vol. 25, 2000.
43. E. Suhir, "Analysis of Interfacial Thermal Stresses in a Tri-Material Assembly," *Journal of Applied Physics*, vol. 89, No. 7, 2001.
44. J-S. Bae, S. Krishnaswamy, "Subinterfacial Cracks in Bimaterial Systems Subjected to Mechanical and Thermal Loading," *Engineering Fracture Mechanics*, vol. 68, No. 9, 2001.
45. J.-S. Hsu, et al., "Photoelastic Investigation on Thermal Stresses in Bonded Structures," *SPIE Congrès Experimental Mechanics*, vol. 4537, 2002.
46. J. Fjelstad, R. Ghaffarian, Y.G. Kim, "Chip Scale Packaging for Modern Electronics," Electrochemical Publications, 2002.
47. E. Suhir, "Accelerated Life Testing (ALT) in Microelectronics and Photonics: Its Role, Attributes, Challenges, Pitfalls, and Interaction with Qualification Tests," *ASME Journal of Electronics Packaging (JEP)*, vol. 124, No. 3, 2002.
48. Y. Wen, C. Basaran, "An Analytical Model for Thermal Stress Analysis of Multi-layered Microelectronics Packaging," 54th ECTC, 2004.
49. N. Zhu, J.D. Van Wyk, Z.X. Liang, "Thermal-Mechanical Stress Analysis in Embedded Power Modules," Proceedings of the 35th Power Electronics Specialists Conference, 2004.
50. N. Noda, R.B. Hetnarski, Y. Tanigawa, *Thermal Stress*, 2nd ed., Taylor and Francis, 2004.
51. D. Sujan, et al., "Engineering Model for Interfacial Stresses of a Heated Bimaterial Structure with Bond Material Used in Electronic Packages," *IMAPS Journal of Microelectronics and Electronic Packaging*, vol. 2, No. 2, 2005.
52. F. Juskey, R. Carson, "DCA on Flex: A Low Cost/Stress Approach," in E. Suhir, et al. (eds.), *Advances in Electronic Packaging*, ASME Press, 1997.
53. D. Shangguan (ed.), *Lead-Free Solder Interconnect Reliability*, ASM, 2005.
54. E. Suhir, "Reliability and Accelerated Life Testing," Semiconductor International, Feb. 2005.
55. N. Vichare, M. Pecht, "Prognostics and Health Management of Electronics," *IEEE Transactions on Components and Packaging Technologies*, vol. 29, No. 1, Mar. 2006.
56. E. Suhir, "Interfacial Thermal Stresses in a Bi-Material Assembly with a Low-Yield-Stress Bonding Layer," *Modeling and Simulation in Materials Science and Engineering*, vol. 14, 2006.
57. E. Suhir, "How to Make a Device into a Product: Accelerated Life Testing It's Role, Attributes, Challenges, Pitfalls, and Interaction with Qualification Testing," in E. Suhir, C.P. Wong, Y.C. Lee, (eds.), *Micro- and Opto-Electronic Materials and Structures: Physics, Mechanics, Design, Packaging, Reliability*, Springer, 2007.
58. J.-H. Zhao, X. Dai, P.S. Ho, "Analysis and Modelling Verification for Thermal-Mechanical Deformation in Flip-Chip Packages," *Proceedings of the 48th Electronic Components and Technology Conference*, Seattle, Washington, May, 1998.

59. E. Suhir, "Thermal Stress Failures in Microelectronics and Photonics: Prediction and Prevention," *Future Circuits International*, No. 5, 1999.
60. Y. Gao, J.-H. Zhao, "A Practical Die Stress Model and Its Applications in Flip-Chip Packages," Proceedings of the 7th Intersociety Conference on Thermal and Thermomechanical Phenomena in Electronic Systems, Las Vegas, Nevada, May, 2000.
61. L. Ceniga, *Analytical Models of Thermal Stresses in Composite Materials*, Nova Science Publishers, 2008.
61. E. Suhir, "Analytical Thermal Stress Modeling in Electronic and Photonic Systems," *ASME Applied Mechanics Reviews*, invited paper, vol. 62, No. 4, 2009.
62. M. Pecht, *Prognostics and Health Management of Electronics*, Wiley-Interscience, 2008.
63. E. Suhir, "Probabilistic Design for Reliability," *Chip Scale Reviews*, vol. 14, No. 6, 2010.
64. E. Suhir, "Thermal Stress Failures in Electronics and Photonics: Physics, Modeling. Prevention," *Journal of Thermal Stresses*, June, 2013.
65. E. Suhir, S. Yi, "Accelerated Testing and Predicted Useful Lifetime of Medical Electronics," IMAPS Conference on Advanced Packaging for Medical Electronics, San Diego, California, Jan. 2018.
66. E. Suhir, "Solder Joint Interconnections in Automotive Electronics: Design-for-Reliability and Accelerated Testing," *Abstracts Proceedings*, SIITME, Jesse, Romania, 2018.
67. P.M. Hall, "Forces, Moments, and Displacements During Thermal Chamber Cycling of Leadless Ceramic Chip Carriers Soldered to Printed Boards," *IEEE CHMT Transactions*, vol. CHMT-7, No. 4, Dec. 1984.
68. P.M. Hall, "Strain Measurements During Thermal Chamber Cycling of Leadless Ceramic Chip Carriers Soldered to Printed Boards," 34th ECTC, New Orleans, Louisiana, May 1984.
69. P.M. Hall, "Creep and Stress Relaxation in Solder Joints in Surface-Mounted Chip Carriers," Proceedings of the 37th ECTC, Boston, Massachusetts, May 1987.
70. E. Suhir, *Applied Probability for Engineers and Scientists*, McGraw-Hill, 1992.
71. E. Suhir, "Failure-Oriented-Accelerated-Testing (FOAT) and Its Role in Making a Viable IC Package into a Reliable Product," Circuits Assembly, 2013.
72. E. Suhir, "The Role of Failure-Oriented-Accelerated-Testing for Field Functional IC Packages," Circuits Assembly, July 2013.
73. E. Suhir, "HALT, FOAT and Their Role in Making a Viable Device into a Reliable Product," Proceedings of the IEEE-AIAA Aerospace Conference, Big Sky, Montana, Mar. 2014.
74. E. Suhir, S. Yi, "Probabilistic Design for Reliability (PDfR) of Medical Electronic Devices (MEDs): When Reliability is Imperative, Ability to Quantify it is a Must," *Journal of SMT*, vol. 30, No. 1, 2017.
75. E. Suhir, R. Ghaffarian, "Probabilistic Palmgren-Miner Rule with Application to Solder Materials Experiencing Elastic Deformations," *Journal of Materials Science: Materials in Electronics*, vol. 28, No. 3, 2017.
76. E. Suhir, "What Could and Should Be Done Differently: Failure-Oriented-Accelerated-Testing (FOAT) and Its Role in Making an Aerospace Electronics Device into a Product," *Journal of Materials Science: Materials in Electronics*, vol. 29, No. 4, 2018.
77. E. Suhir, "Making a Viable Medical Electron Device Package into a Reliable Product," *IMAPS Advancing Microelectronics*, vol. 46, No. 2, 2019.
78. E. Suhir, "Failure-Oriented-Accelerated-Testing (FOAT), Boltzmann-Arrhenius-Zhurkov Equation (BAZ) and Their Application in Microelectronics and Photonics Reliability Engineering," *International Journal of Aeronautical Science and Aerospace Research (IJASAR)*, vol. 6, No. 3, 2019.

79. E. Suhir, "Failure-Oriented-Accelerated-Testing and Its Possible Application in Ergonomics," *Ergonomics International Journal*, vol. 3, No. 2, Apr. 2019.
80. S.N. Zhurkov, "Проблема прочности твёрдых тел" (The Problem of the Strength of Solids), *Vestnik Akad. Nauk SSSR (Bulletin of the USSR Academy of Sciences)*, 11, 1957 (in Russian).
81. S.N. Zhurkov, "Kinetic Concept of the Strength of Solids," *International Journal of Fracture Mechanics*, vol. 1, 1965.
82. S.A. Arrhenius, "Über die Dissociationswärme und den Einfluß der Temperatur auf den Dissociationsgrad der Elektrolyte," *Zeitschrift fur Physikalische Chemie*, vol. 4, 1889 (in German).
83. L. Boltzmann, "Studien über das Gleichgewicht der lebendigen Kraft zwischen bewegten materiellen Punkten," *Wiener Berichte*, vol. 58, 1868 (in German).
84. L. Boltzmann, "The Second Law of Thermodynamics. Populare Schriften," Essay 3, Address to a Formal Meeting of the Imperial Academy of Science, May 29, 1886.
85. E. Suhir, "Boltzmann-Arrhenius-Zhurkov (BAZ) Model in Physics-of-Materials Problems," *Modern Physics Letters B (MPLB)*, vol. 27, Apr. 2013.
86. E. Suhir, S. Kang, "Boltzmann-Arrhenius-Zhurkov (BAZ) Model in Physics-of-Materials Problems," *Modern Physics Letters B (MPLB)*, vol. 27, Apr. 2013.
87. C.E. Shannon, "A Mathematical Theory of Communication," *Bell System Technical Journal*, vol. 27, 1948.
88. L.F. Coffin, "A Study of the Effects of Cyclic Thermal Stresses on a Ductile Metal," *Transactions of ASME, Journal of Applied Mechanics*, vol. 76, 1954.
89. S.S. Manson, "Behavior of Materials Under Conditions of Thermal Stress," NACA Report 1170, Lewis Flight Propulsion Laboratory, 1954.
90. L.F. Coffin, *Internal Stress and Fatigue of Metals*, Elsevier, 1959.
91. F. Garofalo, *Fundamentals of Creep and Creep Fracture of Metals*, McMillan, 1965.
92. L. Anand, "Constitutive Equations for Hot-Working of Metals," *International Journal of Plasticity*, vol. 1, No. 2, 1985.
93. H.D. Solomon, "Fatigue of 60/40 Solder," *IEEE CPMT Transactions*, vol. CHMT-9, No. 4, 1986.
94. S.B. Brown, K.H. Kim, L. Anand, "An Internal Variable Constitutive Model for Hot Working of Metals," *International Journal of Plasticity*, vol. 5, No. 2, 1989.
95. J.U. Akay, Y. Tong, "Thermal Fatigue Analysis of an SMT Solder Joint Using FEM Approach," *Journal of Microcircuits and Electronic Packaging*, vol. 116, 1993.
96. V. Gektin, A. Bar-Cohen, J. Ames, "Coffin-Manson Fatigue Model of Underfilled Flip-Chips," *IEEE CPMT Transactions–Part A*, vol. 20, No. 3, 1997.
97. E. Suhir, "Solder Materials and Joints in Fiber Optics: Reliability Requirements and Predicted Stresses," Proceedings of the International Symposium on "Design and Reliability of Solders and Solder Interconnections," Orlando, Florida, Feb. 1997.
98. E. Suhir, R. Ghaffarian, "Electron Device Subjected to Temperature Cycling: Predicted Time-to-Failure," *Journal of Electronic Materials*, vol. 48, No. 2, 2019.
99. C. Hillman, N. Blattau, M. Lacy, "Predicting Fatigue of Solder Joints Subjected to High Number of Power Cycles," IPC APEX EXPO, 2014.
100. D.W. Peterson, J.N. Sweet, S.N. Burchett, "Validating Theoretical Calculations of Thermomechanical Stress and Deformation Using the ATC4.1 Flip-Chip Test Vehicle," Proceedings of Surface Mount International '97, San Jose, California, Sep. 1997.
101. S. Ling, A. Dasgupta, "A Nonlinear Multi-Domain Thermomechanical Stress Analysis Method for Surface-Mount Solder Joints—Part II: Viscoplastic Analysis," *Journal of Electronic Packaging*, vol. 119, No. 3, 1997.

102. E. Suhir, "Analysis of a Pre-Stressed Bi-Material Accelerated Life Test (ALT) Specimen," *Zeitschrift fur Angewandte Mathematik und Mechanik (ZAMM)*, vol. 91, No. 5, 2011.
103. E. Suhir, "When Reliability is Imperative, Ability to Quantify It is a Must," IMAPS Advanced Microelectronics, Aug. 2012.
104. J. Lau (ed.), *Ball Grid Array Technology*, McGraw-Hill, 1995.
105. J.S. Hwang, *Modern Solder Technology for Competitive Electronics Manufacturing*, McGraw-Hill, 1996.
106. R.J. Iannuzzelli, J.M. Pitarresi, V. Prakash, "Solder Joint Reliability Prediction by the Integrated Matrix Creep Method," *ASME Journal of Electronic Packaging*, vol. 118, 1996.
107. R.A. Riddle, D.R. Lesuer, C.K. Syn, "Damage Initiation and Propagation in Metal Laminates," *ASME PVP*, vol. 342, 1996.
108. E. Suhir, "Flex Circuit vs Regular" Substrate: Predicted Reduction in the Shearing Stress in Solder Joints," Proceedings of the 3rd International Conference on Flex Circuits FLEXCON 96, San Jose, California, Oct. 1996.
109. E.K. Buratynski, "Analysis of Bending and Shearing of Tri-Layer Laminations for Solder Joint Reliability," in E. Suhir et al. (eds.), *Advances in Electronic Packaging 1997*, ASME Press, 1997.
110. V. Gektin, A. Bar-Cohen, S. Witzman, "Coffin-Manson Based Fatigue Analysis of Underfilled DCA's," *IEEE CPMT Transactions–Part A*, vol. 21, No. 4, 1998.
111. J. Wilde, et al., "Rate Dependent Constitutive Relations Based on Anand Model for 92.5Pb5Sn2.5Ag Solder," *IEEE Transactions, Advanced Packaging*, vol. 23, No. 3, Aug. 2000.
112. Z.N. Cheng, G.Z. Wang, L. Chen, J. Wilde, K. Becker, "Viscoplastic Anand Model for Solder Alloys and its Application," *Soldering & Surface Mount Technology*, vol. 12, No. 2, 2000.
113. A. Ramakrishnan, M. Pecht, "A Life Consumption Monitoring Methodology for Electronic Systems," *IEEE Transactions on Components, Packaging Technology*, vol. 26, No. 3, Sep. 2003.
114. R. Ghaffarian, "Qualification Approaches and Thermal Cycle Test Results for CSP/BGA/FCBGA," *Microelectronics Reliability*, vol. 43, No. 5, May 2003.
115. J. Lau, et al., "Thermal Fatigue Life Prediction Equation for Wafer-Level Chip-Scale Package Lead-Free Solder Joints on lead-Free Printed Circuit Board," 54th ECTC, 2004.
116. T.Y. Tee, Z. Zhong, "Integrated Vapor Pressure, Hygroswelling, and Thermo-Mechanical Stress Modelling of QFN Package During Reflow with Interfacial Fracture Mechanics Analysis," *Microelectronics Reliability*, vol. 44, 2004.
117. H.H. Cui, "Accelerated Temperature Cycle Test and Coffin-Manson Model for Electronic Packaging," Reliability and Maintainability Symposium, Proceedings, Jan. 2005.
118. E. Suhir, "Assuring Aerospace Electronics and Photonics Reliability: What Could and Should Be Done Differently," 2013 IEEE Aerospace Conference, Big Sky, Montana, Mar. 2013.
119. E. Suhir, "Could Electronics Reliability Be Predicted, Quantified and Assured?" *Microelectronics Reliability*, No. 53, Apr. 2013.
120. E. Suhir, "Electronics Reliability Cannot Be Assured, if it is not Quantified," *Chip Scale Reviews*, Mar.–Apr. 2014.
121. E. Suhir, A. Bensoussan, "Quantified Reliability of Aerospace Optoelectronics," *SAE International Journal of Aerospace*, vol. 7, No. 1, 2014.

122. E. Suhir, A. Bensoussan, J. Nicolics, L. Bechou, "Highly Accelerated Life Testing (HALT), Failure Oriented Accelerated Testing (FOAT), and Their Role in Making a Viable Device into a Reliable Product," 2014 IEEE Aerospace Conference, Big Sky, Montana, Mar. 2014.
123. E. Suhir, R. Ghaffarian, "Column-Grid-Array (CGA) vs. Ball-Grid-Array (BGA): Board-Level Drop Test and the Expected Dynamic Stress in the Solder Material," *Journal of Materials Science: Materials in Electronics*, vol. 27, No. 11, 2016.
124. E. Suhir, R. Ghaffarian, "Solder Material Experiencing Low Temperature Inelastic Thermal Stress and Random Vibration Loading: Predicted Remaining Useful Lifetime," *Journal of Materials Science: Materials in Electronics*, vol. 28, No. 4, 2017.
125. E. Suhir, "Solder Joint Interconnections in Automotive Electronics: Design-for-Reliability and Accelerated Testing," Abstracts Proceedings, SIITME, Jassy, Romania, 2018.

11 Probabilistic Design for Reliability

"A pinch of probability is worth a pound of perhaps."

—James G. Thurber, American writer and cartoonist

11.1 BACKGROUND/INCENTIVE

Here are some of the major problems envisioned and questions asked in connection with today's practices concerning reliability evaluations and assurances of electronic and photonic devices, products, and systems:

- Electronic and photonic products that underwent highly accelerated life testing (HALT) [1–3] (see also Chapter 10), passed the existing qualifying tests (QT) and survived burn-in testing (BIT) often exhibit premature field failures. Are today's methodologies and practices adequate? [4]. Do industries need new approaches to qualify their products, and if they do, what should be done differently? Could or should the existing practices be improved to an extent that if the product passed the QT, there is a way to assure that it will satisfactorily perform in the field?
- In many applications, such as aerospace, military, long-haul communications, and medical (see, for example, [5]) high reliability of electronics and photonic materials and products is particularly imperative. Could the operational (field) reliability of such products be assured, if it is not predicted, that is, not quantified?
- And if such a quantification is found to be necessary (in many situations it is the case), could it be done on the deterministic, that is, on a nonprobabilistic basis, or are probabilistic risk analyses a must [6–8]?
- Should electronic product manufacturers keep shooting for an unpredictable and, perhaps, unachievable, very long, such as 20 years or so, product lifetime or, considering that every 5 years a new generation of devices appear on the market and that 20-year predictions are rather shaky, should electronic products manufacturers settle for a shorter, but well-substantiated, predictable, and assured lifetime, with a high (actually, adequate) probability of nonfailure?
- How should such a lifetime be related to the acceptable (specified) probability of nonfailure for a particular product and application? It is clear that if a high probability of failure is acceptable, the projected lifetime will be long, and vice versa, the lifetime of a product will be short, if the accepted/specified probability of its failure is low.

- Considering that the principle of superposition does not work in reliability engineering, how to establish the list of the crucial accelerated tests, the adequate physically meaningful stressors and their combinations and levels, and how to bridge the gap between what one "sees" during the accelerated tests [preferably of failure-oriented-accelerated-testing (FOAT) type, geared to a physically meaningful predictive model] and what will they supposedly (most likely) "get" in the field conditions?
- The best engineering product is the best compromise between the requirements for its reliability, cost-effectiveness, and (short-as-possible) time-to-market; it goes without saying that, in order to make any kind of optimization possible, the reliability of such product should also be quantified, but how could such an optimization be done?
- The bathtub curve (BTC), the experimental "reliability passport" of mass-fabricated electronic products, reflects the inputs of two critical irreversible processes: the statistics-of-failure process, statistical failure rate (SFR), that results in a reduced failure rate with time (this trend is particularly strongly pronounced at the initial, infant mortality, portion of the BTC) and physics-of-failure (aging, degradation) process that leads to an increased failure rate with time (this trend is explicitly exhibited by the BTC wear-out portion). Could these two critical processed be separated [9]? The need for that is due to the obvious incentive to minimize, in one way or another, the role and the rate of aging, and this incentive is especially significant for products like lasers, solder joint interconnections, and others, which are characterized by long wear-out portions and when it is economically infeasible to restrict the product's lifetime to the steady-state situation when the two irreversible processes in question compensate each other.
- The last, but, as they say, not the least questions asked have to do with burn-in testing (BIT) [10–13]. BIT is an accepted practice in electronic and photonic manufacturing for detecting and eliminating early failures ("freaks") in newly fabricated products prior to shipping the "healthy" ones that survived BIT to the customer(s). BIT is a costly undertaking: early failures are avoided and the infant mortality portion of the BTC is supposedly eliminated at the expense of the reduced yield. But what is even worse is that the elevated BIT stressors might not only eliminate "freaks," but could cause permanent damage to the main population of the "healthy" products. It is not even clear if BIT is needed at all, the first question is "to BIT or not to BIT?" When highly accelerated life testing (HALT) is relied upon to do the BIT job, it is not even easy to determine whether there exists a decreasing failure rate with time at all. Even if the BIT-related question cannot be given an exhaustive answer, could it at least shed some useful "quantitative light" on the BIT process so that the best strategy possible could be developed?
- In many of today's reliability problems, human factor plays an important role: quite often equipment/instrumentation reliability (both hard- and software) and human performance contribute jointly to the outcome of a particular mission or an extraordinary situation. How could the interaction and integration of a human and a system be considered [14]?

In this chapter, these problems are addressed and the related questions are, at least partially, answered using the probabilistic-design-for-reliability (PDfR) concept [15–20].

11.2 PDFR AND ITS "TEN COMMANDMENTS"

The PDfR concept is an effective means for improving the state of the art in the electronics and photonics reliability field by quantifying, on the probabilistic basis, the operational reliability of a material, a product, or a system. This concept enables predicting the never-zero probability of failure under the given loading conditions and after the given time in operation, and to use this information as a suitable and physically meaningful criterion of the product's expected performance. The following 10 major (governing) principles ("commandments") reflect the rationale behind the PDfR concept:

1. **When reliability of a product is imperative,** and this is typically the case, ability to predict it is a must; reliability cannot be assured if it is not quantified [20].
2. **Nothing is perfect,** and the difference between a highly reliable and an insufficiently reliable product or a system is "merely" the difference in the levels of their never-zero probabilities of failure, and therefore such a quantification, to be consistent and trustworthy, should be done on the probabilistic basis [21, 22]; in effect, reliability is part of applied probability and probabilistic risk management (PRM) bodies of knowledge, and includes the item's (system's) dependability, durability, maintainability, reparability, availability, and testability; that is, probabilities of the corresponding events or characteristics. Each of these characteristics is measured as a certain probability and could be of a greater or lesser importance depending on the particular function and operation conditions of the item or the system, and consequences of failure; applied probability and probabilistic risk management approaches and techniques put the art of Reliability Engineering on a "reliable" ground.
3. **Reliability evaluations cannot be delayed until the product is made;** they should start at the design stage, but, if possible, should be taken care of at all the significant stages of the product's life: at the design stage, when reliability is conceived; at the accelerated testing stage, using electrical, optical, environmental, and mechanical instrumentation; at the production/manufacturing stage, when reliability is implemented; and, if necessary and appropriate, also in the field; then there will be a reason to believe that a "genetically healthy" product is created and its "health" could be maintained by using various prognostics-and-health monitoring (PHM) methods, as well as redundancy, troubleshooting, and other more or less well-established means that could be considered to maintain adequate reliability level, especially if the "genetic health" of the product is not as high as it could and should be [22].
4. **Product's reliability cannot be low, but need not be higher than necessary either: it has to be adequate for the given product and application,** considering its lifetime, environmental conditions, and consequences of failure; overengineered and overrobust products that "never fail" are most

likely more expensive than they could and should be; certainly, the required and, if possible, even specified probability of failure (or safety factor) should be well substantiated, thoroughly quantified and should be different for an electronic or a photonic device or a module installed in a space shuttle, a nuclear warhead, in a passenger aircraft, or in a household product. For example, three classes of electronic products could be suggested:

Class I: The product has to be made as reliable as possible; failure should not be permitted (some military or space objects);

Class II: The product has to be made as reliable as possible, but only up to a certain level of demand; failure is a catastrophe (civil engineering structures, bridges, ships, aircraft, cars);

Class III: The reliability does not have to be very high. Failures are permitted, but should be restricted (consumer products, commercial electronics, and agricultural equipment).

5. **The best product is, as is well known, the best compromise** between the requirements for its reliability, cost-effectiveness and time-to-market; it goes without saying that such a compromise cannot be achieved if reliability is not quantified.
6. **One cannot design a product with quantified, assured and optimized reliability by limiting the effort to the widely used today "black box"—the HALT;** understanding the underlying physics of failure is crucial, and therefore highly cost-effective and highly focused FOAT should be considered and conducted as HALT's possible and natural extension (see Chapter 10).
7. **FOAT, a "white/transparent box," aimed at understanding the physics of failure, should be geared to a limited number of predetermined simple, easy-to-use, and physically meaningful predictive reliability models,** such as the physically meaningful, powerful, and flexible Boltzmann–Arrhenius–Zhurkov (BAZ) model, and is considered as the experimental basis of the PDfR effort (see Chapter 10).
8. **The physically meaningful, easy-to-use, and flexible multiparametric BAZ equation can be used as a suitable model** for the assessment of the remaining "useful" life (RUL) of an electronic product (see Chapter 10). It has been recently demonstrated [23] that this equation is so flexible that that it could also be used in nonelectronic, nonmaterials physics and even in astrobiology.
9. **Predictive modeling,** not limited to FOAT models, and particularly analytical models, confirmed whenever appropriate and possible, by finite element analysis (FEA), is a powerful means to carry out, if necessary, sensitivity analyses (SA) with an objective to quantify and often nearly eliminate failures by making the probability of failure sufficiently low; this principle is referred to in some other areas of engineering (airspace, maritime, civil, military) as "principle of practical confidence."
10. **Consideration of the role of the human factor** is highly desirable and could be naturally and easily incorporated, if there is a need for that, into a PDfR analysis [14]: not only "nothing," but also "nobody" is perfect, and the human role in assessing the likelihood of the adequate performance of a product is getting increasing attention of both system engineers and human psychologists.

11.3 DESIGN FOR RELIABILITY OF ELECTRONIC PRODUCTS: DETERMINISTIC AND PROBABILISTIC APPROACHES

Design for reliability (DfR) is a set of approaches, methods, and best practices that are supposed to be used at the design stage of an electronic or a photonic product to minimize the risk that the fabricated product might not meet the reliability objectives and customer expectations.

When deterministic approach is used, reliability of a product could be based on the belief that sufficient reliability level is assured if a high enough safety factor (SF) is used. The deterministic SF is defined as the ratio $SF = \frac{C}{D}$ of the capacity ("strength") C of the product to the demand ("stress") D. The probabilistic SF could be introduced as the ratio of the mean value $\prec \psi \succ$ of the safety margin $SM = \Psi = C - D$ to its standard deviation, $\hat{s}$, so that the probabilistic SF is evaluated as $SF = \frac{\prec \psi \succ}{\hat{s}}$. When the random time-to-failure (TTF) is of interest, the SF can be found as the ratio of the mean time-to-failure (MTTF) to the standard deviation of the TTF. The use of SF as a measure of the probability of failure is more convenient than the direct use of this probability. This is because the probability of nonfailure is expressed, for highly reliable and, hence, typical electronic or photonic products, by a number, which is very close to one, and, for this reason, even significant changes in the product's design, with an appreciable impact on its reliability, might have a minor effect on the level of the probability of nonfailure, at least the way it appears to and perceived by the user. For a normally distributed variable of interest (such as stress or strain or current or voltage or temperature) the relationship between the probability P that this variable exceeds a certain level $\alpha = SF$ is expressed as

$$P = \frac{1}{2}[1 + \Phi(SF)], \tag{11.1}$$

where

$$\Phi(\alpha) = \frac{2}{\sqrt{\pi}} \int_0^{\alpha} e^{t^2} dt \tag{11.2}$$

is the probability integral (Laplace function). The relationship $\alpha = SF = f(P)$ is shown in Table 11.1:

TABLE 11.1
Safety Factor versus Probability of Nonfailure for a Normally Distributed Random Variable

P	0.999000	0.999900	0.999990	0.999999	1.0
$\alpha = SF$	3.0901	3.7194	4.5255	4.7518	∞

The SF level, whether deterministic or a probabilistic, should be chosen depending on the experience, anticipated operation conditions, possible consequences of failure, acceptable risks, the available and trustworthy information about the capacity and the demand, the accuracy, with which the capacity and the demand are determined, possible costs and social benefits, information on the variability of materials and structural parameters, and fabrication technologies and procedures. It is noteworthy that while the PDfR is new in microelectronics, probabilistic approaches have been used in other areas of engineering to assure safety for at least half a century. It has been established that the probability that the hull of a commercial ocean-going ship will break in half as a result of her continuous navigation for 20 years in the North Atlantic (the most severe region of the world ocean) is between 10^{-7} and 10^{-8} [24]. It would be prudent to assume that considering redundancy, PHM activities, and other measures available in electronics and photonics reliability engineering, the allowable probability of failure of a single device or package could be, perhaps, depending on the device and its application, on the order of 10^{-5} or so (see Section 11.4.1).

11.4 SOME SIMPLE PDFR EXAMPLES

11.4.1 Adequate Heat Sink

Consider a device whose steady-state operation is determined by the Arrhenius equation (10.2). The probability of nonfailure can be found using the exponential law of reliability:

$$P = \exp\left[-\frac{t}{\tau_0}\exp\left(-\frac{U_0}{kT}\right)\right]. \tag{11.3}$$

Solving this equation for the absolute temperature T, we have:

$$T = -\frac{U_0 / k}{\ln\left(-\frac{\tau_0}{t}\ln P\right)}. \tag{11.4}$$

Addressing, for example, surface charge accumulation failure, for which the ratio of the activation energy to the Boltzmann constant is $\frac{U_0}{k} = 11600K$, assuming that the FOAT-predicted time factor τ_0 is, say, $\tau_0 = 2\times10^{-5}$h, that the customer requires that the probability of failure Q at the end of the device's service time of $t = 40{,}000\ h$ is still below $Q = 10^{-5}$, formula (11.4) yields $T = 352.3K = 79.3°C$. Thus, the heat sink should be designed accordingly, and the vendor should be able to deliver such a heat sink, so that the customer of the device manufacturer is satisfied. The situation changes to the worse, if the temperature of the device changes, especially in a random fashion. Such a situation can also be predicted by a simple probabilistic analysis, which is, however, beyond the scope of this book.

11.4.2 Reliable Seal Glass

The maximum interfacial shearing stress in the thin solder glass layer of a Cerdip/Cerquad package design (Figure 11.1) can be computed, based on the concept of interfacial compliance, as [25]

$$\tau_{\max} = kh_g\sigma_{\max}. \tag{11.5}$$

Here,

$$k = \sqrt{\frac{\lambda}{\kappa}} \tag{11.6}$$

is the parameter of the interfacial shearing stress (see also Chapter 2),

$$\lambda = \frac{1-\nu_c}{E_c h_c} + \frac{1-\nu_g}{E_g h_g} \tag{11.7}$$

is the axial compliance of the assembly,

$$\kappa = \frac{h_c}{3G_c} + \frac{h_g}{3G_g} \tag{11.8}$$

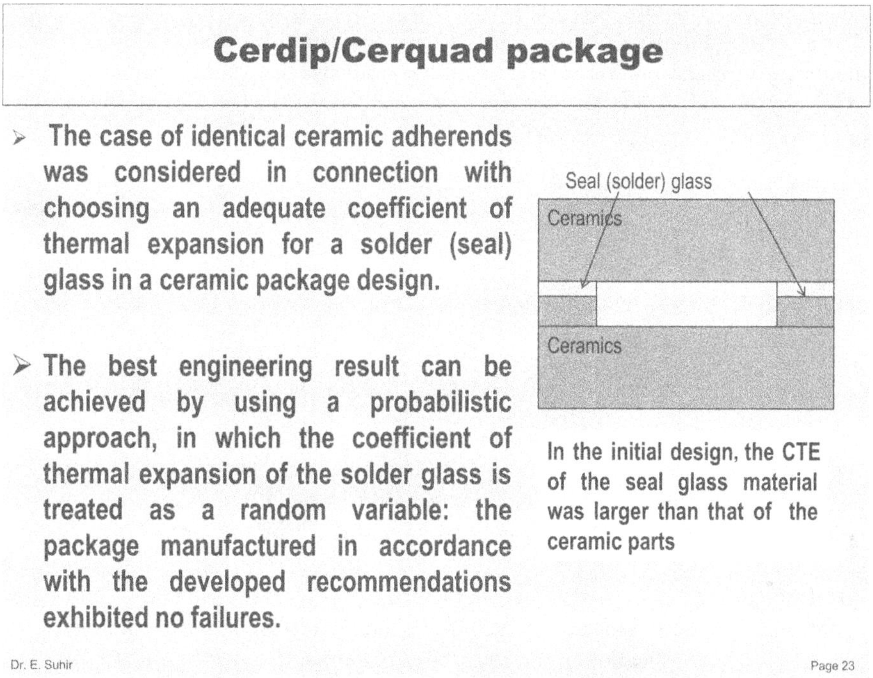
Cerdip/Cerquad package
➤ The case of identical ceramic adherends was considered in connection with choosing an adequate coefficient of thermal expansion for a solder (seal) glass in a ceramic package design.
➤ The best engineering result can be achieved by using a probabilistic approach, in which the coefficient of thermal expansion of the solder glass is treated as a random variable: the package manufactured in accordance with the developed recommendations exhibited no failures.
Seal (solder) glass
Ceramics
Ceramics
In the initial design, the CTE of the seal glass material was larger than that of the ceramic parts
Dr. E. Suhir
Page 23

FIGURE 11.1 Cerdip/Cerquad package (the IC device is located in the space between the two pieces of ceramic)

is its interfacial compliance (the solder glass is a high modulus material and is thin, so that it does not contribute significantly to the interfacial compliance of the assembly),

$$G_c = \frac{E_c}{2(1+\nu_c)}, \quad G_g = \frac{E_g}{2(1+\nu_g)} \tag{11.9}$$

are the shear moduli of the ceramics and glass materials,

$$\sigma_{\max} = \frac{\Delta\alpha\Delta t}{\lambda h_g} \tag{11.10}$$

is the maximum normal stress in the midportion of the glass layer, Δt is the change in temperature from the soldering temperature to the low (room or testing) temperature, $\Delta\alpha = \bar{\alpha}_c - \bar{\alpha}_g$ is the difference in the effective coefficients of thermal expansion (CTEs) of the ceramics and the glass,

$$\bar{\alpha}_{c,g} = \frac{1}{\Delta t}\int_t^{t_0} \alpha_{c,g}(t)dt \tag{11.11}$$

are these coefficients for the given temperature t, t_0 is the annealing (zero stress, setup) temperature, and $\alpha_{c,g}(t)$ are the time dependent CTEs for the materials in question. In an approximate analysis, one could assume that the axial compliance λ of the assembly is due to the thin layer of the seal glass only, so that

$$\lambda \approx \frac{1-\nu_g}{E_g h_g} \tag{11.12}$$

and therefore the maximum normal stress in the solder glass can be evaluated as

$$\sigma_{\max} = \frac{E_g}{1-\nu_g}\Delta\alpha\Delta t. \tag{11.13}$$

While the geometric characteristics of the assembly, the change in temperature, and the elastic constants of the materials can be determined with high accuracy, this is not the case for the difference in the CTEs of the brittle materials of the solder glass and the ceramics. In addition, because of the obvious incentive to minimize this difference, such a mismatch is characterized by a small difference of close and appreciable numbers. This contributes to the uncertainty of the problem in question and justifies the application of the probabilistic approach.

Treating the CTEs of the two materials as normally distributed random variables, we evaluate the probability P that the thermal interfacial shearing stress is compressive (negative) but, in addition, does not exceed a certain allowable level.

This is because it is imperative that the interfacial shearing stresses that result in the compressive normal stresses in the cross sections of the glass layer are not very high and will not result in high shearing stresses and in possible interfacial delamination (cracking). This interfacial shearing stress is proportional to the normal stress in the glass layer, which is, in its turn, proportional to the difference $\Psi = \alpha_c - \alpha_g$ of the CTE of the ceramics and the solder glass materials. So, one wants to make sure that the requirement

$$0 \le \Psi \le \Psi_* = \frac{\sigma_a}{E_g} \frac{1-\nu_g}{\Delta t} \tag{11.14}$$

is fulfilled with a high probability.

For normally distributed random variables α_c and α_g, the variable Ψ is also distributed in accordance with the normal law with the mean value $\prec \psi \succ = \prec \alpha_c \succ - \prec \alpha_g \succ$ and standard deviation $\sqrt{D_\psi} = \sqrt{D_c + D_g}$, where $\prec \alpha_c \succ$ and $\prec \alpha_g \succ$ are the mean values of the materials' CTEs, and D_c, and D_g are their variances. The probability that condition (11.14) takes place is

$$P = \int_0^{\psi*} f_\psi(\psi)d\psi = \Phi_1(\gamma^* - \gamma) - [1 - \Phi_1(\gamma)], \tag{11.15}$$

where

$$\Phi_1(t) = \frac{1}{\sqrt{2\pi}} \int_{-\infty}^{t} e^{-t^2/2} dt \tag{11.16}$$

is the Laplace function (probability integral),

$$\gamma = \frac{\prec \psi \succ}{\sqrt{D_\psi}} \tag{11.17}$$

is the SF for the CTE difference and

$$\gamma^* = \frac{\psi^*}{\sqrt{D_\psi}} \tag{11.18}$$

is the SF for the acceptable level of the allowable stress.

If, for example, the elastic constants of the solder glass are $E_g = 0.66 \times 10^6$ kg/cm^2 and $\nu_g = 0.27$, the sealing (fabrication) temperature is 485°C, the lowest (testing) temperature is –65°C (so that the change in temperature is $\Delta t = 550$°C), the effective CTEs at this temperature are $\bar{\alpha}_g = 6.75 \times 10^{-6}$1/°C and $\bar{\alpha}_c = 7.20 \times 10^{-6}$1/°C, the standard deviations of these CTEs are $\sqrt{D_c} = \sqrt{D_g} = 0.25 \times 10^{-6}$1/°C and the (experimentally obtained) ultimate compressive strength for the glass material is $\sigma_u = 5500$ kg/cm^2, then, with the acceptable SF of, say, 4, the allowable normal stress in compression is $\sigma^* = \sigma_u / 4 = 1375$ kg/cm^2. Actually, it is not this stress, but the maximum shearing stress at the solder glass interface that should be restricted to avoid interfacial damage, but this shearing stress is proportional to the normal compressive stress in the glass.

The corresponding (allowable) level of the random CTE parameter ψ that determines the maximum shearing stress at the glass interface is

$$\psi_* = \frac{\sigma_a}{E_g}\frac{1-\nu_g}{\Delta t} = \frac{1375}{0.66\times 10^6}\frac{0.73}{550} = 2.765\times 10^{-6} 1/°\text{C}.$$

The mean value $\prec \psi \succ$ and the variance D_ψ of the parameter ψ are

$$\prec \psi \succ = \prec \alpha_c \succ - \prec \alpha_g \succ = 0.450\times 10^{-6} 1/°\text{C} \tag{11.19}$$

and

$$D_\psi = D_c + D_g = 0.25\times 10^{-12}(1/°\text{C})^2, \tag{11.20}$$

respectively. Then the SFs for the CTE mismatch and the allowable normal stress in the glass are $\gamma = 1.2726$ and $\gamma^* = 7.8201$, respectively, and the corresponding probability of nonfailure of the seal glass material is, therefore,

$$P = \Phi_1(\gamma^* - \gamma) - [1 - \Phi_1(\gamma)] = 0.898. \tag{11.21}$$

Note that if the standard deviations of the material's CTEs were only

$$\sqrt{D_c} = \sqrt{D_g} = 0.1\times 10^{-6} 1/°\text{C},$$

then the SFs would be much higher: $\gamma = 3.1825$ and $\gamma^* = 19.5559$, and the probability of nonfailure would be as high as $P = 0.999$.

As a result of this analysis, the appropriate solder glass material was selected. No failures were observed in the packages after the original solder glass (that resulted in numerous failures) was replaced in the design with a glass with a reasonably low CTE. Note that the CTE of the initial bonding glass was larger than that of ceramics and all the packages manufactured at the AT&T facility in Allentown, Pennsylvania, using this glass failed during testing.

11.4.3 Extreme Response in Temperature Cycling

Let an electronic device be operated in temperature cycling conditions, and the random amplitude of the induced stress, when a single cycle is applied, is distributed in accordance with the Rayleigh law, so that the probability density function of this amplitude is

$$f(r) = \frac{r}{D_x}\exp\left(-\frac{r^2}{2D_x}\right). \tag{11.22}$$

Our objective is to assess the most likely extreme value of the stress amplitude for a large number n of cycles. The probability distribution density function and the probability distribution function for the extreme value Y_n of the stress amplitude are (see, for example, [18])

$$g(y_n) = n\left\{f(x)[F(x)]^{n-1}\right\}_{x=y_n} \tag{11.23}$$

and

$$G(y_n) = \left\{[F(x)]^n\right\}_{x=\varsigma_n}, \tag{11.24}$$

respectively. Introducing (11.22) into (11.21) we obtain

$$g(y_n) = n\varsigma_n^2 \exp\left(-\frac{\varsigma_n^2}{2}\right)\left[1-\exp\left(-\frac{\varsigma_n^2}{2}\right)\right]^{n-1}, \tag{11.25}$$

where $\varsigma_n = \frac{y_n}{\sqrt{D_x}}$ is the sought dimensionless amplitude. Its maximum value could be determined from equation $g'(y_n) = 0$, which yields

$$\varsigma_n^2\left[n\exp\left(-\frac{\varsigma_n^2}{2}\right)-1\right]-\left[\exp\left(-\frac{\varsigma_n^2}{2}\right)-1\right]=0. \tag{11.26}$$

If the number n is large, the second term in this expression is small compared to the first one and can be omitted, so that

$$n\exp\left(-\frac{\varsigma_n^2}{2}\right)-1=0, \tag{11.27}$$

and

$$y_n = \varsigma_n\sqrt{D_x} = \sqrt{2D_x \ln n}. \tag{11.28}$$

As evident from this result, the ratio of the extreme response y_n, after n cycles are applied, to the maximum response $\sqrt{D_x}$, when a single cycle is applied, is $\sqrt{2\ln n}$. This ratio is 3.2552 for 200 cycles, 3.7169 for 1000 cycles, and 4.1273 for 5000 cycles. Coming back to the adequate heat sink example, we should indicate that the consideration of the effect of the number of cycles of random amplitude could considerably affect the reliability outcome and, hence, the design strategy.

11.5 THE TOTAL COST OF RELIABILITY COULD BE MINIMIZED: ELEMENTARY EXAMPLE

Let us show [26], using rather elementary reasoning, how the total cost of a product associated with reliability (dependability) on one hand and cost-effectiveness on the other could be minimized (Figure 11.2). The rationale behind such an attempt is based on the following reasoning. Assuming that one knows the physics of the possible failures and has a way to prevent them, they could either make the product highly reliable and, hence, highly costly, at the manufacturing stage or to make it reasonably reliable at this stage and wait until reparable failures will actually happen. Figure 11.1 indicates that the minimum total cost could be determined as a reasonable compromise between the "failure cost" (i.e., the cost of repairing the failed products) that decreases with an increase in the reliability level and the "quality-and-reliability cost" that increases with an increase in the reliability level. To quantify this anticipated effect, let us assume that the cost of achieving and improving reliability can be estimated based on the formula

$$C_R = C_R(0)\exp[r(R - R_0)], \tag{11.29}$$

where $R = MTTF$ is the actual level of the MTTF, R_0 is the specified MTTF level, $C_R(0)$ is the cost of achieving the R_0 level of reliability and r is the cost factor associated with reliability improvements. Similarly, let us assume that the failure cost, that is, the cost of reliability repair, can be also assessed by the formula

$$C_F = C_F(0)\exp[-f(R - R_0)], \tag{11.30}$$

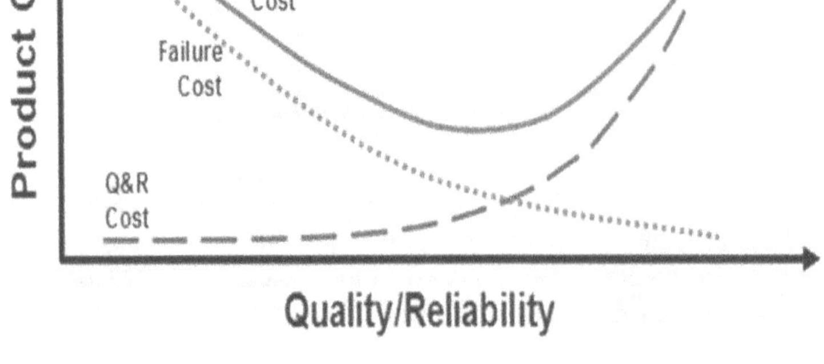

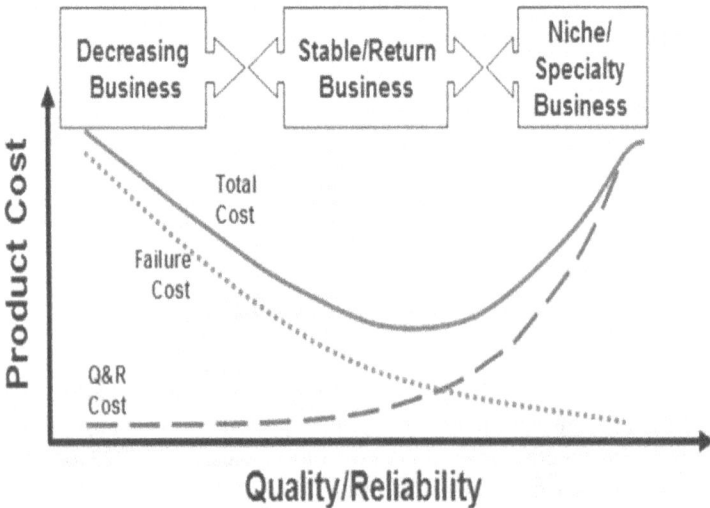

FIGURE 11.2 The total cost of reliability could be minimized.

where $C_F(0)$ is the cost of restoring the product's reliability, and f is the factor of the reliability restoration (repair) cost. The latter formula reflects a natural assumption that the cost of repair is considerably lower for a product of higher reliability. The total cost $C = C_R + C_F$ has its minimum

$$C_{\min} = C_R\left(1+\frac{r}{f}\right) = C_F\left(1+\frac{f}{r}\right), \tag{11.31}$$

when the minimization condition $rC_R = fC_F$ is fulfilled.

Let us further assume that the factor r of the reliability improvement cost is inversely proportional to the MTTF (dependability criterion), and the factor f of the reliability restoration cost is inversely proportional to the mean time to repair MTTR (reparability criterion). Then the minimum total cost is

$$C_{\min} = \frac{C_R}{K} = \frac{C_F}{1-K}, \tag{11.32}$$

where the steady-state availability K (i.e., the probability that the product is sound and is available to the user any time at the steady-state operations) is expressed as

$$K = \frac{1}{1+\dfrac{\prec t_r \succ}{\prec t_f \succ}} = \frac{1}{1+\dfrac{MTTR}{MTTF}}. \tag{11.33}$$

In this formula $\prec t_f \succ = MTTF$ is the mean TTF and $\prec t_r \succ = MTTR$ is the mean time to repair.

The result (11.32) obtained for the total minimum cost establishes, in an elementary way, the relationship between the minimum total cost of achieving and maintaining (restoring) the adequate reliability level and the availability criterion (see also Section 11.6 on the requirements for the restoration time, when reparable failure occurs). The obtained relationship quantifies the intuitively obvious fact that the total cost of the product depends on both the total cost and the availability of the product. The formula

$$\frac{C_F}{C_R} = \frac{1}{K} - 1 \tag{11.34}$$

that follows from (11.32) indicates that if the availability index K is high, the ratio of the cost of repairs to the cost aimed at improved reliability is low. When the availability index is low, this ratio is high. This reasoning can be used, particularly, to interpret the availability index from the cost-effectiveness point of view: the index $K = \dfrac{C_R}{C_{\min}}$ reflects, in effect, the ratio of the cost of improving reliability to the minimum total cost of the product associated with its reliability level. This and similar, even elementary, models can be of help, particularly when there is a need to minimize costs without compromising reliability, such as in various optimization analyses.

11.6 REQUIRED REPAIR TIME TO ASSURE THE SPECIFIED AVAILABILITY

11.6.1 Background/Incentive

One important characteristic of the operational reliability of a complex reparable system, such as air- or spacecraft, motor, railroad, or a maritime vehicle, long-haul communication system, military equipment, an electronic or a photonic product, medical instrumentation, and all kinds of robots, is its availability, which is defined as the probability that the system of importance is in sound condition and is therefore available to the user when needed. For reparable systems, this probability could be kept sufficiently high by assuring high level of the system's reparability characterized, first of all, after failure occurs, by the restoration time [27–36].

The availability level depends on the system's reparability ("serviceability"), which is the ease with which the system can be maintained and, when necessary, repaired. The analysis that follows is motivated by the need to have a simple, easy-to-use, and physically meaningful guideline on how short the repair time should be so that the system's availability would be maintained on the required high enough level. Accordingly, the objective of the analysis is to develop a methodology that enables quantifying, on the probabilistic basis, the role of the random time-to-repair (TTR) versus (also random) time-to-failure (TTF) of the system of importance. The methodology is not restricted to an electronic or a photonic system and is applicable to a wide variety of engineering systems. It is assumed, however, that the failure event always follows the Poisson's distribution and that the restoration time, an exponential distribution. The developed methodology enables assessing the required repair/restoration time, so that the availability is swiftly restored, thereby keeping it on the specified level during the time of operation. The suggested guidelines are presented in the form of Table 11.2 for the (time-dependent) availability function.

The following two inputs are considered: (1) the product of the anticipated failure rate of the system and the time of operation and (2) the ratio of the intensity of the restoration process to the MTTF. This "intensity" is simply a reciprocal to the mean time-to-repair (MTTR). The general concept is illustrated by a practical numerical example. Several extensions of this work are considered. One of them is the role of the human-system interaction ("human-in-the-loop"), when a system's reliability and human performance contribute jointly to the never-100%-failure-free operation process [14].

11.6.2 Analysis

Failures are rare events, and therefore the process of failures and restorations could be characterized by a constant failure rate λ (which is the failure rate at the steady-state portion of the bathtub curve), and the probability of occurrence of n failures during the time t follows the Poisson's distribution (see, for example, [21]):

$$P_n(t) = \frac{(\lambda t)^n}{n!} e^{-\lambda t}. \tag{11.35}$$

TABLE 11.2
Calculated Availability Function $K(t)$

λt	0	0.1	0.2	0.5	1.0	2.0	3.0	4.0	5.0	∞
μ/λ	X	X	X	X	X	X	X	X	X	X
0	1.0	0.9048	0.8187	0.6065	0.3679	0.1353	0.0498	0.0153	0.00674	0
0.2	1.0	0.9058	0.8222	0.6240	0.4177	0.2423	0.1894	0.1735	0.1687	0.1666
0.4	1.0	0.9067	0.8256	0.6404	0.4618	0.3291	0.2964	0.2860	0.2864	0.2857
0.6	1.0	0.9076	0.8288	0.6558	0.5012	0.4005	0.3801	0.3760	0.3752	0.3750
0.8	1.0	0.9085	0.8320	0.6702	0.5362	0.4596	0.4469	0.4448	0.4445	0.4444
1.0	1.0	0.9094	0.8352	0.6839	0.5677	0.5092	0.5012	0.5002	0.5000	0.5000
2.0	1.0	0.9136	0.8496	0.7410	0.7118	0.6675	0.6667	0.6666	0.6666	0.6666
3.0	1.0	0.9176	0.8623	0.7838	0.7546	0.7501	0.7500	0.7500	0.7500	0.7500
4.0	1.0	0.9213	0.8020	0.8164	0.8013	0.8000	0.8000	0.8000	0.8000	0.8000
5.0	1.0	0.9248	0.8835	0.8416	0.8337	0.8333	0.8333	0.8333	0.8333	0.8333
6.0	1.0	0.9281	0.8924	0.8615	0.8572	0.8571	0.8571	0.8571	0.8571	0.8571
10.0	1.0	0.9393	0.9192	0.9095	0.9090	0.9090	0.9090	0.9090	0.9090	0.9090
20.0	1.0	0.9582	0.9531	0.9524	0.9524	0.9524	0.9524	0.9524	0.9524	0.9524
50.0	1.0	0.9805	0.9804	0.9804	0.9804	0.9804	0.9804	0.9804	0.9804	0.9804
100.0	1.0	0.9901	0.9901	0.9901	0.9901	0.9901	0.9901	0.9901	0.9901	0.9901
1000.0	1.0	0.9990	0.9990	0.9990	0.9990	0.9990	0.9990	0.9990	0.9990	0.9990
10000.0	1.0	0.9999	0.9999	0.9999	0.9999	0.9999	0.9999	0.9999	0.9999	0.9999
∞						1.0000				

When restorations are carried out swiftly and the number of the necessary repairs reduces with an increase in their duration, it could be assumed that the restoration time t is an exponentially distributed random variable, so that its probability density distribution function is

$$f(t) = \mu e^{-\mu t}, \tag{11.36}$$

where the intensity $\mu = \frac{1}{\prec t \succ}$ of the restoration process is reciprocal to the mean time $\prec t \succ$ of the process. But how "swiftly" is swiftly enough to keep the system's availability as high as necessary?

Let $K(t)$ be the probability that the product is in the working condition and $k(t)$ is the probability that it is idle. When considering random processes with discrete states and continuous time, it is assumed that the transitions of the system S from the state s_i to the state s_j are defined by transitional probabilities λ_{ij}. If the governing flow of events is of Poisson's type, the random process is a Markovian process (see, for example, [21]). The probability of state $p_i(t) = P\{S(t) = s_i\}, i = 1,2,...,n$ of such a process, that is, the probability that the system S is in the state s_i at the moment t of time, can be found from the Kolmogorov equation (see, for example, [21])

$$\frac{dp_i(t)}{dt} = \sum_{j=1}^{n} \lambda_{jl} p_j(t) - p_i(t) \sum_{j=1}^{n} \lambda_{ij}, \quad i = 1,2,...,n \tag{11.37}$$

Applying this equation to processes (11.35) and (11.36), the following differential equations for the probabilities $K(t)$ and $k(t)$ can be obtained:

$$\frac{dK(t)}{dt} = \mu k(t) - \lambda K(t), \quad \frac{dk(t)}{dt} = \lambda K(t) - \mu k(t). \tag{11.38}$$

The probability normalization condition requires that the relationship $K(t) + k(t) = 1$ takes place for any moment t of time. Then, the probabilities $K(t)$ and $k(t)$ in equations (11.4) can be separated:

$$\frac{1}{\lambda+\mu}\frac{dK(t)}{dt} + K(t) = \frac{\mu}{\lambda+\mu}, \quad \frac{1}{\lambda+\mu}\frac{dk(t)}{dt} + k(t) = \frac{\lambda}{\lambda+\mu}. \tag{11.39}$$

These equations have the following solutions:

$$K(t) = C\exp[-(\lambda+\mu)t] + \frac{\mu}{\lambda+\mu}, \quad k(t) = C\exp[-(\lambda+\mu)t] + \frac{\lambda}{\lambda+\mu}. \tag{11.40}$$

The constant C of integration can be determined from the initial conditions depending on whether the system is in the working or in an idle condition at the initial moment of time. If it is in the working condition, the initial conditions $K(0) = 1$ and $k(0) = 0$ should be used, and then $C = \frac{\lambda}{\lambda+\mu}$. If the item is in the idle condition, then the initial conditions $K(0) = 0$ and $k(0) = 1$ should be applied, and the constant of integration is $C = \frac{\mu}{\lambda+\mu}$. Hence, the availability function is expressed as

$$K(t) = \frac{\mu}{\lambda+\mu} + \frac{\lambda}{\lambda+\mu}\exp[-(\lambda+\mu)t], \tag{11.41}$$

if the system is in the working condition at the initial moment of time, and as

$$K(t) = \frac{\mu}{\lambda+\mu} - \frac{\mu}{\lambda+\mu}\exp[-(\lambda+\mu)t], \tag{11.42}$$

if the system is in the idle condition at the initial moment of time. The constant (steady-state) part

$$K = \frac{\mu}{\lambda+\mu} = \frac{1}{1+\frac{\lambda}{\mu}} = \frac{1}{1+\frac{\langle t_r \rangle}{\langle t_f \rangle}} = \frac{1}{1+\frac{MTTR}{MTTF}} \tag{11.43}$$

of equations (11.41) and (11.42) is known as the availability index. This index determines the percentage of time in which the system is in workable (available) condition. In formula (11.43), $MTTF = \prec t_f \succ = \frac{1}{\lambda}$ is the mean time to failure, and $MTTR = \prec t_r \succ = \frac{1}{\mu}$ is the mean time to repair. If the system consists of many items, formula (11.43) can be generalized as follows:

$$K = \frac{1}{1 + \sum_{l=1}^{n} \frac{\prec t_r \succ_i}{\prec t_f \succ_i}}. \tag{11.44}$$

The relationship (11.41) is tabulated in Table 11.2 and plotted in Figure 11.3. When the failure rate λ is low (the MTTF is high) and the restoration rate μ is high (the MTTR is low), then the ratios $\frac{\mu}{\lambda}$ are significant. This leads to a high steady-state availability index and to a short transition time to it. When the ratio $\frac{\mu}{\lambda}$ is small (long MTTR and short MTTF), the transition period could be significant.

For zero $\frac{\mu}{\lambda}$ ratios (nonrepairable system: MTTR is infinitely long), formula (11.7) yields

$$K(t) = \exp[-(\lambda t)], \tag{11.45}$$

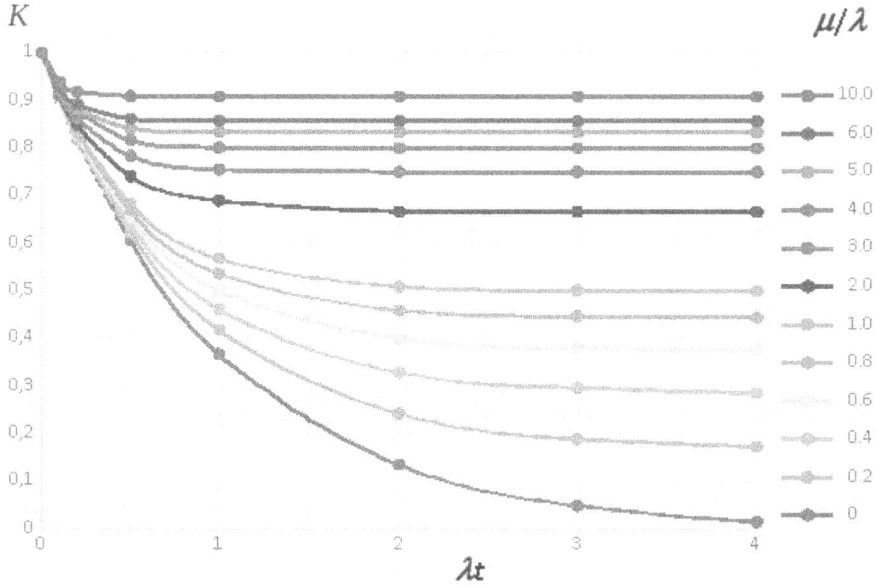

FIGURE 11.3 Availability function $K(t)$ versus the product λt of failure rate λ and time t for different restoration-to-failure-rate ratios μ/λ of the intensity $\mu = \frac{1}{\prec t \succ}$ of the restoration process (this intensity is reciprocal to the mean value $\prec t \succ$ of the restoration time) to the expected failure rate λ. The availability function $K(t)$ is higher and stabilizes faster for high μ/λ ratios, such as for large MTTF and low MTTR.

that is, the availability index is not different from the probability of nonfailure (dependability) of a nonrepairable item. Table 11.2 and Figure 11.3 quantify the role of the $\frac{\mu}{\lambda}$ ratio.

If one intends to determine the ratio $\bar{\mu} = \frac{\mu}{\lambda}$ for the given product λt of the failure rate and time and the availability function $K(t)$ from transcendental equation (11.41), then the following recurrent equation, based on Newton's formula, can be obtained:

$$\bar{\mu}_{n+1} = \bar{\mu}_n - (1+\bar{\mu}_n)\frac{\left[\bar{\mu}_n + \exp\left(-(1+\bar{\mu}_n)\lambda t\right)\right] - K(t)(1+\bar{\mu}_n)}{1-[1+(1+\bar{\mu}_n)\lambda t]\exp\left(-(1+\bar{\mu}_n)\lambda t\right)}, \quad n = 0,1,2,\ldots \tag{11.46}$$

For a steady-state situation, with large λt values, this formula can be simplified as

$$\bar{\mu}_{n+1} = (1+\bar{\mu}_n)^2 K(t) - \bar{\mu}_n^2. \tag{11.47}$$

Let the restoration time τ be a random variable distributed in accordance of the Rayleigh law:

$$f_\tau(\tau) = \frac{\tau}{\tau_0^2}\exp\left(-\frac{\tau^2}{2\tau_0^2}\right). \tag{11.48}$$

Here, τ_0 is the most likely restoration time, which is the maximum value of the function $f_\tau(\tau)$. The rationale behind the suitability of such a distribution is that the most likely restoration time should be short; that the probability of zero and very long restoration times should be zero; that the variable of interest (time) should always be positive, and that the distribution of this variable should be heavily skewed in the direction of short times. The Rayleigh law meets all these physically meaningful requirements. From distribution (11.48), we find that the probability that the random restoration time exceeds a certain level $\hat{\tau}$ can be evaluated as

$$P_\tau(\hat{\tau}) = P(\tau \succ \hat{\tau}) = \exp\left(-\frac{\hat{\tau}^2}{2\tau_0^2}\right). \tag{11.49}$$

Clearly, this probability should be low for low restoration times and high availability of the system.

Using equation (11.41), we obtain the following relationship between the level K_* of the availability function $K(t)$ and the probability $P(K_*)$ that this level is exceeded (reliability criterion):

$$K_* = \frac{1+\tau_* \exp\left(-\frac{1+\tau_*}{\tau_*}\lambda t\right)}{1+\tau_*}. \tag{11.50}$$

Here,

$$\tau_* = \lambda\tau_0\sqrt{-2\ln[P(K_*)]} \tag{11.51}$$

is the level of the corresponding dimensionless restoration time. The threshold τ_* changes from zero to infinity, and the availability threshold K_* and the probability $P(K_*)$ change from one to zero. Thus, high probabilities of nonfailure, as far as high enough availability is concerned, are characterized by low thresholds τ_* that should be kept low for high availability. Formula (11.51) indicates that lower products of the failure rate and the most likely TTR, and high $P(K_*)$ values keep the threshold τ_* of the TTF low.

If the product λt is significant (steady-state situation, long time t), formula (11.50) can be simplified:

$$K_* = \frac{1}{1+\tau_*}. \tag{11.52}$$

By putting $\frac{\mu}{\lambda} = \frac{1}{\tau_*}$ in Table 11.2 and in Figure 11.3, one can use this table and figure to assess the K_* level for the given failure rate λ, the most likely restoration time τ_0, the probability $P(K_*)$ and the time t in operation.

11.6.3 Numerical Example

Let, for example, the steady-state failure rate of the system defined by the bathtub curve be $\lambda = 10^{-3}$ 1/h, and the reduced, but still acceptable, level of the availability function in steady-state operations, because of the detected malfunction in the system, should not be below $K(t) = 0.95$. Table 11.2 and Figure 11.3 indicate that the corresponding steady-state $\bar{\mu} = \frac{\mu}{\lambda}$ ratio should be as high as about 20. This means that the corresponding MTTR should not exceed $MTTR = \frac{1}{\mu} = \frac{1}{\bar{\mu}\lambda} = \frac{1}{20\times 10^{-3}} = 50$ hours. A more accurate number for the $\bar{\mu} = \frac{\mu}{\lambda}$ ratio of the restoration-to-failure rates could be obtained using formula (11.13) and $\bar{\mu} = 20$ value as zero approximation:

$$\bar{\mu}_1 = (1+\bar{\mu}_0)^2 K(t) - \bar{\mu}_0^2 = 21^2 \times 0.95 - 20^2 = 18.95,$$

$$\bar{\mu}_2 = (1+\bar{\mu}_1)^2 K(t) - \bar{\mu}_1^2 = 19.95^2 \times 0.95 - 18.95^2 = 18.95 = 19.00,$$

$$\bar{\mu}_3 = (1+\bar{\mu}_2)^2 K(t) - \bar{\mu}_2^2 = 20.0^2 \times 0.95 - 19.00^2 = 19.00.$$

Then the more accurate prediction, as far as the required MTTR is concerned, is

$$MTTR = \frac{1}{\mu} = \frac{1}{\bar{\mu}\lambda} = \frac{1}{19\times 10^{-3}} = 52.6\text{ h}.$$

Using this number as the most likely MTTR τ_0 and formula (11.52) with $K_* = 0.95$, we conclude that the nondimensional threshold value (the lower the better) of the most likely MTTR should be

$$\tau_* = \frac{1}{K_*} - 1 = \frac{1}{0.95} - 1 = 0.052632$$

for high enough availability. Then, the corresponding probability $P(K_*)$ can be found as

$$P(K_*) = \exp\left[-\frac{1}{2}\left(\frac{\tau_*}{\lambda\tau_0}\right)^2\right] = \exp\left[-\frac{1}{2}\left(\frac{0.052632}{10^{-3}\times 19}\right)^2\right] = 0.02156,$$

and is not high at all. This should be attributed to the application of the Rayleigh-based distribution. The situation might be different with the other distributions, such as the Weibull distribution for the TTR.

11.6.4 Conclusion

The following conclusions are drawn from the analysis:

- A simple and easy-to-use methodology is suggested to quantify, on the probabilistic basis, the role of the random TTR in connection with the (also random) TTF to keep the availability of the system on the acceptable (and high enough) level.
- The methodology is not restricted to a particular repairable system and, in the authors' opinion, is applicable to a wide variety of engineering systems and applications.
- Several extensions of this work are considered, and particularly the role of the human–system interaction. Since it is assumed in this analysis that the failure event always follows the Poisson's distribution and the restoration time always follows an exponential distribution, possible deviations from these distributions and the effect of such deviations on the obtained data is considered as future work. Because in practical applications of the suggested methodology it might not be easy to determine the failure rate and the restoration time as the major input parameters, possible approaches of handling this circumstance within the framework of the conducted analysis are also considered as future work.

11.7 BURN-IN TESTING OF ELECTRONIC AND PHOTONIC PRODUCTS: TO BIT OR NOT TO BIT?

11.7.1 Background/Initiative

Burn-in testing (BIT) [10–13] is an accepted practice in electronic manufacturing for detecting and eliminating early failures ("freaks") in newly fabricated products prior to shipping the "healthy" ones that survived BIT to the customer(s). BIT can be based on temperature cycling, elevated temperatures, voltage, current, humidity, random vibrations, and/or, since the principle of superposition does not work in reliability engineering, on the appropriate combination of these stressors.

BIT is a costly undertaking: early failures are avoided and the infant mortality portion (IMP) of the bathtub curve (BTC) is supposedly eliminated at the expense of the reduced yield. But what is even worse is that the elevated BIT stressors might not only eliminate "freaks," but could cause permanent damage to the main population of the "healthy" products. This kind of testing should be therefore well understood, thoroughly planned, and carefully executed. This is not the current situation though.

It is even unclear whether BIT is always needed (to BIT or not to BIT is always a question?), nor to what extent the current BIT practices are adequate and effective. HALT that is currently employed as a BIT vehicle is, as has already been indicated, is a "black box" that tries "to kill many birds with one stone." HALT is therefore unable to provide any trustworthy information on what this testing does. It remains unclear what is actually happening during, and as a result of, the HALT-based BIT and how to effectively eliminate "freaks," while minimizing the testing time, reducing BIT cost and avoiding damaging the sound devices. When HALT is relied upon to do the BIT job, it is not even easy to determine whether there exists a decreasing failure rate with time at the initial, infant mortality portion, of the BTC. There is, therefore, an obvious incentive to develop ways, in which the BIT process could be better understood, trustworthily quantified, effectively monitored and possibly optimized. Accordingly, in the subsequent analysis, some important BIT aspects are addressed for a packaged electronic or a photonic product comprised of numerous mass-produced components. We intend to shed some quantitative light on the BIT process, and, since nothing is perfect (the difference between a highly reliable process or a product and an insufficiently reliable one is merely in the levels of their never-zero probability of failure, is it not?), such a quantification should be done on a probabilistic basis. Particularly, we intend to come up with a suitable criterion to answer the fundamental "to burn-in or not to burn-in" question, and, in addition, if BIT is decided upon, to find a way to quantify its outcome using our physically meaningful and flexible BAZ model.

11.7.2 Objective

In our analysis, the role and significance of the following important factors that affect the testing time and the stress level are addressed:

1. the random statistical failure rate (SFR) of mass-produced components that the product of interest is comprised of;
2. the way to assess, from the highly focused and highly cost-effective FOAT, the activation energy of the "freak" population more or less typical for the given manufacturing technology;
3. the role of the applied stressor(s); and, most importantly,
4. the probabilities of the "freak" failures depending on the duration of the BIT loading, and a way to assess, using the BAZ equation, these probabilities as functions of the duration and level of the BIT, as well as the variance of the random SFR of the mass-produced components that the product of interest is comprised of.

We intend to show that the BTC-based time-derivative of the failure rate at the initial moment of time (the beginning of the IMP portion) can be considered as a suitable criterion of whether BIT for a packaged IC device should be, or does not have to be, conducted. We intend to also show that this criterion is, in effect, the variance of the random SFR of the mass-produced components that the manufacturer of the given product received from numerous vendors. The commitments of these vendors to reliability of their mass-produced components are usually unknown, and therefore the random statistical failure rated (SFR) of these components might vary significantly, from zero to infinity. Based on the developed general formula for the nonrandom SFR of a product comprised of such components, the solution for the case of normally distributed random SFR of the constituent components has been obtained. This information enables answering the "to burn-in or not to burn-in" question in electronics manufacturing.

If BIT is decided upon, the BAZ model can be employed for the assessment of its required duration and level. Our analyses sheds light on the role and significance of important factors that affect the testing time and stress level: the random SFR of mass-produced components that the product of interest is comprised of; the way to assess, from the highly focused and highly cost-effective FOAT, the activation energy of the "freak" population; the role of the applied stressor(s); and, most importantly, the probabilities of the "freak" failures depending on the duration of the BIT effort. These factors should be considered when there is an intent to quantify and, eventually and hopefully, to optimize the BIT's procedure. This is addressed using two mutually complementary and independent analyses:

1. the analysis of the configuration of the infant mortality portion (IMP) of a BTC obtained for a more or less well-established manufacturing technology of interest; and
2. the analysis of the role of the random SFR of the mass-produced components that the product of interest is comprised of, as far as the effect of this SFR on the nonrandom initial SFR of the product is concerned.

11.7.3 Information Based on the Available BTC

The desirable steady-state portion of the BTC commences at the BIT's end as a result of the interaction of two major irreversible time-dependent processes: the "favorable" statistical SFR process that results in a decreasing failure rate with time, and the "unfavorable" physics-of-failure (PoF)–related process resulting in an increasing failure rate. The first process dominates at the IMP of the BTC and is considered here. The IMP of a typical BTC, the "reliability passport" of a mass-produced electronic product using a more or less well-established manufacturing technology, can be approximated as

$$\lambda(t)=\lambda_0+(\lambda_1-\lambda_0)\left(1-\frac{t}{t_1}\right)^{n_1},\quad 0\le t\le t_1. \tag{11.53}$$

Here, λ_0 is BTC's steady-state ordinate, λ_1 is its initial (highest) value at the beginning of the IMP, t_1 is the IMP duration, the exponent n_1 is $n_1 = \frac{\beta_1}{1-\beta_1}$, and β_1 is the fullness of the BTC's IMP. This fullness is defined as the ratio of the area below the BTC to the area $(\lambda_1 - \lambda_0)t_1$ of the corresponding rectangular. The exponent n_1 changes from zero to one, when the β_1 changes from zero to 0.5. The time derivative of the failure rate at the IMP's initial moment of time ($t = 0$) is

$$\lambda'(0) = \frac{\lambda_1 - \lambda_0}{t_1} \frac{\beta_1}{1-\beta_1}. \tag{11.54}$$

If this derivative is zero or next-to-zero, this means that the IMP of the BTC is parallel to the time axis (so that there is, in effect, no IMP at all), that no BIT is needed to eliminate this portion, and "not to burn-in" is the answer to the basic question: the initial value λ_1 of the BTC is not different from its steady-state λ_0 value. What is less obvious is that the same result takes place for $\frac{\beta_1}{t_1} = 0$.

This means that although the BIT *is* needed, the testing could be short and low level, because there are not too many "freaks" in the manufactured population and because, although these "freaks" exist, they are characterized by very low probabilities of nonfailure, so that the planned BIT process could be a next-to-an-instantaneous one. The maximum value of the fullness β_1 is $\beta_1 = 0.5$. This corresponds to the case when the IMP of the BTC is a straight line connecting the initial, λ_1, and the steady-state, λ_0, BTC ordinates. The derivative $\lambda'(0)$ is

$$\lambda'(0) = \frac{d\lambda(t)}{dt} = -\frac{\lambda_1 - \lambda_0}{t_1} \tag{11.55}$$

in this case, and this seems to be the case, when the BIT is mostly needed. It has been found [37] (see also Appendix A) that the expression

$$\lambda_{ST}(t) = \frac{\int_0^\infty \lambda \exp(-\lambda t) f(\lambda) d\lambda}{\int_0^\infty \exp(-\lambda t) f(\lambda) d\lambda} \tag{11.56}$$

for the nonrandom time dependent SFR can be obtained from the probability density distribution function $f(t)$ for the random SFR λ for the components obtained from the vendors. When this rate is normally distributed, such as

$$f(\lambda) = \frac{1}{\sqrt{2\pi D}} \exp\left(-\frac{(\lambda - \bar{\lambda})^2}{2D}\right), \tag{11.57}$$

formula (11.56) yields

$$\lambda_{ST}(t) = \sqrt{2D}\varphi[\tau(t)]. \tag{11.58}$$

The "time function" $\varphi[\tau(t)]$ in this formula depends on the dimensionless "physical" (effective) time

$$\tau = t\sqrt{\frac{D}{2}} - s, \tag{11.59}$$

where

$$s = \frac{\bar{\lambda}}{\sqrt{2D}} \tag{11.60}$$

value ("safety factor") can be interpreted as a measure of the degree of uncertainty of the random SFR. The time derivative with respect to the actual (real) time is

$$\lambda'_{ST}(t) \text{ is } \lambda'_{ST}(t) = \sqrt{2D}\frac{d\varphi[\tau(t)]}{dt} = \sqrt{2D}\frac{d\varphi}{d\tau}\frac{d\tau}{dt} = D\varphi'(\tau). \tag{11.61}$$

It can be shown that the derivative $\varphi'(\tau)$ at the initial moment of time ($t = 0$) is equal to -1.0, so that

$$\lambda'_{ST}(0) = \lambda'_1 = -D. \tag{11.62}$$

This result explains the physical meaning of this derivative: it is the variance (with a "minus" sign, of course) of the random SFR of the constituent components.

As to the use of the kinetic BAZ model (see Chapter 10) in the problem in question, this model suggests a simple, easy-to-use, highly flexible, and physically meaningful way to evaluate of the probability of failure of a material or a device after the given time in testing or operation at the given temperature and under the given stress or stressors. Using this model, the probability of nonfailure during the BIT can be sought as

$$P = \exp\left[-\gamma_t DI_* t \exp\left(-\frac{U_0 - \gamma_\sigma \sigma}{kT}\right)\right]. \tag{11.63}$$

Here, D is the variance of the random SFR of the mass-produced components, I is the measured/monitored signal (such as leakage current, whose agreed-upon high value I_* could be considered as an indication of failure; or an elevated electrical resistance, particularly suitable as an indication of failure of solder joint interconnections), t is time, σ is the "external" stressor, U_0 is the activation energy (unlike in the original BAZ model, this energy may or may not be affected by the level of the external stressor), T is the absolute temperature, γ_σ is the stress sensitivity factor for the applied stress and γ_t is the time/variance sensitivity factor.

The distribution (11.63) makes physical sense. Indeed, the probability P of nonfailure decreases with an increase in the variance D, the time t, the level I_* of the leakage current at failure and the temperature T, and increases with an increase in the activation energy U_0. As has been shown, the maxima of the entropy $H(P)$ and the probability P of nonfailure take place at the moment of time

$$t = \frac{1}{\gamma_t D I_*} \exp\left(\frac{U_0 - \gamma_\sigma \sigma}{kT}\right), \tag{11.64}$$

accepted in the BAZ model as the MTTF.

There are three unknowns in expression (11.64): the product $\rho = \gamma_t D$ of the time related stress-sensitivity factor γ_t and the variance D, the stress-sensitivity factor γ_σ and the activation energy U_0. These unknowns, as has been previously demonstrated, could be determined from a two-step FOAT.

At the first step, testing should be carried out for two temperatures, T_1 and T_2, but for the same effective activation energy $U = U_0 - \gamma_\sigma \sigma$. Then the relationships

$$P_{1,2} = \exp\left[-\rho I_* t_{1,2} \exp\left(-\frac{U_0 - \gamma_\sigma \sigma}{kT_{1,2}}\right)\right] \tag{11.65}$$

for the probabilities of nonfailure can be obtained. Here $t_{1,2}$ are the corresponding times, at which the failures have been detected, and I_* is the agreed upon leakage current value at failure. Since the numerator $U = U_0 - \gamma\sigma$ in relationships (11.65) is kept the same in the conducted tests, the amount $\rho = \gamma_t D$ can be found as

$$\rho = \exp\left[\frac{1}{\theta - 1}\left(\frac{n_2^\theta}{n_1}\right)\right], \tag{11.66}$$

where the notations

$$n_{1,2} = -\frac{\ln P_{1,2}}{I_* t_{1,2}}, \quad \theta = \frac{T_2}{T_1} \tag{11.67}$$

are used.

The second step of testing is aimed at the evaluation of the stress sensitivity factor γ_σ and should be conducted at two stress levels, σ_1 and σ_2 (say, temperatures or voltages). If the stresses σ_1 and σ_2 are thermal stresses determined for the temperatures T_1 and T_2, they could be evaluated using a suitable stress model (see, for example, [38, 39]). Then,

$$\lambda_\sigma = k \frac{T_1 \ln n_1 - T_2 \ln n_2 + (T_2 - T_1) \ln \rho}{\sigma_1 - \sigma_2}. \tag{11.68}$$

If, however, the external stress is not a thermal stress, then the temperatures at the second step tests should preferably be kept the same. Then the ρ value will not affect the factor γ_σ, which could be found as

$$\lambda_\sigma = \frac{kT}{\sigma_1 - \sigma_2} \ln\left(\frac{n_1}{n_2}\right), \tag{11.69}$$

Where T is the testing temperature. Finally, after the product ρ and the factor γ_σ are determined, the activation energy U_0 can be determined as

$$U_0 = -kT_1 \ln\left(\frac{n_1}{\rho}\right) + \gamma\sigma_1 = -kT_2 \ln\left(\frac{n_2}{\rho}\right) + \gamma\sigma_2. \tag{11.70}$$

The TTF can be obviously determined as

$$TTF = MTTF(-\ln P), \tag{11.71}$$

where the MTTF has been previously defined.

Let, for example, the following data were obtained at the first step of FOAT:

1. After $t_1 = 14\ h$ of testing at the temperature of $T_1 = 60°\text{C} = 333\ \text{K}, 90\%$ of the tested devices reached the critical level of the leakage current of $I_* = 3.5\ \mu A$ and, hence, failed, so that the recorded probability of nonfailure is $P_1 = 0.1$; the applied stress is elevated voltage $\sigma_1 = 380\ V$;
2. After $t_2 = 28\ h$ of testing at the temperature of $T_2 = 85°\text{C} = 358\ \text{K}, 95\%$ of the samples failed, so that the recorded probability of nonfailure is $P_2 = 0.05$. The applied stress is still elevated voltage $\sigma_1 = 380\ V$. Then the parameters $n_{1,2} = -\frac{\ln P_{1,2}}{I_* t_{1,2}}$ are

$$n_1 = -\frac{\ln P_1}{I_* t_1} = -\frac{\ln 0.1}{3.5 \times 14} = 4.6991 \times 10^{-2} \mu A^{-1} h^{-1};$$

$$n_2 = -\frac{\ln P_2}{I_* t_2} = -\frac{\ln 0.05}{3.5 \times 28} = 3.0569 \times 10^{-2} \mu A^{-1} h^{-1};$$

With the temperature ratio $\theta = \frac{T_2}{T_1} = \frac{358}{333} = 1.0751$, we have

$$\rho = \exp\left[\frac{1}{\theta - 1}\left(\frac{n_2^\theta}{n_1}\right)\right] = \exp\left[\frac{1}{0.0751}\left(\frac{0.030569^{1.0751}}{0.046991}\right)\right] = 785.3197 \mu A^{-1} h^{-1}.$$

At the second step of FOAT, one can use, without conducting additional testing, the information from the first step, its duration, and outcome, and let the second step of testing has shown that after $t_2 = 36h$ of testing at the same temperature of

$T = 60°C = 333K$, 98% of the tested samples failed, so that the predicted probability of nonfailure is $P_2 = 0.02$. If the stress σ_2 is the elevated voltage $\sigma_2 = 220V$, then the parameter n_2 becomes

$$n_2 = -\frac{\ln P_2}{I_* t_2} = -\frac{\ln 0.02}{3.5 \times 36} = 3.1048 \times 10^{-2} \mu A^{-1} h^{-1},$$

and the sensitivity factor γ_σ for the applied stress is

$$\gamma_\sigma = kT \frac{\ln\left(\frac{n_1}{n_2}\right)}{\sigma_1 - \sigma_2} = 8.61733 \times 10^{-5} \times 333 \frac{\ln\left(\frac{4.6991 \times 10^{-2}}{3.1048 \times 10^{-2}}\right)}{380 - 220}$$

$$= 4326 \times 10^{-5} eV \times V^{-1}.$$

The zero-stress activation energies calculated for these parameters n_1 and n_2, and the stresses σ_1 and σ_2 is

$$U_0 = -kT \ln\left(\frac{n_1}{\rho}\right) + \gamma_\sigma \sigma_1 =$$

$$= -8.61733 \times 10^{-5} \times 333 \ln\left(\frac{4.6991 \times 10^{-2}}{785.3197}\right) + 4326 \times 10^{-5} \times 380$$

$$= 0.2790 + 0.0282 = 0.3072 eV.$$

To make sure that there was no calculation error, the zero-stress activation energy can also be found as

$$U_0 = -kT \ln\left(\frac{n_2}{\rho}\right) + \gamma_\sigma \sigma_2 =$$

$$= -8.61733 \times 10^{-5} \times 333 \ln\left(\frac{3.1048 \times 10^{-2}}{785.3197}\right) + 4326 \times 10^{-5} \times 220$$

$$= 0.2909 + 0.0164 = 0.3072 eV.$$

No wonder that these values are considerably lower than the activation energies of "healthy" products. Many manufacturers consider as a sort of "rule of thumb" that the level of 0.7eV can be used as an appropriate tentative number for the activation energy of healthy electronic products.

In this connection, it should be indicated that when the BIT process is monitored and the activation energy U_0 is being continuously calculated based on the number of the failed devices, the BIT process should be terminated, when the calculations, based on the observed and recorded FOAT data, indicate that the stress-free activation energy U_0 starts to increase.

TABLE 11.3
TTF versus Probability-of-Nonfailure

P	0.0050	0.0075	0.0100	0.0500
TTF, h	85.745	79.183	74.528	48.481

The MTTF can be computed as

$$t = MTTF = \frac{1}{\rho I_*}\exp\left(\frac{U_0 - \gamma_\sigma \sigma}{kT}\right) = \frac{1}{785.3197 \times 3.5}\exp\left(\frac{0.3072 - 7.4326 \times 10^{-5}}{8.61733 \times 10^{-5} \times 333}\right)$$
$$= 16.1835h,$$

The TTF, however, depends on the probability of nonfailure. Its values calculated as $TTF = MTTF \times (\ln P)$ are shown in Table 11.3.

Clearly, the probabilities of nonfailure for successful BITs should be low enough. It is clear also that the BIT process should be terminated when the calculated probabilities of nonfailure and the activation energy U_0 start rapidly increasing. Although our BIT analyses do not suggest any straightforward and complete way of how to optimize BIT, they nonetheless shed useful and insightful light on the significance of some important factors that affect the BIT's need, and, if decided upon, its required time and stress level for a packaged product comprised of mass-produced components.

11.8 CONCLUSION

The application of FOAT, the PDfR concept, and particularly the multiparametric BAZ model enables one to improve dramatically the state of the art in the field of the electronic products reliability prediction and assurance. Since FOAT cannot do without simple, easy-to-use, and physically meaningful predictive modeling, the role of such modeling, both computer-aided and analytical (mathematical), is making the suggested new approach to QT practical and successful. It is imperative that the reliability physics that underlies the mechanisms and modes of failure is well understood. Such an understanding can be achieved only provided that flexible, powerful, and effective PDfR efforts are implemented.

APPENDIX A RELIABILITY OF AN ELECTRONIC PRODUCT COMPRISED OF MASS-PRODUCED COMPONENTS

A.1 SUMMARY

Two major irreversible processes, statistics-of-failure-related (SFR) and physics-of-failure-related (PFR), take place during the operation of an electronic product. The SFR process results in a decreasing failure rate with time, while the PFR process leads to an increasing failure rate. A way to quantify the favorable effect of the SFR process is suggested for a product comprised of mass-produced components. The SFR failure rate of the components the product is made of is assumed to be a random variable distributed between zero and infinity. Normal and Rayleigh distributions of this failure rate are considered and the corresponding constitutive equations are obtained. Calculations indicate that the probability distribution of the mass-produced components comprising the product of interest has a significant effect on its nonrandom SFR and that the favorable SFR effect may or may not have an appreciable impact on the total probability of nonfailure. The PFR probabilities of nonfailure become very low at the wear-out portion of the product's life, while the total probability of nonfailure is determined, for the given moment of time, as the product of the probabilities of nonfailure associated with both SFR and PFR processes. It is concluded that the favorable role of the decrease in the product's SFR, owing to the random nature of the SFR of its mass-produced components, should always be assessed in a thorough reliability analysis, and that such an assessment could be done on the basis of the developed methodology. An incentive for that might be significant for products not prone to extensive degradation, which suppresses, to a greater or lesser extent, the favorable SFR effect in question.

A.2 BACKGROUND/INCENTIVE

It is usually assumed that the governing FOAT model, such as Miner-Palmgren, Coffin–Manson, stress–strength interference, and the BAZ equation [A.1] (see also Chapter 10), is applicable also to the field conditions. One just has to consider the actual, much lower, operation loading (stress) level. This naturally leads to a significantly longer expected lifetime of the product. The more favorable loading is, however, not the only difference between the field and the accelerated test conditions. Another favorable effect has to do with the significantly larger number of mass-produced components that the actual product of interest is comprised of. In the analysis that follows, a physically meaningful and trustworthy way to account for this circumstance is suggested, that is, a way to assess the effect of the mass-produced components comprising a product on its propensity to failure. The analysis is based on the major assumption that the nonrandom SFR of the product can be evaluated from the known (assumed or determined) random SFR failure rate of its mass-produced components fabricated by different and numerous vendors, and manufactured using different technologies. Normal (Gaussian) or Rayleigh distributions of the SFR failure of the constituent components have been considered.

A.3 ANALYSIS

A.3.1 Analytical Bathtub Curve (Diagram)

The typical bathtub curve (BTC), the "reliability passport" of a mass-produced product (Figure A.1 [A.2]) can be approximated as follows:

$$\lambda(t) = \begin{cases} \lambda_0 + (\lambda_1 - \lambda_0)\left(1 - \dfrac{t}{t_1}\right)^{n_1} = \lambda_0\left[1 + \left(\dfrac{\lambda_1}{\lambda_0} - 1\right)\left(1 - \dfrac{t}{t_1}\right)^{n_1}\right], 0 \le t \le t_1 \\ \lambda_0 + (\lambda_2 - \lambda_0)\left(\dfrac{t - t_1}{t_2}\right)^{n_2} = \lambda_0\left[1 + \left(\dfrac{\lambda_2}{\lambda_0} - 1\right)\left(\dfrac{t}{t_2} - \dfrac{t_1}{t_2}\right)^{n_2}\right], t_1 \le t \le t_1 + t_2 \end{cases} \tag{A.1}$$

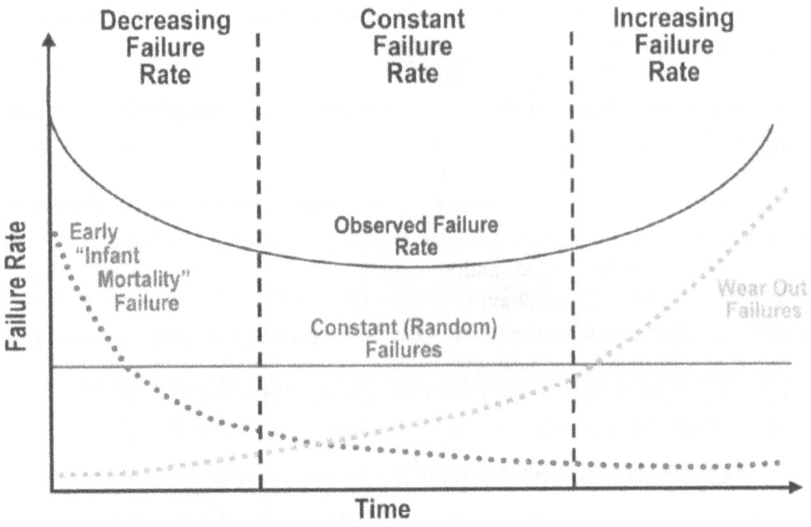

FIGURE A.1 Typical BTC "reliability passport" of an electronic technology. (From E. Wyrwas, L. Condra, A. Hava, "Accurate Quantitative Physics-of-Failure Approach to Integrated Circuit Reliability," IPC APEX EXPO Technical Conference, 2011.)

Here, $\lambda(t)$ is time-dependent failure rate, λ_0 is its "steady-state" minimum, λ_1 is its initial (high) value at the beginning of the infant mortality portion, t_1 is the duration of this portion, λ_2 is the final value of the failure rate at the end of the wear-out portion, t_2 is the duration of this portion, and the exponents n_1 and n_2 are expressed through the fullnesses β_1 and β_2 of the BTC infant-mortality and the wear-out portions as $n_{1,2} = \frac{\beta_{1,2}}{1-\beta_{1,2}}$. These fullnesses are defined as the ratios of the areas above the BTC to the areas $(\lambda_1 - \lambda_0)t_1$ and $(\lambda_2 - \lambda_0)t_2$ of the corresponding rectangular. The exponents n_1 and n_2 change from 1 to infinity, when the fullnesses β_1 and β_2 change from 0.5 (triangle) to 1 (rectangular). The lowest $\lambda(t)$ values can be achieved in the case of the largest β_1 and β_2 (or n_1 and n_2) values.

A.3.2 Statistical Failure Rate

Let a product manufacturer receive the components for this product from n independent vendors. The components are made by different and numerous vendors, using different technologies and different reliability levels for the components. The probability of nonfailure of the product, assuming that the exponential law of reliability is applicable, can be sought as

$$P(t) = \sum_{k=1}^{n} p_k \exp(-\lambda_k t), \tag{A.2}$$

where p_k is the component fraction received from the kth vendor, $\exp(-\lambda_k t)$ is the probability of nonfailure of the kth component, λ_k is its (random) failure rate, and t is time. The sum in formula (A.2) can be substituted, for a large number n of mass-produced components, by the integral

$$P(t) = \int_0^\infty \exp(-\lambda t)\, dF(\lambda) = \int_0^\infty \exp(-\lambda t) \frac{dF(\lambda)}{d\lambda} d\lambda = \int_0^\infty \exp(-\lambda t) f(\lambda) d\lambda, \quad \text{(A.3)}$$

where $F(\lambda)$ is the cumulative probability distribution function and $f(\lambda) = \frac{dF(\lambda)}{d\lambda}$ is the probability density distribution function of the continuously distributed random failure rate λ of the components. The nonrandom SFR of the product can be determined as the ratio of the current rate $\frac{dN_f(t)}{dt}$ of the number $N_f(t)$ of products that failed by the time t to the number $N_s(t)$ of products that still remained sound by this time:

$$\lambda_{ST}(t) = \frac{\frac{dN_f(t)}{dt}}{N_s(t)}. \quad \text{(A.4)}$$

Substituting the number of the sound items with the probability $P(t)$ of nonfailure and the number of the failed items with the probability $Q(t) = 1 - P(t)$ of failure, this formula can be written as

$$\lambda_{ST}(t) = \frac{\frac{d[1 - P(t)]}{dt}}{P(t)} = -\frac{1}{P(t)} \frac{dP(t)}{dt}, \quad \text{(A.5)}$$

or, considering (A.3), as

$$\lambda_{ST}(t) = \frac{\int_0^\infty \lambda \exp(-\lambda t) f(\lambda) d\lambda}{\int_0^\infty \exp(-\lambda t) f(\lambda) d\lambda}. \quad \text{(A.6)}$$

A.3.3 The Case When Random SFR is Normally Distributed

Using the probability density distribution function

$$f(\lambda) = \frac{1}{\sqrt{2\pi D}} \exp\left(-\frac{(\lambda - \bar{\lambda})^2}{2D}\right) \quad \text{(A.7)}$$

for the random failure rate of the component received from an arbitrary vendor (here $\bar{\lambda}$ is the mean value of the random failure rate λ, and D is its variance), introducing (A.7) into formula (A.6) and using [A.3] information, the following expression for the non-random SFR of the product can be obtained:

$$\lambda_{ST}(t)=\frac{\int_0^\infty \lambda \exp\left(-\frac{(\lambda-\bar{\lambda})^2}{2D}-t\lambda\right)d\lambda}{\int_0^\infty \exp\left(-\frac{(\lambda-\bar{\lambda})^2}{2D}-t\lambda\right)d\lambda}=\sqrt{2D}\varphi(\xi). \tag{A.8}$$

The function

$$\varphi(\xi)=-\xi+\frac{1}{\bar{\Phi}(\xi)} \tag{A.9}$$

(see Table A.1) depends on the dimensionless time

$$\xi=\frac{Dt-\bar{\lambda}}{\sqrt{2D}} \tag{A.10}$$

and so do the auxiliary function

$$\bar{\Phi}(\xi)=\sqrt{\pi}\exp(\xi^2)[1-\Phi(\xi)]\approx\frac{1}{\xi}\left[1+\sum_{k=1}^{\infty}(-1)^k\frac{1x3x...(2k-1)}{2^k\xi^{2k}}\right]\approx$$
$$\approx\frac{1}{\xi}\left(1-\frac{1}{2\xi^2}+\frac{3}{4\xi^4}-\frac{15}{8\xi^6}+\frac{105}{16\xi^8}...\right) \tag{A.11}$$

and the probability integral (Laplace function)

$$\Phi(\xi)=\frac{2}{\sqrt{\pi}}\int_0^\xi \exp\left(-\eta^2\right)d\eta. \tag{A.12}$$

TABLE A.1
The Governing Function $\varphi(\xi)$ of the Effective Dimensionless Time ξ

ξ	**–3.0**	**–2.5**	**–2.0**	**–1.5**	**–1.0**	**–0.5**	**–0.25**	**0**	**0.25**
$\varphi(\xi)$	3.0000	2.5005	2.0052	1.5302	1.1126	0.7890	0.6652	0.5642	0.4824
ξ	**0.5**	**1.0**	**1.5**	**2.0**	**2.5**	**3.0**	**3.5**	**4.0**	**4.5**
$\varphi(\xi)$	0.4163	0.3194	0.2541	0.2080	0.1618	0.1456	0.1300	0.1166	0.1053
ξ	**5.0**	**6.0**	**7.0**	**8.0**	**9.0**	**10.0**	**11.0**	**12.0**	**13.0**
$\varphi(\xi)$	0.0958	0.0809	0.0699	0.0615	0.0549	0.0495	0.0451	0.0414	0.0391
ξ	**15.0**	**20.0**	**30.0**	**50.0**	**100.0**	**200.0**	**500.0**	**1000.0**	**1500.0**
$\varphi(\xi)$	0.0332	0.0249	0.0166	0.0100	0.0050	0.0025	0.0010	0	0

Formula (A.9) indicates that the "physical" (effective) time ξ depends not only on the "chronological" (actual) time t, but also on the mean and variance of the distribution of the components' random SFR, and that the rate of changing of the dimensionless effective time ξ with the change in the actual time t is $\frac{d\xi}{dt}=\sqrt{\frac{D}{2}}$: the effective time ξ changes the faster the larger the standard deviation $\sqrt{D}$ of the random SFR is.

Expansion (A.11) can be used for large ξ values, exceeding, say, 2.5, and has been employed, when calculating the Table A.1 data for the function $\varphi(\xi)$. The function $\bar{\Phi}(\xi)$ changes from infinity to zero, when the effective time ξ changes from $-\infty$ to ∞. For effective times ξ below –2.5 the function $\bar{\Phi}(\xi)$ is large, the second term in (A.9) becomes small compared to the first term, and the function $\varphi(\xi)$ coincides with the time ξ itself, with an opposite sign though.

At the initial moment of time ($t = 0$), Formulas (A.10), (A.11), and (A.8) yield

$$\xi=-\frac{\bar{\lambda}}{\sqrt{2D}},\quad \bar{\Phi}(\xi)=\sqrt{\pi}\exp\left(\frac{\bar{\lambda}^2}{2D}\right)\left[1+\Phi\left(\frac{\bar{\lambda}}{\sqrt{2D}}\right)\right],\quad \lambda_{ST}=\bar{\lambda}+\sqrt{\frac{2D}{\pi}}\frac{\exp\left(-\frac{\bar{\lambda}^2}{2D}\right)}{1+\Phi\left(\frac{\bar{\lambda}}{\sqrt{2D}}\right)} \tag{A.13}$$

With the product's initial SFR value $\lambda_{ST}=\lambda_1$ (the degradation failure rate λ_{DG} is obviously zero at the initial moment of time), the last formula in (A.13) yields

$$\frac{\lambda_1}{\sqrt{2D}}=\frac{\bar{\lambda}}{\sqrt{2D}}+\frac{1}{\sqrt{\pi}}\frac{\exp\left(-\frac{\bar{\lambda}^2}{2D}\right)}{1+\Phi\left(\frac{\bar{\lambda}}{\sqrt{2D}}\right)}. \tag{A.14}$$

This relationship, tabulated in Table A.2, indicates that the effective initial value $\lambda_1/\sqrt{2D}$ of the BTC can be established based on the mean value and the variance of the distribution (A.2).

When the ratio $\frac{\bar{\lambda}}{\sqrt{2D}}$ increases from zero to infinity, the ratio $\frac{\lambda_1}{\sqrt{2D}}$ increases from $\frac{1}{\sqrt{\pi}}=0.5642$ to infinity. The initial failure rate can be put equal to its mean

TABLE A.2
The Initial SFR versus Its Mean Value

$\bar{\lambda}/\sqrt{2D}$	0	0.5000	1.0000	1.5000	2.0000	2.5000	2.7500	2.8000	2.9000	3.0000
$\lambda_1/\sqrt{2D}$	0.5642	0.7890	1.1126	1.5302	2.0052	2.5005	2.7501	2.8001	2.9001	3.0000

value, if the ratio $\frac{\bar{\lambda}}{\sqrt{2D}}$ exceeds 2.5. This is usually indeed the case in an actual situation, since the accepted normal distribution, when applied to a random variable that cannot be negative, should be characterized by a significant ratio of its mean value to the standard deviation, so that the negative values of such a distribution are meaningless and do not contribute appreciably to the sought information.

The illustrative numerical example in Table A.3 is carried out with the following input data: mean (initial) value factor of the SFR of the component population: $\frac{\bar{\lambda}}{\sqrt{2D}} = \frac{\lambda_1}{\sqrt{2D}} = 3.0$; standard deviation of the random failure rate of the component population: $\sqrt{D} = 2x10^{-4}h^{-1}$; initial nonrandom failure rate of the product: $\lambda_1 = 8.4853 \times 10^{-4}h^{-1}$; the lowest failure rate of the product: $\lambda_0 = 9.6000 \times 10^{-4}h^{-1}$; the highest (allowable) failure rate of the product: $\lambda_2 = 19.8 \times 10^{-4}$; the duration of the infant mortality portion for the product: $t_1 = 48h$; the duration of the wear-out portion of the product's BTC: $t_2 = 39{,}952h$ (obtained as the difference between the total time of operation of $40{,}000h$ and the duration of the infant mortality portion); the "fullness" of the infant mortality portion of the product's BTC:$\beta_1 = 0.8$ ($n_1 = 4$); the "fullness" of its wear-out portion: $\beta_2 = 0.75$ ($n_2 = 3$). Assuming that the SFR and PFR are statistically independent, the PFR are calculated, for each moment of time, as the difference between the BTC ordinates obtained using formulas (A.1) and the predicted SFR $\lambda_{ST}(t)$. The PFR probabilities P_{DG} of nonfailure are calculated based on formula (A.2). The SFR probabilities P_{ST}^* of nonfailure are calculated using the

TABLE A.3
Calculated Probabilities of Nonfailure in the Case of the Normally Distributed Failure Rate of the Mass-Produced Components

1	t,h	**0**	**48**	**100**	**250**	**1000**	**5000**
2	$\xi = t\sqrt{D/2} - \bar{\lambda}/\sqrt{2D}$	−3.000	−2.993	−2.986	−2.965	−2.859	−2.293
3	$\varphi(\xi) = -\xi + \frac{1}{\bar{\Phi}(\xi)}$	3.000	2.993	2.986	2.965	2.859	2.347
4	$\lambda_{ST}(t) = \sqrt{2D}\varphi(\xi)x10^5h^{-1}$	84.85	84.67	84.45	83.862	80.864	64.937
5	$P_{ST} = \exp(-\lambda_{ST}t)$	1.000	0.960	0.919	0.8109	0.4455	0.0389
6	$P_{ST}^* = \exp(-\lambda_{ST}^*t)$	1.000	0.960	0.919	0.8089	0.4280	0.0144
7	P_{DG}	1.000	0.9998	0.9995	0.9993	0.9942	0.0767
8	$P = P_{DG}P_{ST}$	1.000	0.9598	0.9185	0.8099	0.4429	2.984E-3
1	t,h	**7500**	**10000**	**20000**	**25000**	**30000**	**40000**
2	$\xi = t\sqrt{D/2} - \bar{\lambda}/\sqrt{2D}$	−1.9394	−1.5858	−0.1716	0.5355	1.2426	2.6568
3	$\varphi(\xi) = -\xi + \frac{1}{\bar{\Phi}(\xi)}$	1.9460	1.6089	0.6319	0.4073	0.2822	0.1567
4	$\lambda_{ST}(t) = \sqrt{2D}\varphi(\xi)x10^5h^{-1}$	55.041	45.506	17.874	11.521	7.982	4.4327
5	$P_{ST} = \exp(-\lambda_{ST}t)$	0.0161	0.0106	0.0280	0.0561	0.0912	0.1698
6	$P_{ST}^* = \exp(-\lambda_{ST}^*t)$	1.72E-3	2.10E-4	4.26E-8	6.13E-10	8.80E-12	1.82E-15
7	P_{DG}	1.911E-3	2.37E-4	5.62E-8	8.66E-10	1.33E-11	3.16E-15
8	$P = P_{DG}P_{ST}$	3.077E-5	2.512E-6	1.574E-9	4.86E-11	1.21E-12	5.37E-16

initial (time-independent) SFR. The calculated data indicate that the statistical probability of nonfailure (based on the nonrandom SFR of the product and the exponential law of reliability) decreases with time at the rather significant portion of the product's life despite the decrease in its SFR.

At some moment of time (beginning, in the carried out example, after about 10,000 hours in operation), the effect of the decreasing SFR starts to prevail, and the SFR related probability of nonfailure begins to increase with time. This circumstance does not play, however, an important role here, because the degradation (reliability physics) related failure rates become significant and suppress the increase in the probability of nonfailure associated with the decreased SFR. It should be pointed out, however, that although in the carried out example, the increase in the SFR-related probability of nonfailure turned out to be small, the situation might be quite different with other input data, and particularly for highly physically reliable materials and products.

It is noteworthy that for short times at the beginning of the infant mortality (burn-in) process, the function $\bar{\Phi}(\xi)$ is significant, and the second term in formula (A.4) is small compared to the first term. In such a situation, the linear formula $\lambda_{ST} = \lambda_1 - Dt$ can be used to evaluate the SFR. Indeed, this simplified formula predicts the SFR at the end of the infant mortality time as $\lambda_{ST} = 84.834 \times 10^{-5} h^{-1}$. The exact number $\lambda_{ST} = 84.668 \times 10^{-5} h^{-1}$ is only 0.2% lower.

A.3.4 The Case When Random SFR is Distributed in Accordance with the Rayleigh Law

Assuming that the components' failure rate is distributed in accordance with the Rayleigh law

$$f(\lambda) = \frac{\lambda}{D} \exp\left(-\frac{\lambda^2}{2D}\right) \tag{A.15}$$

Formula (A.8) yields

$$\lambda_{ST}(t) = \frac{\int_0^\infty \lambda^2 \exp\left(-\frac{\lambda^2}{2D} - t\lambda\right) d\lambda}{\int_0^\infty \lambda \exp\left(-\frac{\lambda^2}{2D} - t\lambda\right) d\lambda} = \sqrt{\frac{\pi D}{2}} \varphi(\xi). \tag{A.16}$$

Here,

$$\varphi(\xi) = \frac{2}{\sqrt{\pi}} \frac{\left(\frac{1}{2} + \xi^2\right)\bar{\Phi}(\xi) - \xi}{1 - \xi\bar{\Phi}(\xi)} \tag{A.17}$$

is a function of the dimensionless time

$$\xi = t\sqrt{\frac{D}{2}}, \tag{A.18}$$

and so are the auxiliary function

$$\bar{\Phi}(\xi) = \sqrt{\pi}\exp(\xi^2)[1-\Phi(\xi)] \approx \frac{1}{\xi}\left[1+\sum_{k=1}^{\infty}(-1)^k \frac{1\times 3\times ...(2k-1)}{2^k \xi^{2k}}\right]$$
$$\approx \frac{1}{\xi}\left(1-\frac{1}{2\xi^2}+\frac{3}{4\xi^4}-\frac{15}{8\xi^6}+\frac{105}{16\xi^8}...\right) \tag{A.19}$$

and the probability integral $\Phi(\xi)$ is defined by formula (A.12). The calculations using the same input data as in the case of the normally distributed failure rate of mass-produced components are carried out in Table A.4.

TABLE A.4
Calculated Probabilities of Nonfailure in the Case of the SFR Distributed in Accordance with the Rayleigh Law

1 t, h	**0**	**48**	**100**	**250**	**1000**	**5000**
2 $\xi = t\sqrt{\frac{D}{2}}$	0	0.6788	1.4142	3.53	14.1	70.7
3 $\varphi(\xi) = \frac{2}{\sqrt{\pi}}\frac{\left(\frac{1}{2}+\xi^2\right)\bar{\Phi}(\xi)-\xi}{1-\xi\bar{\Phi}(\xi)}$	1.000	0.9998	0.9995	0.9988	0.9951	0.9780
4 $\lambda_{ST}(t) = \sqrt{\frac{\pi D}{2}}\varphi(\xi)$	25.07	25.06	25.05	25.036	24.943	24.513
5 $P_{ST} = \exp(-\lambda_{ST}t)$	1.000	0.9880	0.9753	0.9393	0.7792	0.2936
6 P_{DG}	1.000	0.9998	0.9995	0.9993	0.9942	0.0767
7 $P = P_{DG}P_{ST}$	1.000	0.9878	0.9748	0.9386	0.7747	0.0225
1 t, h	**7500**	**10000**	**20000**	**25000**	**30000**	**40000**
2 $\xi = t\sqrt{\frac{D}{2}}$	106.1	141.4	282.8	353.6	424.3	567.7
3 $\varphi(\xi) = \frac{2}{\sqrt{\pi}}\frac{\left(\frac{1}{2}+\xi^2\right)\bar{\Phi}(\xi)-\xi}{1-\xi\bar{\Phi}(\xi)}$	0.9695	0.9369	0.8763	0.8563	0.8258	0.7784
4 $\lambda_{ST}(t) = \sqrt{\frac{\pi D}{2}}\varphi(\xi)$	24.301	23.484	21.965	21.464	20.699	19.511
5 $P_{ST} = \exp(-\lambda_{ST}t)$	0.1616	0.0955	1.24E-2	4.67E-3	2.01E-3	0.
6 P_{DG}	1.911E-3	2.37E-4	5.62E-8	8.66E-10	1.33E-11	3.16E-15
7 $P = P_{DG}P_{ST}$	3.088E-5	2.263E-5	6.969E-10	4.044E-12	2.673E-15	0

TABLE A.5
Calculated Probabilities of Nonfailure for the Normal and Rayleigh Distributions of the Random Failure Rates of an Electronic Product Components

t, h	**0**	**48**	**100**	**250**	**1000**	**5000**
P_N	1.000	0.9598	0.9185	0.8099	0.4429	2.984E-3
P_R	1.000	0.9878	0.9748	0.9386	0.7747	0.0225
t, h	**7500**	**10000**	**20000**	**25000**	**30000**	**40000**
P_N	3.077E-5	2.512E-6	1.574E-9	4.86E-11	1.21E-12	5.37E-16
P_R	3.088E-5	2.263E-5	6.969E-10	4.044E-12	2.673E-15	0

Probabilities of nonfailure computed using normal and Rayleigh distributions of the random SFR are summarized in Table A.5. The normal distribution of the random SFR failure rates results, for the major portion of the BTC, in appreciably lower probabilities P_N of nonfailure than the probabilities P_R of nonfailure computed on the basis of the Rayleigh law, and should be preferred therefore in approximate and conservative engineering analyses, unless more reliable SFR information becomes available.

A.3.5 Probability of Nonfailure

The second segment of a typical bathtub curve for an electronic or an optoelectronic product is known to be steady state. The failure rate is assumed to be more or less constant at this segment, and this is confirmed by numerous experiments. It has been established also that the steady-state segment of the BTC is characterized primarily by instantaneous random failures. Their occurrence could be adequately described by the one-parametric exponential law of reliability:

$$P(t) = \exp(-t / \tau_1) = \exp(-\lambda_1 t). \tag{A.20}$$

This law establishes the probability of nonfailure of a product at the moment t of operation. In formula (A.20), τ_1 is the MTTF, and $\lambda_1 = \frac{1}{\tau_1}$ is the steady-state failure rate. The MTTF can be defined as the moment of time $t = \tau_1$, when the entropy $H(P) = -P \ln P$ of the distribution (A.20) reaches its maximum value $H_{\max} = \frac{1}{e} = 0.3679$. The wear-out portion of the BTC, on the other hand, is characterized primarily by continuous and accelerated physical degradation (aging) of the product.

The wear-out failures are described by the two-parametric normal distribution

$$P(t) = \frac{1 - \Phi\left(\frac{t - \tau_2}{\sqrt{2D_\sigma}}\right)}{1 + \Phi\left(\frac{\tau_2}{\sqrt{2D_\sigma}}\right)}. \tag{A.21}$$

Here, $\tau_2 = \tau_2(t)$ is the MTTF, D_σ is the variance of the applied load σ, and

$$\Phi(\alpha) = \frac{2}{\sqrt{\pi}} \int_0^\alpha \exp(-\xi^2)\, d\xi \tag{A.22}$$

is the Laplace function. Clearly, the log-normal law of reliability can also be used in this case. The MTTF τ_1, when the failures are instantaneous and random, is typically assumed to be time independent, while the MTTF $\tau_2 = \tau_2(t)$, when the failures are caused by the degradation process, is time dependent and decreases, when time progresses. Generally, both types of failures could possibly occur in a particular product at any stage of its operation. The only difference is the likelihood of a particular failure mode: aging related failures are less likely at the steady-state period of the product's operation, while instantaneous failures are less likely at the wear-out portion of the BTC. When both modes could possibly occur and could be assumed statistically independent, the probability of nonfailure can be determined as

$$P(t) = \exp\left(-\frac{t}{\tau}\right) \frac{1-\Phi\left(\frac{t-\tau}{\sqrt{2D_\sigma}}\right)}{1+\Phi\left(\frac{\tau}{\sqrt{2D_\sigma}}\right)} \tag{A.23}$$

A.4 CONCLUSIONS

When conducting accelerated testing and predicting on its basis the operational/field reliability of an electronic product comprised of mass-produced components, one should consider not only less severe loading conditions in the field, but also that many more devices are operated in the field than the number of tested specimens. This circumstance also has a favorable effect on the field probability of nonfailure, and this analysis shows how this could be considered in advance.

The product's nonrandom SFR is dependent on the probability distribution of the random SFR failure rate of its mass-produced components. Future work should therefore include statistical analysis of the most realistic distributions of the random failure rates for the most vulnerable mass-produced components obtained by the particular manufacturer from its major vendors.

REFERENCES

1. J.T. Duane, et al., "Accelerated Test Methods for Reliability Prediction," *IEEE Transactions Aerospace*, vol. 2, 1964.
2. W.B. Nelson, *Accelerated Testing: Statistical Models, Test Plans, and Data Analysis*, John Wiley & Sons, 1990.
3. E. Suhir, A. Bensoussan, J. Nicolics, L. Bechou, "Highly Accelerated Life Testing (HALT), Failure Oriented Accelerated Testing (FOAT), and Their Role in Making a Viable Device into a Reliable Product," IEEE Aerospace Conference, Big Sky, Montana, Mar. 2014.

4. E. Suhir, R. Mahajan, "Are Current Qualification Practices Adequate?" Circuit Assembly, Apr. 2011.
5. E. Suhir, S. Yi, "Accelerated Testing and Predicted Useful Lifetime of Medical Electronics," IMAPS Conference on Advanced Packaging for Medical Electronics, San Diego, California, Jan. 2018.
6. E. Suhir, "Quantifying the Unquantifiable in Electronics and Aerospace Engineering: Review", Journal of Aerospace Engineering and Mechanics, 2020, in print.
7. E. Suhir, R. Mahajan, A. Lucero, L. Bechou, "Probabilistic Design for Reliability (PDfR) and a Novel Approach to Qualification Testing (QT)," 2012 IEEE/AIAA Aerospace Conference, Big Sky, Montana, 2012.
8. E. Suhir, "Could Electronics Reliability Be Predicted, Quantified and Assured?" *Microelectronics Reliability*, No. 53, Apr. 2013.
9. E. Suhir, "Aerospace Electronics Reliability Must Be Quantified to Be Assured: Application of the Probabilistic Design for Reliability Concept", *International Journal of Aeronautical Science and Aerospace Research*, 2020, in print.
10. E. Suhir, "Burn-in: When, For How Long and at What Level?" *Chip Scale Reviews*, Oct. 2019.
11. E. Suhir, "To Burn-In, or Not to Burn-in: That's the Question," *Aerospace*, vol. 6, No. 3, 2019.
12. E. Suhir, "Is Burn-in Always Needed?" *International Journal of Advanced Research in Electrical, Electronics and Instrumentation Engineering (IJAREEIE)*, vol. 9, No. 1, Jan. 2020.
13. E. Suhir, "For How Long Should Burn-in Testing Last?" *Journal of Electrical & Electronic Systems (JEES)*, vol. 2, 2020.
14. E. Suhir, *Human-in-the-Loop: Probabilistic Modeling of an Aerospace Mission Outcome*, CRC Press, 2018.
15. E. Suhir, "Probabilistic Design for Reliability," *Chip Scale Reviews*, vol. 14, No. 6, 2010.
16. E. Suhir, "When Reliability is Imperative, Ability to Quantify It is a Must," IMAPS Advanced Microelectronics, Aug. 2012.
17. E. Suhir, "Electronics Reliability Cannot Be Assured, if it is not Quantified," *Chip Scale Reviews*, Mar.–Apr. 2014.
18. E. Suhir, "Probabilistic Design for Reliability of Electronic Materials, Assemblies, Packages and Systems: Attributes, Challenges, Pitfalls," Plenary Lecture, MMCTSE, Cambridge, UK, Feb. 2017.
19. E. Suhir, "The Outcome of an Engineering Undertaking of Importance Must Be Quantified to Assure its Success and Safety: Review," *Reliability Engineering and System Safety Journal*, 2020, submitted.
20. E. Suhir, "Could Electronics Reliability Be Predicted, Quantified and Assured?" *Microelectronics Reliability*, No. 53, Apr. 2013.
21. E. Suhir, *Applied Probability for Engineers and Scientists*, McGraw-Hill, 1997.
22. E. Suhir, "Remaining Useful Lifetime (RUL): Probabilistic Predictive Model," *International Journal of Prognostics and Health Management*, vol. 2, No. 2, 2011.
23. E. Suhir, "Survivability of Species in Different Habitats: Application of Multi-Parametric Boltzmann-Arrhenius-Zhurkov Equation," Acta Astronautica, 2020.
24. E. Suhir, Y. Raskin, A. Tunik, "Russian Strength Standards for Commercial Ships," American Bureau of Shipping, 1982.
25. E. Suhir, B. Poborets, "Solder Glass Attachment in Cerdip/Cerquad Packages: Thermally Induced Stresses and Mechanical Reliability," Procedures of the 40th ECTC, Las Vegas, Nevada, May 1990.
26. E. Suhir, L. Bechou, "Availability Index and Minimized Reliability Cost," Circuit Assemblies, Feb. 2013.

28. D. Kececioglu, *Reliability Engineering Handbook, Vol. 2,* Prentice Hall, 1991.
29. E. Elsayed, *Reliability Engineering*, Addison Wesley, 1996.
30. W.Q. Meeker, L.A. Escobar, *Statistical Methods for Reliability Data*, Wiley and Sons, 1998.
31. U.S. Department of Defense, *MIL-HDBK-338B, Electronic Reliability Design Handbook,* Oct. 1, 1998.
32. W.R. Blischke, D.N. Prabhakar Murthy, *Reliability Modeling, Prediction, and Optimization*, Wiley and Sons, 2000.
33. U.S. Department of Defense, *Guide for Achieving Reliability, Availability, and Maintainability*, Sept. 11, 2011.
34. U.S. Department of Defense, *MIL-HDBK-189C, Reliability Growth Management,* Jun. 2011.
35. M. Grimwade, *Managing Operational Risk: New Insights and Lessons Learnt*, The Institute of Operational Risk, 2016.
36. P. McConnell, *Systemic Operational Risk: Theory, Case Studies and Regulation*, Risk Books, 2016.
37. E. Suhir, "Statistics- and Reliability-Physics-Related Failure Processes," *Modern Physics Letters B (MPLB)*, vol. 28, No. 13, 2014.
38. E. Suhir, "Thermal Stress Failures in Electronics and Photonics: Physics, Modeling. Prevention," *Journal of Thermal Stresses*, vol. 36, No. 6, Jun. 2013.
39. E. Suhir, "Analytical Thermal Stress Modeling in Electronic and Photonic Systems," *ASME Applied Mechanics Reviews*, invited paper, vol. 62, No. 4, 2009.

APPENDIX REFERENCES

A.1. E. Suhir, A. Bensoussan, "Application of Multi-Parametric BAZ Model in Aerospace Optoelectronics," IEEE Aerospace Conference, Big Sky, Montana, Mar. 2014.
A.2. E. Wyrwas, L. Condra, A. Hava, "Accurate Quantitative Physics-of-Failure Approach to Integrated Circuit Reliability," IPC APEX EXPO Technical Conference, 2011.
A.3. I.S. Gradshteyn, I.M. Ryzhik, *Tables of Integrals, Series, and Products*, Academic Press, 1980.

12 Fiber Optics Systems and Reliability of Solder Materials

"Say not 'I have found the truth,' but rather, 'I have found a truth."

—Kahlil Gibran, Lebanese American artist, poet, and writer

12.1 BACKGROUND/OBJECTIVE

In this chapter, it is shown how methods of analytical modeling in structural analyses of fiber optics systems can be effectively employed to evaluate stresses in, and provide recommendations for, the rational structural design of these systems. The emphasis is on stress analyses and reliability of solder materials—the reliability bottleneck in optical engineering. This undertaking actually forms a new direction—Fiber Optics Structural Analysis (FOSA) [1–8]. All the findings based on analytical modeling were confirmed, for the provided numerical examples, by finite element analysis (FEA) (see, for example, [9]). The developed analytical FOSA models can be or, actually, have been used to assess the reliability of the fiber optics materials and structures (see, for example, [10–13]). The Bell Labs researchers involved in reliability engineering of microelectronic and photonic systems used to say that "you could have the best electronic or a photonic chip in the world, but if you put it on a piece of junk that is called a substrate or a submount, you end up with a piece of junk." Therefore, it was argued [14] that the selection and evaluation of the materials and structures in the emerging technologies in electronics and photonics engineering is critical: nothing happens to an electron or a photon per se, but it is the methods and approaches of materials science, structural analyses, and reliability engineering that deal with the "pieces of junk" that are needed to assure the failure-free performance of the electronic and photonic materials, structures, and systems of the future.

The contents of this chapter are mostly based on the author's research conducted at Bell Labs, Physical Sciences and Engineering Research Division, Murray Hill, New Jersey, USA, during his approx. 20-year tenure with this company and, to a lesser extent, on his recent work in the field. In the past, we have addressed coated fibers [15–34]; low-temperature micro-bending of fibers intended for undersea communication systems [35–40]; mechanical behavior of optical fiber interconnects treated as flexible beams [41–57]; accelerated testing ("proof-testing") of optical fibers [58–65]; free vibrations of fused bi-conical taper lightwave couplers (FBTLCs) [66–68]; strain-free planar optical waveguides [69]; fibers soldered into ferrules [70]; apparatus and method for thermostatic compensation of temperature sensitive, mostly optical and fiber optics–based, devices [71]; fiber “pigtails” bent on cylindrical surfaces [72, 73]; elastic stability of glass fibers in a micro-machined fiber-optic switch [74]; application of the probabilistic approach in thermal stress modeling [75] and in the interpretation of the accelerated testing results [76]; the attributes of aerospace optoelectronics [77, 78]; application of the Boltzmann–Arrhenius–Zhurkov (BAZ) model [79] and use of nanotechnologies to provide moisture resistant nanoparticle based material for an overclad in fiber optics [80–85] as well as elevated interfacial compliance in micro- and optoelectronic systems [86]. In these publications, optical fibers subjected to thermal and/or mechanical loading (stresses) in bending, tension, compression, or to combinations of such loadings were addressed, and the considered structural elements included optical fibers of finite length (bare, jacketed and dual-coated fibers); fibers experiencing thermal loading; the roles of geometric and material nonlinearity; dynamic response of fiber systems to shocks and vibrations; and possible applications of nanomaterials in new generations of fiber optics coating and cladding systems. The objective of this chapter is to address, using the probabilistic-design-for-reliability (PDfR) concept and BAZ model, to quantify, on a probabilistic basis, the likelihood of operational failures of solder materials employed in fiber optics [87–95].

12.2 FIBER OPTICS STRUCTURAL ANALYSIS (FOSA) IN FIBER OPTICS ENGINEERING: ROLE AND ATTRIBUTES

The discipline of FOSA employs methods and approaches of reliability physics and structural analysis to evaluate stresses, strains, and displacements in fiber optics structures, to carry out physical (mechanical) design of these structures, and assess and assure short- and long-term reliability of fiber optics products. FOSA treats these products as structures, in which the materials interaction, their size and configuration, and the applied loads are as important as the properties and characteristics of the employed materials. As such, FOSA naturally complements fiber optics materials and optical fiber communication sciences.

The application of methods and approaches of FOSA enables one to design, fabricate, and operate viable and reliable fiber optics products. Like other branches of Structural Analysis (civil, aircraft, space, maritime, automotive), FOSA considers the specifics, associated with the properties of the materials used, typical structures employed, and the nature, magnitude and variability of the applied loads. Typical structures in fiber optics engineering are bare or composite (coated) rods and beams of various lengths and flexural rigidities. These rods and beams could be soldered into ferrules, adhesively bonded into capillaries, or embedded into various materials and media. Typical materials are, of course, silica glasses, but also polymers (coatings, adhesives, and even polymer light-guides) and solders, “hard” (e.g., gold-tin) or “soft” (e.g., silver-tin). Typical loads include internal thermal loads caused by dissimilar materials in the structure and/or by temperature gradients, and high- or low temperature environments (temperature extremes); external (“mechanical”) loads due to the inevitable or imposed deformations; possible dynamic loads due to shocks, vibrations and/or acoustic noise. High voltage, elevated electric current, ionizing radiation and/or extensive light output from a powerful laser source should also be considered, in accordance with the multiparametric BAZ equation, as loads (stressors, stimuli).

FOSA pursues, but is not limited to, the following major objectives:

1. Determine (and, to an extent possible, idealize, for the sake of the theoretical analysis) the most likely loading conditions;
2. Evaluate the stresses, strains, displacements, and, when methods and approaches of fracture mechanics are pursued, also fracture characteristics of the fiber optics materials and structures;
3. Assure, typically on a probabilistic basis, that the acceptable strength and reliability criteria will remain, during the lifetime of the optoelectronic product, within the limits that are allowable from the standpoint of its structural integrity, elastic stability, dependability, availability, and normal operation, both physical (structural) and functional (optical). While an optical engineer is and should be concerned, first of all, with the functional (optical) performance of an optoelectronic product, an adequate performance of such a product cannot be sustained and assured if the product's ability to withstand elevated stresses (physical reliability) and to exhibit adequate environmental durability (ability to withstand degradation and aging) is not taken care of.

12.3 FIBERS SOLDERED INTO FERRULES

Solder materials and joints are as important in photonics and, particularly, in fiber optics, as they are in microelectronics. There are, however, specific requirements for the solder materials and joints used in photonics: ability to achieve high alignment, requirement for a very low creep, and so on. Thermally induced stresses in optical fibers soldered into various ferrules were addressed in [70] (Figure 12.1). It has been shown that low expansion enclosures with good thermal expansion (contraction) match with the silica fiber is not always the right choice from the standpoint of the thermally induced stresses in the metalized optical fibers soldered into ferrules, as well as, and first of all, from the standpoint of the stresses in the solder material itself. Indeed, the low expansion enclosures result, at low temperatures, in tensile radial stresses in the solder ring, and could lead to the delamination of the metallization from the fiber and/or to the excessive radial deformations in the solder material.

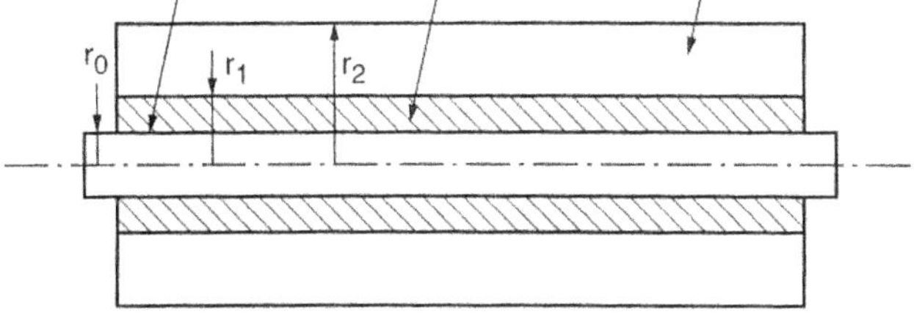

FIGURE 12.1 Optical glass fiber soldered into a ferrule.

On the other hand, high expansion (contraction) enclosures might result in high compressive stresses in the solder material, and, because of that, in low-cycle fatigue conditions during temperature cycling of the joint. The most feasible material of the enclosure can be found based on the developed model for the given thickness of the solder ring and solder material.

12.4 THERMAL STRESSES IN A CYLINDRICAL SOLDERED TRIMATERIAL BODY WITH APPLICATION TO OPTICAL FIBERS

12.4.1 Background/Incentive

Ability to predict thermal stress failures in electronics and photonics is of obvious importance. Numerous predictive models for the evaluation of thermal stresses in cylindrical bi-and trimaterial bodies were developed during the last decade in application to dual-coated or jacketed optical fibers. Despite the success in understanding the physics of thermal stresses in various electronics and photonics assembles, there still exists an incentive for the development of practically useful and physically meaningful engineering predictive models for various particular applications, such as silicon photonics. This technology offers numerous novel solutions in different areas of optical communications, optical computer interconnects, sensing, bio-applications, and so on. It has been recognized, however, that the reliability of materials in silicon-photonics structures is not always satisfactory. Accordingly, the objective of the analysis that follows is to develop a simple engineering model for the evaluation of the thermal stresses in trimaterial cylindrical bodies of finite length, with application to silicon photonics and other optical fiber–based technologies.

12.4.2 Analysis

The radial displacements in a circular ring subjected to the internal, p_0, and to external, p_1, pressures applied to its inner and the outer boundaries of the radii r_0 and r_1, can be evaluated, using plane strain approximation, by formula [18]

$$u(r) = -\frac{1+\nu}{E}\frac{(1-2\nu)(p_1r_1^2 - p_0r_0^2)r + r_0^2r_1^2(p_1 - p_0)\frac{1}{r}}{r_1^2 - r_0^2}. \tag{12.1}$$

Here, r is the current radius, and E and ν are the elastic constants of the material. This formula enables one to seek the interfacial radial displacements in a trimaterial body subjected to the change ΔT in temperature as follows:

$$u_0 = \left[-\alpha_0(1+\nu_0)\Delta T + \frac{(1+\nu_0)(1-2\nu_0)}{E_0}\sigma_0\right]r_0, \tag{12.2}$$

$$u_{10} = \left[-\alpha_1(1+\nu_1)\Delta T - \frac{1+\nu_1}{E_1}\frac{[1+(1-2\nu_1)\gamma_0^2]\sigma_0 - 2(1-\nu_1)\sigma_1}{1-\gamma_0^2}\right]r_0, \tag{12.3}$$

$$u_{12} = \left[-\alpha_1(1+\nu_1)\Delta T - \frac{1+\nu_1}{E_1} \frac{2(1-\nu_1)\gamma_0^2\sigma_0 - (\gamma_0^2+1-2\nu_1)\sigma_1}{1-\gamma_0^2} \right] r_1, \quad (12.4)$$

$$u_2 = \left[-\alpha_2(1+\nu_2)\Delta T - \frac{1+\nu_2}{E_2} \frac{1+(1-2\nu_2)\gamma_1^2}{1-\gamma_1^2} \right] r_1. \quad (12.5)$$

Here, $E_i, i = 0,1,2$, are Young's moduli of the materials, $\nu_i, i = 0,1,2$, are their Poisson's ratios, $\alpha_i, i = 0,1,2$, are the materials CTEs, $r_i, i = 0,1,2$, are the outer radii of the cylinders, $\gamma_i = \frac{r_i}{r_{i+1}}, i = 0,1$, are the radii ratios, u_0 are the radial interfacial displacements of the zero ("core") component, u_{10} are the displacements of intermediate component #1 ("bond") at its interface with the zero component, u_{12} are the displacements of component #1 at its interface with outer component #2, u_2 are the displacements of component #2 at its interface with component #1, σ_0 is the stress at the zero interface (between the zero and component #1), and σ_1 is the stress at the interface #1 (between components #1 and #2). The stresses σ_0 and σ_1 are considered positive if they are tensile, that is, if they are directed inward at the inner boundary of the intermediate material, and outward at its outer boundary. The conditions $u_0 = u_{10}$, $u_2 = u_{12}$ of the displacement compatibility result in equations:

$$-\delta_{11}\sigma_0 + \delta_{12}\sigma_1 = \Delta_1, -\delta_{21}\sigma_0 + \delta_{22}\sigma_1 = \Delta_2, \quad (12.6)$$

where

$$\delta_{11} = \frac{(1+\nu_0)(1-2\nu_0)}{E_0} + \frac{1+\nu_1}{E_1} \frac{1+(1-2\nu_1)\gamma_0^2}{1-\gamma_0^2}, \quad \delta_{12} = 2\frac{1-\nu_1^2}{E_1} \frac{1}{1-\gamma_0^2},$$

$$\delta_{22} = \frac{1+\nu_1}{E_1} \frac{\gamma_0^2+1-2\nu_1}{1-\gamma_0^2} + \frac{1+\nu_2}{E_2} \frac{1+(1-2\nu_1)\gamma_1^2}{1-\gamma_1^2}, \quad \delta_{21} = 2\frac{1-\nu_1^2}{E_1} \frac{\gamma_0^2}{1-\gamma_0^2},$$

$$\Delta_1 = [\alpha_1(1+\nu_1) - \alpha_0(1+\nu_0)]\Delta T, \quad \Delta_2 = [\alpha_1(1+\nu_1) - \alpha_2(1+\nu_2)]\Delta T. \quad (12.7)$$

From (12.6), we obtain

$$\sigma_0 = \frac{\delta_{12}\Delta_2 - \delta_{22}\Delta_1}{D}, \quad \sigma_1 = \frac{\delta_{11}\Delta_2 - \delta_{21}\Delta_1}{D}, \quad (12.8)$$

where $D = \delta_{11}\delta_{22} - \delta_{12}\delta_{21}$ is the determinant of equations (12.6). The radial, σ_r, and the tangential (circumferential), σ_t, normal stresses are equal to σ_0 throughout the zero component and are expressed as

$$\sigma_r^{(1)} = \frac{1}{1-\gamma_0^2} \left[-\gamma_0^2\sigma_0 + \sigma_1 + (\sigma_0 - \sigma_1)\frac{r_0^2}{r^2} \right],$$

$$\sigma_t^{(1)} = \frac{1}{1-\gamma_0^2} \left[-\gamma_0^2\sigma_0 + \sigma_1 - (\sigma_0 - \sigma_1)\frac{r_0^2}{r^2} \right] \quad (12.9)$$

for the intermediate component #1 (bond). At the outer surface of component #2, the radial normal stress is zero, and the tangential stress is

$$\sigma_t^{(2)} = -2\frac{\gamma_1^2}{1-\gamma_1^2}\sigma_1. \tag{12.10}$$

This stress fades away rapidly with an increase in the component's thickness. When the ratio of the outer radius of outer component #2 to its inner radius is equal to 5, the tangential stress at the outer surface, where the radial stress is zero, is only 8.33% of the tangential stress at the component's inner surface. The axial forces, $F_i, i = 0,1,2$, can be determined based on the condition that the thermally induced strain

$$\varepsilon = -\alpha_i(1+\nu_i)\Delta T + \lambda_i F_i, i = 0,1,2 \tag{12.11}$$

is the same for all the components of the body. In this equation, λ_i are the axial compliances of the components. These compliances can be evaluated by formulas

$$\lambda_0 = \frac{1+\nu_0}{\pi E_0 r_0^2}, \quad \lambda_1 = \frac{1+\nu_1}{\pi E_1 r_1^2}\frac{1}{1-\gamma_0^2}, \quad \lambda_2 = \frac{1+\nu_2}{\pi E_2 r_2^2}\frac{1}{1-\gamma_1^2}. \tag{12.12}$$

From condition (12.11), we have:

$$F_i = \frac{\varepsilon + \alpha_i(1+\nu_i)\Delta T}{\lambda_i}, \quad i = 0,1,2 \tag{12.13}$$

The equilibrium condition $\sum_{i=0}^{2} F_i = 0$ results in the following formula for the induced strain: $\varepsilon = -\alpha_e \Delta T$, where

$$\alpha_e = \frac{\sum_{l=0}^{2}\frac{\alpha_i(1+\nu_i)}{\lambda_i}}{\sum_{l=0}^{2}\frac{1}{\lambda_i}} \tag{12.14}$$

is the effective coefficient of thermal expansion (CTE) of the body. Formula (12.13) yields

$$F_i = \frac{\alpha_i(1+\nu_i) - \alpha_e}{\lambda_i}\Delta T, \quad i = 0,1,2, \tag{12.15}$$

so that

$$
\begin{aligned}
F_0 &= \frac{\alpha_0(1+\nu_0)(\lambda_1+\lambda_2)-\alpha_1(1+\nu_1)\lambda_2-\alpha_2(1+\nu_2)\lambda_1}{\lambda_0\lambda_1+\lambda_1\lambda_2+\lambda_2\lambda_0}\Delta T,\\
F_1 &= \frac{\alpha_1(1+\nu_1)(\lambda_0+\lambda_2)-\alpha_0(1+\nu_0)\lambda_2-\alpha_2(1+\nu_2)\lambda_0}{\lambda_0\lambda_1+\lambda_1\lambda_2+\lambda_2\lambda_0}\Delta T,\\
F_2 &= \frac{\alpha_2(1+\nu_2)(\lambda_0+\lambda_1)-\alpha_0(1+\nu_0)\lambda_1-\alpha_1(1+\nu_1)\lambda_0}{\lambda_0\lambda_1+\lambda_1\lambda_2+\lambda_2\lambda_0}\Delta T.
\end{aligned}
\tag{12.16}
$$

The axial stresses can be found as

$$
\begin{aligned}
\sigma_z^{(0)} &= \frac{F_0}{\pi r_0^2}+\nu_0\left(\sigma_r^{(0)}+\sigma_t^{(0)}\right)=\frac{F_0}{\pi r_0^2}+2\nu_0\sigma_0\\
\sigma_z^{(1)} &= \frac{F_1}{\pi(r_1^2-r_0^2)}+\nu_1\left(\sigma_r^{(1)}+\sigma_t^{(1)}\right)=\frac{1}{1-\gamma_0^2}\left[\frac{F_1}{\pi}+2\nu_1(\sigma_1-\gamma_0^2\sigma_0)\right]\\
\sigma_z^{(2)} &= \frac{F_2}{\pi(r_2^2-r_1^2)}+\nu_2\left(\sigma_r^{(2)}+\sigma_t^{(2)}\right)=\frac{1}{1-\gamma_1^2}\left(\frac{F_2}{\pi}-2\nu_2\gamma_1^2\sigma_1\right)
\end{aligned}
\tag{12.17}
$$

No appreciable longitudinal interfacial shearing stresses occur in the midportion of the body. These stresses might be significant, however, at its end portions. We use the concept of interfacial compliance (see Chapter 2) to address these stresses. In accordance with this concept, the interfacial longitudinal displacements of the body components can be sought as

$$
\begin{aligned}
w_0(z) &= -\alpha_0(1+\nu_0)\Delta Tz+\lambda_0\int_0^z F_0(\varsigma)d\varsigma-\kappa_0\tau_0(z),\\
w_{10}(z) &= -\alpha_1(1+\nu_1)\Delta Tz-\lambda_1\int_0^z [F_0(\varsigma)+F_2(\varsigma)]d\varsigma+\kappa_{10}\tau_0(z),\\
w_{12}(z) &= -\alpha_1(1+\nu_1)\Delta Tz-\lambda_1\int_0^z [F_0(\varsigma)+F_2(\varsigma)]d\varsigma+\kappa_{12}\tau_1(z)\\
w_2(z) &= -\alpha_2(1+\nu_2)\Delta Tz+\lambda_2\int_0^z F_2(\varsigma)d\varsigma-\kappa_2\tau_1(z).
\end{aligned}
\tag{12.18}
$$

Here, $w_0(z)$ are the longitudinal interfacial displacements of the inner (zero) component at its interface with intermediate component #1; $w_{10}(z)$ are the displacements of component #1 at its interface with the zero component; $w_{12}(z)$ are the displacements of component #1 at its interface with outer component #2; $w_2(z)$ are the displacements of outer component #2 at its interface with intermediate component #1; α_0, α_1 and α_2 are the CTE's of the materials; λ_0, λ_1 and λ_2 are the axial compliances of the components; ΔT is the change in temperature; $F_0(z)$, $F_1(z) = -[F_0(z) + F_2(z)]$ and $F_2(z)$ are the axial forces; κ_0 is the longitudinal interfacial compliance of the zero component, κ_{10} is the compliance of component #1 at its interface with the zero component, κ_{10} is the compliance of component #1 at its interface with outer component #2, κ_2 is the compliance of component #2 at its interface with component #1; and $\tau_0(z)$ and $\tau_1(z)$ are the interfacial shearing stresses. They are related to the axial forces $F_0(z)$ and $F_2(z)$ as follows:

$$F_0(z) = \int_{-l}^{z} \tau_0(\varsigma)d\varsigma, \quad F_2(z) = \int_{-l}^{z} \tau_1(\varsigma)d\varsigma. \tag{12.19}$$

Here, l is half the body's length. The origin of the longitudinal z axis is at the body's mid-cross section. The longitudinal interfacial compliances can be evaluated by formulas [18]

$$\kappa_0 = \frac{r_0}{E_0}, \quad \kappa_{10} = -\frac{r_0}{G_1}\ln\gamma_0, \quad \kappa_{12} = -\frac{r_1}{G_1}\ln\gamma_0,$$

$$\kappa_2 = \frac{r_1}{4E_2}\frac{(1+3\gamma_1)(1-\gamma_1) + (1+\nu_2)\gamma_1 \ln\dfrac{1+\gamma_1}{2\gamma_1}}{1-\gamma_1^2}. \tag{12.20}$$

Here, $G_1 = \dfrac{E_1}{2(1+\nu_1)}$ is the shear modulus of the intermediate material. The conditions $w_0(z) = w_{10}(z)$, $w_2(z) = w_{12}(z)$ of the compatibility of the longitudinal interfacial displacements result in the following equations for the interfacial shearing stresses $\tau_0(z)$ and $\tau_1(z)$:

$$(\kappa_0 + \kappa_{10})\tau_0(z) - (\lambda_0 + \lambda_1)\int_0^z F_0(\varsigma)d\varsigma - \lambda_1 \int_0^z F_2(\varsigma)d\varsigma$$

$$= [\alpha_1(1+\nu_1) - \alpha_0(1+\nu_0)]\Delta T z$$

$$(\kappa_2 + \kappa_{12})\tau_1(z) - (\lambda_1 + \lambda_2)\int_0^z F_2(\varsigma)d\varsigma - \lambda_1 \int_0^z F_0(\varsigma)d\varsigma$$

$$= [\alpha_1(1+\nu_1) - \alpha_2(1+\nu_2)]\Delta T z \tag{12.21}$$

Differentiating these equations twice with respect to the coordinate z and considering relationships (12.19), we obtain the following equations for the shearing stress functions $\tau_0(z)$ and $\tau_1(z)$:

$$(\kappa_0+\kappa_{10})\tau_0''(z)-(\lambda_0+\lambda_1)\tau_0(z)-\lambda_1\tau_1(z)=0,$$
$$(\kappa_2+\kappa_{12})\tau_1''(z)-\lambda_1\tau_0(z)-(\lambda_1+\lambda_2)\tau_1(z)=0. \tag{12.22}$$

The shearing stresses are antisymmetric with respect to the mid-cross section of the body and could be sought as

$$\tau_0(z)=C_0\sinh kx,\quad \tau_0(z)=C_1\sinh kx. \tag{12.23}$$

Introducing these solutions into equations (22), we obtain the following two homogeneous equations:

$$[(\kappa_0+\kappa_{10})k^2-(\lambda_0+\lambda_1)]C_0-\lambda_1C_1=0,$$
$$-\lambda_1C_0+[(\kappa_2+\kappa_{12})k^2-(\lambda_1+\lambda_2)]C_0=0. \tag{12.24}$$

The requirement that the determinant of these equations should be equal to zero for nonzero constants C_0 and C_1 results in the following equation for the parameter k of the interfacial shearing stress:

$$k^4-(k_{10}^2+k_{12}^2)k^2+\lambda k_{10}^2k_{12}^2=0, \tag{12.25}$$

where

$$k_{10}=\sqrt{\frac{\lambda_0+\lambda_1}{\kappa_0+\kappa_{10}}},\quad k_{12}=\sqrt{\frac{\lambda_2+\lambda_1}{\kappa_2+\kappa_{12}}},\quad \lambda=\frac{\lambda_0\lambda_1+\lambda_1\lambda_2+\lambda_0\lambda_2}{(\lambda_0+\lambda_1)(\lambda_1+\lambda_2)}. \tag{12.26}$$

Biquadratic equation (12.25) has the following solution:

$$k=\sqrt{\frac{k_{10}^2+k_{12}^2}{2}\left[1+\sqrt{1-\lambda\left(\frac{2k_{10}k_{12}}{k_{10}^2+k_{12}^2}\right)^2}\right]}. \tag{12.27}$$

We seek the axial forces at the body's ends in the form

$$F_i(z)=F_i\left(1-\frac{\cosh kz}{\cosh kl}\right),\quad i=0,1,2 \tag{12.28}$$

that satisfies the zero boundary conditions for the forces at the body's ends. By differentiation, we obtain

$$\tau_0(z) = F_0'(z) = -kF_0 \frac{\sinh kz}{\cosh kl}, \quad \tau_1(z) = F_2'(z) = -kF_2 \frac{\sinh kz}{\cosh kl}. \tag{12.29}$$

Comparing these formulas with formulas (12.23), we find the following expressions for the constants C_0 and C_1:

$$C_0 = -\frac{kF_0}{\cosh kl}, \quad C_1 = -\frac{kF_2}{\cosh kl}. \tag{12.30}$$

Then, formulas (12.23) yield

$$\tau_0(z) = -kF_0 \frac{\sinh kz}{\cosh kl}, \quad \tau_1(z) = -kF_2 \frac{\sinh kz}{\cosh kl}. \tag{12.31}$$

The maximum stresses occur at the body's ends:

$$\tau_{0,\max} = \tau_0(l) = -kF_0 \tanh kl, \quad \tau_{1,\max} = \tau_1(l) = -kF_2 \tanh kl. \tag{12.32}$$

For long (large l) bodies with stiff (large k) interfaces, when the product kl is larger than, say, 3, the maximum shearing stresses become body length independent:

$$\tau_{0,\max} = \tau_0(l) = -kF_0, \quad \tau_{1,\max} = \tau_1(l) = -kF_2. \tag{12.33}$$

The distribution of the interfacial shearing stresses, in the case of long and/or stiff bodies, can be found from (12.31):

$$\tau_0(z) = -kF_0 e^{-k(l-z)}, \quad \tau_1(z) = -kF_2 e^{-k(l-z)}. \tag{12.34}$$

Thus, the interfacial stresses concentrate at the body's ends.

12.4.3 Numerical Example

Input data

Consider a trimaterial body comprised of 0.125 mm dia glass fiber ($E_0 = 7400\ \text{kg/mm}^2$, $\nu_0 = 0.2$, $\alpha_0 = 0.5 \times 10^{-6}\,1/°\text{C}$, $r_0 = 0.0625$ mm) as the zero ("core") component; 2 mil = 0.0508 mm thick ring of the silver-tin solder eutectic (melting point $T_{melt} = 221°\text{C}$, $E_1 = 650\ \text{kg/mm}^2$, $\nu_1 = 0.35$, $\alpha_1 = 29.0 \times 10^{-6}\,1/°\text{C}$, $r_1 = 0.1133$ mm) as the intermediate," bonding," component #1; and the silicon "enclosure"

($E_2 = 12300$ kg/mm^2, $\nu_2 = 0.27$, $\alpha_2 = 2.6\times10^{-6}$1/°C) as the outer (#2) material. The "enclosure" is considered thick, so that its axial compliance λ_2 can be put equal to zero.

Calculated data

$$\delta_{11'} = 33.5521\times10^{-4}\ \text{mm}^2/\text{kg},\quad \delta_{12'} = 38.8098\times10^{-4}\ \text{mm}^2/\text{kg},$$
$$\delta_{21'} = 11.8098\times10^{-4}\ \text{mm}^2/\text{kg},\quad \delta_{22'} = 19.0731\times10^{-4}\ \text{mm}^2/\text{kg},$$
$$D = 181.6248\times10^{-8}\ \text{mm}^4/\text{kg}^2,$$
$$\Delta_1 = 77.100\times10^{-4},\quad \Delta_2 = 71.696\times10^{-4},$$
$$\sigma_0 = 7.2235\ \text{kg/mm}^2,\quad \sigma_1 = 8.2317\ \text{kg/mm}^2.$$

Because both the interfacial radial stresses are positive, the solder ring is in tension, and the radial stress at its interface with silicon is about 14% higher than the radial stress at its interface with the glass. One has to make sure that these stresses are still low enough in order not to cause excessive inelastic deformations in the solder and/or delaminations at its interface.

The radial stress in the solder at its inner interface is

$$\sigma_r^{(1)} = \sigma_0 = 7.2235\ \text{kg/mm}^2;$$

the tangential stress in the solder at its interface with the fiber is

$$\sigma_t^{(1)} = 10.1219\ \text{kg/mm}^2,$$

which is about 40% higher than the radial stress; the radial and tangential stresses in the solder at its interfaces with silicon are

$$\sigma_r^{(1)} = \sigma_1 = 8.2317\ \text{kg/mm}^2 \quad \text{and} \quad \sigma_t^{(1)} = 9.1137\ \text{kg/mm}^2,$$

respectively (the latter stress is by about 11% higher than the radial stress); the radial and the tangential stresses in the silicon at its interface with the solder ring are

$$\sigma_r^{(2)} = \sigma_1 = 8.2317\ \text{kg/mm}^2 \quad \text{and} \quad \sigma_t^{(2)} = -8.23177\ \text{kg/mm}^2,$$

respectively.

Thus, the highest lateral normal stresses are the tangential (circumferential) stresses at the solder–fiber interface.

The computed axial forces and he normal stresses are:

$$F_0 = -0.040908\ \text{kg},\quad F_1 = 0.096847\ \text{kg},\quad F_2 = -0.055939\ \text{kg}$$

(their sum is certainly zero) and

$$\sigma_z^{(0)} = -0.4441 \text{ kg/mm}^2, \quad \sigma_z^{(1)} = 6.1152 \text{ kg/mm}^2, \quad \sigma_z^{(2)} = -0.0178 \text{ kg/mm}^2.$$

Thus, while the axial normal stresses in the glass are low, and the axial stress in the silicon is only about 4% of the stress in the glass, the axial stress in the solder is significant. It is of the same order of magnitude as the lateral stresses are.

In order to evaluate the interfacial shearing stresses one has to compute first the longitudinal interfacial compliances. These are:

$$\kappa_0 = 8.4459 \times 10^{-6} \text{ mm}^3/\text{kg}, \quad \kappa_{10} = 154.4536 \times 10^{-6} \text{ mm}^3/\text{kg},$$
$$\kappa_{12} = 787.0955 \times 10^{-6} \text{ mm}^3/\text{kg}, \quad \kappa_2 = 128.4248 \times 10^{-6} \text{ mm}^3/\text{kg}.$$

The partial parameters of the interfacial shearing stresses areh

$$k_{10} = 23.1418 \text{ mm}^{-1}, \quad k_{12} = 8.9923 \text{ mm}^{-1},$$

and the computed parameter of the interfacial stress for the entire body is $k = 24.6085 \text{ mm}^{-1}$. Assuming that the body is long enough, so that the product kl exceeds considerably the $kl = 3$ value (indeed, with such a significant k value, a body of the length $2l = \frac{6}{k} = 0.244$ mm is long enough to be considered long), the computed maximum shearing stresses are

$$\tau_0 = -kF_0 = 1.0067 \text{ kg/mm}^2, \quad \tau_1 = -kF_2 = 1.3766 \text{ kg/mm}^2.$$

These stresses are substantially lower than the lateral or the axial normal stresses, but, still on the same order of magnitude, and should be considered in the stress–strain and reliability analyses of the bodies of the type in question.

12.4.4 Conclusion

Thermal stresses in a bonded elongated elastic cylindrical trimaterial body of finite length are predicted based on the developed simple, easy-to-use, and physically meaningful analytical (mathematical) model. The numerical example is carried out in application to structures, in which a glass fiber is soldered into a silicon chip (silicon photonics technology) using a "soft" silver-tin solder. It is concluded that the appropriate solder material and its thickness should be selected based on the calculations using the developed model.

REFERENCES

1. E. Suhir, *Structural Analysis in Microelectronics and Fiber Optic Systems*, Van-Nostrand, 1991.
2. E. Suhir, "Mechanical Behavior of Materials in Microelectronic and Fiber Optic Systems: Application of Analytical Modeling-Review," *MRS Symposia Proceedings*, vol. 226, 1991.

3. E. Suhir, "Structural Analysis in Fiber Optics," in J. Menon, (ed.), *Trends in Lightwave Technology*, Council of Scientific Information, 1995.
4. E. Suhir, "Fiber Optic Structural Mechanics–Brief Review," *Editor's Note, ASME Journal of Electronics Packaging (JEP)*, Sept. 1998.
5. E. Suhir, "The Future of Microelectronics and Photonics and the Role of Mechanics and Materials," *ASME Journal of Electronics Packaging*, Mar. 1998.
6. E. Suhir, "Microelectronic and Photonic Systems: Role of Structural Analysis," ASME InterPack 2005 Conference, San Francisco, California, Jul. 2005.
7. E. Suhir, "Mechanical Behavior of Optical Fibers and Interconnects: Application of Analytical Modeling," in H. Altenbach, (ed.), *Encyclopedia of Continuum Mechanics*, Springer, 2019.
8. E. Suhir, "Analytical Predictive Modeling in Fiber Optics Structural Analysis: Review and Extension," SPIE, San Francisco, California, Feb. 2015.
9. B. Welker, M. Uschitsky, E. Suhir, S. Kher, G. Bubel, "Finite Element Analysis of Optical Fiber Structures," in E. Suhir, (ed.), *Structural Analysis in Microelectronics and Fiber Optics*, Symposium Proceedings, ASME Press, 1996.
10. E. Suhir, "Elastic Stability, Free Vibrations, and Bending of Optical Glass Fibers: The Effect of the Nonlinear Stress-Strain Relationship," *Applied Optics*, vol. 31, No. 24, 1992.
11. E. Suhir, "Fiber Optics Engineering: Physical Design for Reliability," *Facta Universitatis: series Electronics and Energetics*, vol. 27, No 2, Jun. 2014.
12. E. Suhir, M. Fukuda, C.R. Kurkjian, "Reliability of Photonic Materials and Structures," *Materials Research Society Symposia Proceedings*, vol. 531, 1998.
13. M. Uschitsky, E. Suhir, S. Kher, G. Bubel, "Epoxy Bonded Optical Fibers: the Effect of Voids on Stress Concentration in the Epoxy Material," in E. Suhir, (ed.), *Structural Analysis in Microelectronic and Fiber Optic Systems*, Symposium Proceedings, ASME Press, 1995.
14. E. Suhir, "Microelectronics and Photonics–the Future," *Microelectronics Journal*, vol. 31, No. 11–12, 2000.
15. E. Suhir, "Stresses in Dual-Coated Optical Fibers," *ASME Journal of Applied Mechanics*, vol. 55, No. 10, 1988.
16. E. Suhir, "Calculated Stresses in Dual-Coated Optical Fibers," *Polymer Engineering & Science*, vol. 30, 1990.
17. E. Suhir, "Stresses in a Coated Fiber Stretched on a Capstan," *Applied Optics*, vol. 29, No. 18, 1990.
18. E. Suhir, "Can the Curvature of an Optical Glass Fiber be Different from the Curvature of Its Coating?" *International Journal of Solids and Structures*, vol. 30, No. 17, 1993.
19. E. Suhir, "Approximate Evaluation of the Interfacial Shearing Stress in Circular Double Lap Shear Joints, with Application to Dual-Coated Optical Fibers," *International Journal of Solids and Structures*, vol. 31, No. 23, 1994.
20. E. Suhir, "Critical Strain and Postbuckling Stress in Polymer Coated Optical Fiber Interconnect: What Could Be Gained by Using Thicker Coating?" International Workshop on Reliability of Polymeric Materials and Plastic Packages of IC Devices, Paris, France, Nov.–Dec. 1998.
21. E. Suhir, "Thermal Stress in a Polymer Coated Optical Glass Fiber with a Low Modulus Coating at the Ends," *Journal of Materials Research*, vol. 16, No. 10, 2001.
22. E. Suhir, "Thermal Stress Modeling in Micro- and Opto-Electronics: Review and Extension," Invited Presentation, ASME Symposium Dedicated to Dr. Richard Chu, IBM, Washington, DC, Nov. 2003.
23. E. Suhir, "Polymer Coated Optical Glass Fibers: Review and Extension," Proceedings of the POLYTRONIK'2003, Montreux, Switzerland, Oct. 2003.

24. E. Suhir, V. Ogenko, D. Ingman, "Two-Point Bending of Coated Optical Fibers," Proceedings of the PhoMat 2003 Conference, San Francisco, California, Aug. 2003.
25. E. Suhir, "Modeling of Thermal Stress in Microelectronic and Photonic Structures: Role, Attributes, Challenges and Brief Review," Special Issue, *ASME Journal of Electronic Packaging (JEP)*, vol. 125, No. 2, Jun. 2003.
26. E. Suhir, "Polymer Coated Optical Glass Fiber Reliability: Could Nano-Technology Make a Difference?" Polytronic '04, Portland, Oregon, Sept. 2004.
27. E. Suhir, "Mechanics of Coated Optical Fibers: Review and Extension," ECTC 2005, Orlando, Florida, 2005.
28. E. Suhir, "Coated Optical Glass Fiber," U.S. Patent #6,647,195, 2003.
29. E. Suhir, "Thermal Stress in Electronics and Photonics: Prediction and Prevention," Keynote presentation, Therminic, Budapest, Hungary, Sept. 2012.
30. E. Suhir, L. Bechou, "Saint-Venant's Principle and the Minimum Length of a Dual-Coated Optical Fiber Specimen in Reliability (Proof) Testing," ESREF, Arcachon, France, 2013.
31. E. Suhir, "Thermal Stress Failures in Electronics and Photonics: Physics, Modeling. Prevention," *Journal of Thermal Stresses*, vol. 36, No. 6, Jun. 2013.
32. E. Suhir, "Elastic Stability of a Dual-Coated Fiber," SPIE Paper #8621-37, Photonics West, Feb. 2013.
33. E. Suhir, "Compressed Cantilever Beam on an Elastic Foundation, with Application to a Dual-Coated Fiber-Optic Connector," *International Journal of Engineering Sciences*, vol. 83, Oct. 2014.
34. E. Suhir, S. Yi, "Elastic Stability of a Dual-Coated Fiber-Optic Connector," 2017 SPIE Photonics West, San Francisco, California, Feb. 2017. (Published in the SPIE Digital Library as part of the proceedings of the Silicon Photonics XII conference.)
35. E. Suhir, "Effect of Initial Curvature on Low Temperature Microbending in Optical Fibers," *IEEE/OSA Journal of Lightwave Technology*, vol. 6, No. 8, 1988.
36. E. Suhir, "Spring Constant In the Buckling of Dual-Coated Optical Fibers," *IEEE/OSA Journal of Lightwave Technology*, vol. 6, No. 7, 1988.
37. E. Suhir, "Calculated Stresses in Microelectronic and Fiber-Optic Structures," Proceedings of the 1st Pan American Congress of Applied Mechanics, American Academy of Mechanics, Rio de Janeiro, Brazil, Jan. 1989.
38. E. Suhir, "Mechanical Approach to the Evaluation of the Low Temperature Threshold of Added Transmission Losses in Single-Coated Optical Fibers," *IEEE/OSA Journal of Lightwave Technology*, vol. 8, No. 6, 1990.
39. S.T. Shiue, "The Spring Constant in the Buckling of Tightly Jacketed Double-Coated Optical Fibers," *Journal of Applied Physics*, vol. 81, No. 8, 1997.
40. E. Suhir, J.J. Vuillamin, Jr, "Effects of the CTE and Young's Modulus Lateral Gradients on the Bowing of an Optical Fiber: Analytical and Finite Element Modeling," *Optical Engineering*, vol. 39, No. 12, 2000.
41. E. Suhir, "Bending Performance of Clamped Optical Fibers: Stresses Due to the End Off-Set," *Applied Optics*, vol. 28, No. 3, Feb. 1989.
42. E. Suhir, "Bending of a Partially Coated Glass Fiber Subjected to the Ends Off-Set," *IEEE CPMT Transactions*, Jun. 1997.
43. E. Suhir, "Coated Optical Fiber Interconnect Subjected to the Ends Off-Set and Axial Loading," *International Workshop on Reliability of Polymeric Materials and Plastic Packages of IC Devices*, Paris, France, Nov.–Dec. 1998.
44. E. Suhir, "Bending Stress in an Optical Fiber Interconnect Experiencing Significant Ends Off-Set," *MRS Symposia Proceedings*, vol. 531, 1998.
45. E. Suhir, "Optical Fiber Interconnect Subjected to a Not-Very-Small Ends Off-Set: Effect of the Reactive Tension," *MRS Symposia Proceedings*, vol. 531, 1998.

46. E. Suhir, "Optical Fiber Interconnect with the Ends Offset and Axial Loading: What Could Be Done to Reduce the Tensile Stress in the Fiber?" *Journal of Applied Physics*, vol. 88, No. 7, 2000.
47. E. Suhir, "Optical Fiber Interconnect with the Ends Offset and Axial Loading: What Could Be Done to Reduce the Tensile Stress in the Fiber?" *Journal of Applied Physics*, vol. 88, No. 7, 2000.
48. E. Suhir, "Silica Optical Fiber Interconnects: Design for Reliability," Proceedings of the Annual Conference of the American Ceramic Society, St. Louis, Missouri, May, 2000.
49. E. Suhir, "Optical Fiber Interconnects Having Offset Ends with Reduced Tensile Strength and Fabrication Method," U.S. Patent #6,606,434, 2003.
50. E. Suhir, "Predicted Curvature and Stresses in an Optical Fiber Interconnect Subjected to Bending," *IEEE/OSA Journal of Lightwave Technology*, vol. 14, No. 2, 1996.
51. E. Suhir, "Method of Improving the Performance of Optical Fiber, which is Interconnected Between Two Misaligned Supports," U.S. Patent #6,314,218, 2001.
52. E. Suhir, "Interconnected Optical Devices Having Enhanced Reliability," U.S. Patent #6,327,411, 2001.
53. E. Suhir, "Method for Determining and Optimizing the Curvature of a Glass Fiber for Reducing Fiber Stress," U.S. Patent #6,016,377, 2000.
54. E. Suhir, "Elastic Stability of a Dual-Coated Optical Fiber of Finite Length," *Journal of Applied Physics*, vol. 102, No. 5, 2007.
55. E. Suhir, "Elastic Stability of a Dual-Coated Optical Fiber with a Stripped Off Coating at Its End," *Journal of Applied Physics*, vol. 102, No. 4, 2007.
56. E. Suhir, "Lateral Compliance of a Compressed Cantilever Beam, with Application to Micro-Electronic and Fiber-Optic Structures," *Journal of Applied Physics D*, vol. 41, No. 1, 2008.
57. E. Suhir, "Optical Fiber Interconnects: Design for Reliability," *Society of Optical Engineers (SPIE)*, Proceedings of the SPIE, vol. 7607, 2010.
58. E. Suhir, "The Effect of the Nonlinear Behavior of the Material on Two-Point Bending in Optical Glass Fibers," *ASME Journal of Electronics Packaging (JEP)*, vol. 114, No. 2, 1992.
59. E. Suhir, "Effect of the Nonlinear Stress-Strain Relationship on the Maximum Stress in Silica Fibers Subjected to Two-Point Bending," *Applied Optics*, vol. 32, No. 9, 1993.
60. E. Suhir, "Analytical Modeling of the Interfacial Shearing Stress During Pull-Out Testing of Dual-Coated Lightguide Specimens," *Applied Optics*, vol. 32, No. 7, 1993.
61. E. Suhir, "Pull Testing of a Glass Fiber Soldered into a Ferrule: How Long Should the Test Specimen Be?" *Applied Optics*, vol. 33, No. 19, 1994.
62. E. Suhir, "Accelerated Life Testing (ALT) in Microelectronics and Photonics: Its Role, Attributes, Challenges, Pitfalls, and Interaction With Qualification Tests," *ASME Journal of Electronics Packaging (JEP)*, vol. 124, No. 3, 2002.
63. E. Suhir, "Method and Apparatus for Proof-testing Optical Fibers," U.S. Patent #6,119,527, 1998.
64. E. Suhir, "Design-for-Reliability and Accelerated-Testing of Solder Joint Interconnections," *Chip Scale Reviews*, Nov.–Dec. 2019.
65. E. Suhir, "Failure-Oriented-Accelerated-Testing (FOAT), Boltzmann-Arrhenius-Zhurkov Equation (BAZ) and Their Application in Microelectronics and Photonics Reliability Engineering," *International Journal of Aeronautical Science and Aerospace Research (IJASAR)*, vol. 6, No. 3, 2019.
66. E. Suhir, "Free Vibrations of a Fused Biconical Taper Lightwave Coupler," *International Journal of Solids and Structures*, vol. 29, No. 24, 1992.
67. E. Suhir, "Vibration Frequency of a Fused Biconical Taper (FBT) Lightwave Coupler," *IEEE/OSA Journal of Lightwave Technology*, vol. 10, No. 7, 1992.

68. E. Suhir, "Predicted Stresses and Strains in Fused Biconical Taper Couplers Subjected to Tension," *Applied Optics*, vol. 32, No. 18, 1993.
69. E. Suhir, "Strain Free Planar Optical Waveguides," U.S. Patent #6,389,209, 2002.
70. E. Suhir, "Thermally Induced Stresses in an Optical Glass Fiber Soldered into a Ferrule," *IEEE/OSA Journal of Lightwave Technology*, vol. 12, No. 10, 1994.
71. E. Suhir, "Apparatus and Method for Thermostatic Compensation of Temperature Sensitive Devices," U.S. Patent #6,337,932, 2002.
72. E. Suhir, "Optical Glass Fiber Bent on a Cylindrical Surface," *MRS Symposia Proceedings*, vol. 531, 1998.
73. E. Suhir, "Optimized Configuration of an Optical fiber "Pigtail" Bent on a Cylindrical Surface," in T. Winkler, A. Schubert, (eds.), *Materials Mechanics, Fracture Mechanics, Micromechanics*, An Anniversary Volume in Honor of B. Michel's 50th Birthday, Fraunhofer IZM, 1999.
74. E. Suhir, "Elastic Stability of the Glass Fibers in a Micromachined Fiber-Optic Switch Packaged into a Dual-in-Line Ceramic Package," ECTC, 1999.
75. E. Suhir, "Thermal Stress Modeling in Microelectronics and Photonics Packaging, and the Application of the Probabilistic Approach: Review and Extension," *IMAPS International Journal of Microcircuits and Electronic Packaging*, vol. 23, No. 2, 2000 (invited paper).
76. E. Suhir, "How to Make a Photonic Device Into a Product: Role of Accelerated Life Testing," Keynote Address at the International Conference of Business Aspects of Microelectronic Industry, Hong Kong, Jan. 2003.
77. E. Suhir, A. Bensoussan, "Quantified Reliability of Aerospace Optoelectronics," *SAE International Journal of Aerospace*, vol. 7, No. 1, 2014.
78. E. Suhir, A. Bensoussan, "Application of Multi-Parametric BAZ Model in Aerospace Optoelectronics," 2014 IEEE Aerospace Conference, Big Sky, Montana, Mar. 2014.
79. E. Suhir, "Static Fatigue Lifetime of Optical Fibers Assessed Using Boltzmann-Arrhenius-Zhurkov (BAZ) Model," *Journal of Materials Science: Materials in Electronics*, vol. 28, No. 16, 2017.
80. D. Ingman, V. Ogenko, E. Suhir, A. Glista, "Moisture Resistant Nano-Particle Material and Its Applications," U.S. Patent #7,321,714B2, 2008.
81. D. Ingman, E. Suhir, "Optical Fiber with Nano-Particle Overclad," U.S. Patent, #7,162,138 B2, 2007.
82. E. Suhir, "Fiber-Optics Structural Mechanics and Nano-Technology Based New Generation of Fiber Coatings: Review and Extension," in E. Suhir, C.P. Wong, Y.C. Lee, (eds.), *Micro- and Opto-Electronic Materials and Structures: Physics, Mechanics, Design, Packaging, Reliability*, 2 volumes, Springer, 2008.
83. E. Suhir, "New Nano-Particle Material (NPM) for Micro- and Opto-Electronic Packaging Applications," IEEE Workshop on Advanced Packaging Materials, Irvine, California, Mar. 2005.
84. E. Suhir, "Fiber Optics Structural Mechanics, and a New Generation of Nano-Technology Based Optical Fiber Cladding and Coating," Invited talk at the Photonics West Conference, San Jose, California, Jan. 2006.
85. E. Suhir, "Polymer Coating of Optical Silica Fibers, and a Nanomaterial-Based Coating System," Keynote Presentation, Polytronic'2007, Proceedings of the International Conference on Polymeric Materials for Micro- and Opto-Electronics Applications, Tokyo, Japan, Jan. 2007.
86. E. Suhir, D. Ingman, "Highly Compliant Bonding Material and Structure for Micro- and Opto-Electronic Applications," ECTC '06 Proceedings, San Diego, California, May 2006.
87. E. Suhir, "Solder Materials and Joints in Fiber Optics: Reliability Requirements and Predicted Stresses," Proceedings of the International Symposium on Design and Reliability of Solders and Solder Interconnections," Orlando, Florida, Feb. 1997.

88. E. Suhir, "Predicted Thermal Mismatch Stresses in a Cylindrical Bi-Material Assembly Adhesively Bonded at the Ends," *ASME Journal of Applied Mechanics*, vol. 64, No. 1, 1997.
89. E. Suhir, "Bi-Material Assembly Adhesively Bonded at the Ends and Fabrication Method," U.S. Patent #6,460,753, 2002.
90. T. Reinikainen, E. Suhir, "Novel Shear Test Methodology for the Most Accurate Assessment of Solder Material Properties," IEEE ECTC, 2009.
91. E. Suhir, T. Reinikainen, "Interfacial Stresses and a Lap Shear Joint (LSJ): The 'Transverse Groove Effect' (TGE)," *JSME Journal of Solid Mechanics and Materials Engineering (JSMME)*, vol. 3, No. 6, 2009.
92. E. Suhir, T. Reinikainen, "Interfacial Stresses in a Lap Shear Joint (LSJ): The 'Transverse Groove Effect' (TGE)," *JSME Journal of Solid Mechanics and Materials Engineering (JSMME)*, vol. 4, No. 8, 2010.
93. E. Suhir, S. Kang, J. Nicolics, C. Gu, A. Bensoussan, L. Bechou, "Predicted Thermal Stresses in a Cylindrical Tri-Material Body, with Application to Optical Fibers Embedded into Silicon," *Journal of Electrical and Control Engineering*, vol. 3, No. 6, Dec. 2013.
94. E. Suhir, S. Yi, "Thermal Stress in an Optical Silica Fiber Embedded (Soldered) into Silicon," 2017 SPIE Photonics West, San Francisco, California, Jan. 2017.
95. E. Suhir, "Solder Joint Interconnections in Aerospace Electronics: Design-for-Reliability and Accelerated Testing," IEEE MetroAeroSpace Conference, Torino, Italy, Jun. 2019.

Index

Note: Locators in *italics* represent figures and **bold** indicate tables in the text.

J

K

L

M

N

O

P

T

U

V

W

X

Y

Z

For Product Safety Concerns and Information please contact our EU
representative GPSR@taylorandfrancis.com
Taylor & Francis Verlag GmbH, Kaufingerstraße 24, 80331 München, Germany

www.ingramcontent.com/pod-product-compliance
Lightning Source LLC
Chambersburg PA
CBHW060751220326
41598CB00022B/2401
9780367635886